Lecture Notes in Computer Science 994
Edited by G. Goos, J. Hartmanis and J. van Leeuwen

Springer
Berlin
Heidelberg
New York
Barcelona
Budapest
Hong Kong
London
Milan
Paris
Santa Clara
Singapore
Tokyo

Martial Hebert Jean Ponce
Terry Boult Ari Gross (Eds.)

Object Representation in Computer Vision

International NSF-ARPA Workshop
New York City, NY, USA, December 5-7, 1994
Proceedings

Series Editors

Gerhard Goos, Karlsruhe University, Germany

Juris Hartmanis, Cornell University, NY, USA

Jan van Leeuwen, Utrecht University, The Netherlands

Volume Editors

Martial Hebert
The Robotics Institute, Carnegie Mellon University
5000 Forbes Avenue, Pittsburgh, PA 15213, USA

Jean Ponce
Beckmann Institute, University of Illinois
Urbana, IL 61801, USA

Terry Boult
Lehigh University, EECS Department, Packard Laboratory
19 Memorial Drive West, Bethlehem, PA 18015, USA

Ari Gross
Department of Computer Science, Graduate Center and Queens College
CUNY, Flushing, New York 11367, USA

Cataloging-in-Publication data applied for

Die Deutsche Bibliothek - CIP-Einheitsaufnahme

Object representation in computer vision : proceedings / International NSF ARPA Workshop, New York City, NY, USA, December 5 - 7, 1994. Martial Hebert ... (ed.). - Berlin ; Heidelberg ; New York ; Barcelona ; Budapest ; Hong Kong ; London ; Milan ; Paris ; Tokyo : Springer, 1995
(Lecture notes in computer science ; Vol. 994)
ISBN 3-540-60477-4
NE: Hebert, Martial [Hrsg.]; International NSF ARPA Workshop <1994, New York, NY>; GT

CR Subject Classification (1991): I.5.4, I.3, I.2.9-10, J.2

ISBN 3-540-60477-4 Springer-Verlag Berlin Heidelberg New York

Typesetting: Camera-ready by author
SPIN 10485977 06/3142 – 5 4 3 2 1 0 Printed on acid-free paper

Preface

Three-dimensional object representation is paramount in computer vision. Algorithms for recognition, inspection, tracking, etc., are only as good as the underlying representation. Research in this area is very active and healthy, but relatively uncoordinated: the past few years have witnessed a proliferation of approaches, ranging from well-established ideas, based on generalized cylinders for example, to emerging new concepts, such as invariant-based techniques. Although each technique has been demonstrated on a specific set of examples, the research lacks a systematic study of the strengths and weaknesses of the approaches and a consistent method for evaluating them. Maybe more importantly, it is not clear how today's representation schemes will scale up to more complex geometries, and how they will handle large model databases or the need to describe object classes.

We believe that constructing the next generation of model-based vision systems will require a critical review of the current representational schemes and a concerted effort from the computer vision community to define specific requirements for object representation. Making a first step in this direction was the rationale for holding a workshop on three-dimensional object representation for computer vision. This three-day international workshop was funded by a joint grant of the National Science Foundation and the Advanced Research Projects Agency, and it took place at the CUNY Graduate School and University Center in New York City on December 5-7, 1994, with participants chosen among the recognized experts in the field.

This book is the outcome of the workshop. It contains the complete set of papers presented at the workshop. In addition, the first paper is a report on the workshop in which the panel discussions as well as the conclusions and recommendations reached by the workshop participants are summarized. This report, along with abstracts of the contributed papers, is available through the World-Wide Web at: //www.ius.cs.cmu.edu/usr/users/hebert/www/workshop/report.html.

We wish to thank Prof. Stanley Habib, executive officer of the CUNY Graduate School and University Center, for providing a wonderful place where to hold the workshop, and Joseph Driscoll, also from the CUNY Graduate School and University Center, for taking care of the local arrangements. We also want to thank Marie Elm, from Carnegie-Mellon University, for helping us with the workshop organization. We acknowledge the valuable comments on early drafts of the workshop report from Seth Hutchinson, Dave Kriegman, and Alison Noble.

Finally, we wish to thank the following organizations which provided partial support for the authors during the preparation of the workshop and the editing of this book: Jean Ponce was partially supported by the Department of Mechanical Engineering of the California Institute of Technology, the Center for Advanced Study of the University of Illinois at Urbana-Champaign, and the Beckman Institute for Advanced Science and Technology; Ari Gross was supported by the CUNY Research Foundation; and Terry Boult was supported in part by an NSF PYI Grant and by industry grants from AT&T, IBM, Siemens, and TI.

August 1995

Martial Hebert
Jean Ponce
Terry Boult
Ari Gross

CONTENTS

M. Hebert, *Carnegie Mellon University*
J. Ponce, *University of Illinois*
T. Boult, *Lehigh University*
A. Gross, *City University of New York*
Report on the 1995 Workshop on 3-D Object Representations in Computer Vision 1

J.L. Mundy, *General Electric*
Object Recognition: The Search for Representation 19

L.G. Shapiro, M. S. Costa, *University of Washington* 51
Appearance-Based 3D Object Recognition

P. Gros, *INRIA*
Using Quasi-Invariants for Automatic Model Building and Object Recognition: an Overview 65

G. Stockman, *Michigan State University*
Object Representation for Recognition-By-Alignment 77

D. Forsyth, *University of California at Berkeley*
A. Zisserman, *Oxford University*
J. Malik, *University of California at Berkeley*
Distinctive Representations for the Recognition of Curved Surfaces Using Outlines and Markings 89

R.S. Weiss, *University of Massachusetts*
The Epipolar Parametrization 101

D.Huttenlocher, *Cornell University and Xerox*
Using Two-Dimensional Models to Interact with the Three-Dimensional World 109

D. Kriegman, *Yale University*
J. Ponce, *University of Illinois*
Representations for Recognizing Complex Curved 3D Objects 125

D.B. Cooper, Z. Lei, *Brown University*
On Representation and Invariant Recognition of Complex Objects Based on Patches and Parts 139

S. Petitjean, *CRIN*
Algebraic Geometry and Object Representation in Computer Vision 155

G. Taubin, *IBM*
Discrete Surface Signal Processing: The Polygon as the Surface Element 167

V. Koivunen, R. Bajcsy, *University of Pennsylvania*
Spline Representations in 3-D Vision 177

P.J. Besl, *General Motors*
Triangles as a Primary Representation 191

Tom O. Binford, *Stanford University*
Body-Centered Representation and Perception 207

M. Zerroug and G. Medioni, *University of Southern California*
The Challenge of Generic Object Recognition 217

Dimitri Metaxas, *University of Pennsylvania*
A Physics-Based Framework for Segmentation, Shape and Motion Estimation 233

A. Pentland, *MIT*
S. Sclaroff, *Boston University*
Modal Representations 249

C. Nastar, *MIT*
N. Ayache, *INRIA*
Time Representation of Deformations: Combining Vibration Modes and Fourier Analysis 263

T. Boult, *Lehigh University*
S. D. Fenster, T. O'Donnell, *Columbia University*
Physics in a Fantasy World vs. Robust Statistical Estimation 277

O. I. Camps, *Pennsylvania State University*
Towards a Robust Physics-Based Object Recognition System 297

A. Gross, *City University of New York*
L. Latecki, *University of Hamburg*
Toward Non-Parametric Digital Shape Representation and Recovery 313

K. Ikeuchi, M. Hebert, *Carnegie Mellon University*
Spherical Representations: from EGI to SAI 327

D. Terzopoulos, *University of Toronto*
From Physics-Based Representation to Functional Modeling of Highly Complex Objects 347

Report on the 1995 Workshop on 3-D Object Representations in Computer Vision

Martial Hebert[1], Jean Ponce[2], Terrance Boult[3] and Ari Gross[4]

[1] Robotics Institute
Carnegie Mellon University
5000 Forbes Avenue, Pittsburgh PA15213
[2] Beckman Institute University of Illinois
Urbana, IL 61801
[3] EECS Dept.
Lehigh University
304 Packard Laboratory
19 Memorial Drive West, Bethlehem, PA 18015
[4] University Graduate Center and Queens College
City University of New York
Flushing NY 11367

1 Introduction

Computer vision programs are designed to use some form of visual data to accomplish a specific task. Like any other computer programs, they must possess some internal representation of the task and of the data. Here we are interested in tasks which involve reasoning about (and therefore representing) three-dimensional objects. Examples include object recognition, robot navigation, medical image analysis and the construction of virtual reality models.

We believe that computer vision is at a stage where identifying the qualities that make a good representation is the key to progress. We also believe that, although the choice of representation obviously depends on both the task and the form of data, there are general principles (to be discovered and/or articulated) that will dictate the form of object representation useful for large classes of tasks and data.

This motivated us, first to organize two panels, held respectively at the 1994 IEEE workshop on CAD-based vision and at the 1994 IEEE conference on computer vision and pattern recognition, then to organize an international workshop on 3D object representation for computer vision. This workshop was held in New York City on Dec. 5-7, 1994, under joint sponsorship from the National Science Foundation, the Advanced Research Projects Agency and the CUNY Graduate School and University Center.

The workshop brought together about thirty computer vision researchers. Its goal was to evaluate the current representation schemes and to identify important issues and promising research directions.

This document describes the format of the workshop, summarizes its panel sessions, and reports its findings. In addition, we have included a short essay by David Forsyth which addresses the current limitations of recognition systems as related to the underlying representation schemes. The need to address these limitations clearly emerged from the panel discussions and this essay is motivated by some of the views expressed during the panels. Finally, an appendix contains the list of the workshop participants and panelists, as well as extended abstracts of the papers presented at the workshop.

2 Workshop Format

The core of the workshop was a set of three half-day sessions corresponding to the broad classes of three-dimensional object representations: appearance-, primitive-, and physics-based approaches.

Each session consisted of eight oral presentations, each one lasted for twenty five minutes including questions, and was followed by a two-hour panel discussion. The speakers were encouraged to present their views on the current state and future of object representation as it relates to their own research. Each panel started with a brief summary report of the issues identified during the technical presentations. This report was delivered by the panel moderator and was used as a basis for the general panel discussion.

A final panel included an overall report presentation and a discussion of the findings of the workshop, including recommendations for future research.

The next four sections are summaries of the panel discussions. It should be noted that we have deliberately chosen to summarize the ideas that emerged from the panel discussion rather than transcribing the detailed discussions and attributing to each panelist his or her comments.

3 Panel 1: Appearance-Based Approaches

The goal of this first panel was to assess the performance and future of view-based object representation schemes that may include photometric as well as purely geometric information. The panelists were (alphabetically) David Forsyth, Patrick Gros, Dan Huttenlocher, Shree Nayar, Jean Ponce (moderator), George Stockman and Richard Weiss.

3.1 General Comments

There is a traditional dichotomy between object- and viewer-centered representations in computer vision: while object-centered representations usually store an explicit three-dimensional description of the object of interest in some world coordinate frame, viewer-centered schemes generally rely on (possibly abstractions of) collections of images of an object as the implicit description of its shape. Aspect graphs are a classical instance of the latter type of representation, focusing mostly on image geometry. Because recent new approaches may also store photometric, color, or texture information, we will refer to the whole body of view-based techniques as "appearance-based" methods. An example is Nayar's work, which permits the compression of a collection of images (including pose and illumination variations) into a compact spline description.

Appearance-based approaches may be classified according to the degree of structure that is extracted from the image data. At one extreme, some techniques store explicit collections of pictures as object models, and use this representation to recognize the object by either correlating an input image directly with the model images or by comparing the input image with the principal components (in the SVD sense) of the model data set. Huttenlocher's work on image comparison using the Hausdorff distance and Nayar's parametric representation of image sets are two examples of this approach.

At the other end of the spectrum are techniques that impose more structure on the data by first extracting features from the images and then organizing these features into aspects (or "view classes"), or by explicitly identifying invariants of those features. Recognition using this type of representation is achieved by indexing the model using the features detected in the input image. These approaches differ in the type of features used and the criteria utilized for grouping the features into invariant sets. For example, Gros' use of affine invariants computed from sets of views and Shapiro's and Stockman's work on recognizing objects from aspects all fall into this category.

Although the various appearance-based approaches to computer vision may differ substantially, they are all based on the premise that object-centered three-dimensional object descriptions are not required. Accordingly, the panel discussion focussed on the constraints imposed on the visual processes by this fundamental premise and identified three main issues that must be confronted by all appearance-based techniques: To what extent do they depend on effective segmentation? How can large model databases be handled in an appearance-based scheme? Should 3D information be stored along with the underlying description in terms of views? These issues are addressed in the rest of this section.

3.2 Segmentation

Constructing an object description from a 2D image requires solving two difficult problems: first, the object of interest must be isolated from the background (figure/ground discrimination). Second, reliable features must be found and grouped into significant structures (grouping). The overall process is called image segmentation.

Because they cannot exploit explicit three-dimensional information to help the segmentation process in a top-down manner, appearance-based techniques rely fairly heavily on robust bottom-up feature grouping and figure/ground discrimination. Bottom-up segmentation is intrinsically difficult, and at least at the current level of image segmentation technology, this may limit the general applicability of appearance-based methods. It should be noted however that recent work has demonstrated that these techniques can be extremely successful in the controlled environments (and accordingly easier segmentation) afforded by certain tasks, e.g., Pentland's face recognition work, Huttenlocher's tracking algorithm for indoor robot navigation, or Nayar's illumination planning and recognition work.

Also, it is sometimes possible to guide the segmentation process within the appearance- based paradigm: First, the view information stored in a model can be used to predict typical configuration of features (this is of course also true for object-centered representations, as discussed later). Second, the coherence of successive images in stereo or motion data sets may be used to guide figure/ground separation. Third, the segmentation process may be viewed as a feedback loop in which an initial segmentation is refined using predictions from particular segmentation events. A simple example is using the collinearity of edge elements to predict new features and refining the segmentation.

3.3 Large Model Databases

Assuming that appropriate segmentation tools have been developed, the question of the tractability of the models remains, in particular in the case of large object libraries. A possible objection to appearance-based representations is that each object is itself described by a potentially very large number of views. How is it possible then to efficiently store, organize and index model databases that may contain thousands of objects? Certainly, the views of an object are quite redundant, and remarkably compact object descriptions can be extracted from a large number of views (see again the work of Nayar mentioned earlier), allowing efficient indexing of the model database through table look-up. However, this still does not solve the problem of organizing appearance-based representations, which may be necessary to index really large model databases or handle object classes.

An intuitive way of structuring an object description is to divide it into hierarchical descriptions of its parts. Besides allowing (potentially) faster indexing of the model library, this is better suited to handling classes of objects by allowing the use of symbolic relational representations.

Giving an operational, computable definition of parts is by itself a very difficult problem, and it is not clear at this point how appearance-based representations can be structured into part- based descriptions. Much more work is needed in this area.

3.4 2D vs. 3D Information

The fundamental tenet of appearance-based approaches is that a set of views of an object captures all the information relevant to the target tasks, eliminating the need for explicitly describing the object's 3D geometry. The most radical approaches actually do not store any three-dimensional information at all. It is likely, however, that such information will be useful in certain cases, for example to understand occlusion. Therefore the question arises as to when is it appropriate to use 2D information and at what point does 3D information become necessary.

There are two main arguments in favor of using view-based representations. The first one is that shape information is not sufficient for object recognition; object models must include appearance information as well. In fact, appearance is often a more powerful discriminator than geometry. Appropriate use of appearance information does require a sensor model in order to discriminate between variations due to object appearance, which should be retained, and variations due to sensor pose, which should be discarded. The second argument is that a sequences of 2D images does in fact, albeit implicitly, capture the object's 3D geometry. This is supported by recent work by Faugeras, Hartley, Mohr and others on projective shape reconstruction from discrete sets of images, by Weiss' technique for smooth surface reconstruction from continuous image sequences, and by the work in aspect graphs which shows that the object appearance can be completely captured by a finite set of characteristic views.

The question still remains as to under which conditions should 3D geometry be made explicit. Two cases were identified by the panel. The first one occurs when the target task requires the vision system to operate with complex scenes, that is, scenes with a high degree of occlusion and clutter. In particular, an explicit description of the spatial relationship between objects must be used in order to reason about occlusion. The second case is when the goal is to describe object classes as opposed to single object instances. The intuition here is that models based solely on 2D information may be hard to generalize, whereas 3D shape can be generalized into classes by using appropriate parameterizations or by using part decompositions, for example.

4 Panel 2: Primitive-Based Approaches

With the second panel, the discussion shifted to object-centered representations, with a focus on techniques based on a set of primitive shapes. The panelists were: Paul Besl, David Cooper, Martial Hebert (moderator), David Kriegman, Gérard Medioni, Sylvain Petitjean and Gabriel Taubin.

4.1 General Comments

Most object-centered approaches rely on one type or another of shape primitives: thirty years ago, Roberts used simple polyhedral models as a basis for object recognition; since then, primitives as diverse as parametric surface or volume patches, generalized cylinders, and point or curve sets with various invariant properties have been used to support recognition and other visual tasks.

These approaches can be classified according to the degree of generality of the underlying primitives. At one end of the spectrum, primitive shapes are explicitly defined by numerical parameters. As more and more parameters are used, it takes fewer and fewer primitives to describe complicated objects: for example, it may take hundreds of triangles or quadric patches to faithfully capture a shape reasonably well approximated by a handful of, say, superquadrics or algebraic surfaces.

It is clear that, however large the number of parameters, there will always be common- place objects that cannot be described by a single (or even a few) parametric primitives. The next step is to define primitive shapes by precise generative rules instead of numerical parameters: solids of revolution and straight homogeneous generalized cylinders (SHGCs) are instances of this approach, the latter for example are obtained by sweeping a 2D cross-section along a straight line, preserving the shape of the cross-section but scaling it as necessary.

Primitives based on rigid generative rules may in fact also be seen as parametric representations: for example, SHGCs are parameterized by their cross-section shape and scaling function. More generality can be achieved by relaxing these rules (as in Binford's original definition of generalized cylinders) so they apply to a wide variety of shapes while still capturing the intended object structure. A more extreme position is to abandon generative rules altogether and to base the representation solely on the viewpoint-invariant (or quasi-invariant) properties that its primitives may hold. Kriegman's work on invariant curves in an instance of this approach.

Generality and invariance are not the only factors that affect the choice of a primitive set: certain computations may be particularly simple for certain primitives (e.g., rendering of triangular meshes, or formal manipulation of algebraic surfaces). Besides, primitives may be chosen because they capture volume information (hence also the spatial relationship between surface elements) instead of surface information only. Examples include generalized cylinders and to a certain degree various types of implicit surfaces (such as superquadrics) with well-defined inside-outside functions.

Whatever approach is used, it must support robust decomposition of complex shapes into the appropriate catalogue of primitives. This problem is addressed in the next section. As in the panel, we explore in the following sections two connected problems: first, the logical next step in the quest for more and more general primitives is the construction of object classes; how should these classes be defined? From a more pragmatic point of view, how does the level of generality of the primitives affect the performance for a given task?

4.2 Decomposing a Shape into Primitives

As remarked earlier, no single primitive can capture the shape of a complex object. Hence a critical prerequisite for the applicability of primitive-based approaches is that they support stable decomposition into primitive parts. Such decompositions were proposed in the first panel as a mechanism for tackling the combinatorial problems associated with indexing a large model database. Here the motivation is different: part decomposition is an integral part of primitive-based approaches. Without it, these techniques would be limited to the toy world formed by primitive instances.

For low-level parametric primitives (e.g., polygons or quadric patches), "part decomposition" is essentially a matter of fitting and searching: split-and-merge techniques have indeed been used quite successfully for extracting stable decompositions of range images of complex objects into polygonal or quadric patches (see for example Faugeras' and Hebert's early work). Unfortunately, this approach does not appear to scale up very well to more complex parametric primitives, such as superquadrics and algebraic surfaces. In this case, researchers have so far mostly relied on external processes, which are essentially independent of the underlying primitive class, for part decomposition (e.g., Cooper's work).

Such processes can be guided by prior higher-level knowledge: in fact, functional considerations may in many cases be more appropriate than purely geometrical ones. Separating the handle from the body of a teapot is an example in which function is the main reason for the decomposition. Recent results (e.g., Stark's work) show that functional information can be used successfully in some recognition tasks. How well would current techniques perform in the context of part decomposition is not clear, and there is a definite need for more research in this area.

Somewhere in the middle ground, high-level generative object models such as generalized cylinders may offer a primitive-based approach to part decomposition which is related to functional considerations. In this case, parts can be defined by primitive instances separated by discontinuities in their parameters. Although this is intuitively appealing (e.g., it seems possible to decompose the duck used by Kriegman in his recognition experiments into a small set of elongated shapes and to separate for example the neck from the body because the cross-section area is discontinuous at the junction), it is not clear how to find the primitives before having done the decomposition (a duck-and-egg problem, so to speak). View-invariant and quasi-invariant properties should prove useful in this context. We will come back to this issue in later sections.

4.3 Object Classes

Part decompositions in general, and primitive-based decompositions in particular, lead naturally to the concept of object class. More precisely, they offer an intuitive approach to measuring shape similarity at both the primitive and overall part structure levels. Like part decompositions, object classes can in principle be used to add an extra hierarchical layer to the model database and are thus attractive from a purely combinatorial point of view. Another pragmatic reason for constructing object class models is that answering the question "How many cars are there in this parking lot?" may be much more appropriate than answering the question "How many 1985 Chevy Sprint are there in this parking lot?", at least for certain applications.

A practical problem is to define a precise and computable notion of primitive shape similarity. Low-level parametric models such as triangulations do not appear particularly well suited to defining object classes. For medium-level parametric representations such as cylinders, cones and superquadrics, there is a natural notion of shape similarity (or equivalently, shape difference) based on the distance between primitive instances in parameter space.

In the case of more complex parametric representations (e.g., algebraic surfaces) and rigid generative models (e.g., solids of revolution or SHGCs), shape similarity measures can be based on certain abstractions (invariants) of the primitive parameters. In both cases, it is not clear how well these similarity measures capture the intuitive notion of class (e.g., two objects with very different parameters may look very similar).

In general, it is likely that, as the generality of the primitive representation increases (e.g., non-parametric generalized cylinders), the overall part structure will become more important than the actual primitive parameters, so that rough, qualitative shape similarity measures may be appropriate.

4.4 Generality vs. Effectiveness

Although the representations discussed in this section are all based on some notion of shape primitives, they operate at widely varying levels of abstraction. Therefore, it is important to understand which level should be used for specific applications.

Object recognition from a large library of models probably does require object classes based on relatively general shape primitives. Medioni's work on recognizing objects using SHGCs is an example of such a situation. However, a number of obstacles must be overcome before such systems become practical. It is therefore important to recognize the merits of less general representations whenever possible. For example, less structured shape representations have been demonstrated successfully for single-model recognition and pose determination. Kriegman's work on recognizing objects by comparing the descriptions of their HOT curves is one example. Cooper's work on comparing algebraic invariants computed from the polynomial approximation of patches on the object's surface is another. Those techniques are maturing quickly and are likely to be practical in the near future.

Finally, many applications do not require general, unconstrained recognition. These include registration problems in which the goal is to compute the precise pose of the object assuming complete knowledge of the object model. They also include applications such as surface checking, part mating and quality control. In this case, the goal is to compute a quantitative measure of shape similarity, not to identify an object. For those applications, low-level representations based on, for example, surface triangulations have been shown to be extremely effective, and they are at the core of many of the industrial applications of computer vision (see Besl's work). Moreover, successful attempts at building systematic algorithms for manipulating digital surfaces, such as Taubin's work on surface smoothing, have emerged in support of those surface representations.

5 Panel 3: Physics-Based Approaches

The third panel discussed the role of physics-based approaches in object representation and computer vision. The panelists were Terry Boult (moderator), Octavia Camps, Ari Gross, Dimitri Metaxas, Chahab Nastar, Alex Pentland and Demetri Terzopoulos.

5.1 General Comments

Physics-based approaches typically model shape as a mechanical system submitted to forces that reflect material properties as well as smoothness and image constraints. These methods have been used with a great deal of success for modeling very complex objects whose shape may vary over time (e.g., anatomical surfaces, see Nastar's work).

Further, the mechanical interaction between objects and their environment, and even some modeling of perception and action, can be captured by physics-based techniques (e.g., Terzopoulos' work on fish simulation). In fact, the computational tools used in these approaches can be used in visual tasks without necessarily modeling any actual physical process (e.g., Pentland's use of finite element methods for object recognition).

Accurate reflectance models which go beyond the classical Lambertian model also fall under the physics-based umbrella (e.g., Nayar's generalized Lambertian model). Thus it is clear that the galaxy of physics-based approaches has a lot to offer to computer vision, and, as reflected by the panel discussion, it is important to clarify the interplay between the different aspects of these approaches in order to take full advantage of their potential.

Accordingly, we examine in the next section the relationship between methods whose goal is to actually model a physical process and techniques which use physical analogies simply for the computational leverage they offer. We then address the issue of model complexity, which must be assessed before physics-based approaches find a wide use in recognition tasks.

5.2 Physical Analogies vs. Actual Physical Processes

In the early days of physics-based vision research (e.g., in Terzopoulos' Ph.D. thesis), there was no attempt at (and of course no claim of) modeling actual physical phenomena: instead, it was realized that physical analogies (e.g., representing surface points as massive particles submitted to image forces) allowed the use of the very powerful technology developed to support the modeling of physical processes. In particular, physical analogies give a natural framework for merging various types of constraints (e.g., smoothness and image constraints). Furthermore, efficient computational techniques, such as finite-element methods, have been developed in numerical analysis to solve the corresponding variational problems. These techniques are typically much more powerful than the gradient-descent least-squares techniques commonly used in the vision community.

As mentioned earlier, this pragmatic approach to physics-based vision has been extremely successful in modeling very complex surfaces. Examples of applications include medical imaging, but also modeling of man-made objects (e.g., Metaxas' work) and natural ones (e.g., Pentland's and Sclaroff's work on modeling fish, and Ikeuchi's and Hebert's work on modeling rocks, faces, etc.).

Recently there has also been a push toward actually modeling physical processes. An obvious motivation is the idea that the more accurately the model reflects reality, the better. This is certainly true from a simulation point of view:

to model the expression of a face for example, it is useful to model the geometry, material properties and dynamics of the skin and muscle layers. Another example is Terzopoulos' simulation of schools of fish, that swim (the interaction between the muscle actions, fin motion, gravity, water reaction, etc. is modeled), see (forming images through ray tracing then interpreting them with vision algorithms), follow each other, eat, etc.

As shown by this example, physics-based approaches are by no means limited to modeling object shape. In fact, an obvious domain of application of physics-based methods is in modeling the appearance of objects. More precisely, physics-based techniques may be used to predict the appearance of an object under different illumination conditions. As mentioned earlier, modeling appearance in addition to geometry is key to building robust vision systems.

Here again, there are two main approaches. The first one uses ad-hoc models developed for pragmatic reasons. For example, Lambert's and Phong's models are simple approximations of matte and specular surface reflectance. These models cannot handle interreflections, shadows, or surface roughness. In limited domains however (convex objects, say, illuminated with a single distant light source for example), both models have been used with some success.

The second approach is based on modeling the physical interaction between light, surfaces, and bodies. This approach has been used successfully for modeling smooth and rough specular surfaces (see the work of Nayar, Ikeuchi and Kanade, and Healey), smooth matte surfaces with interreflections (e.g., Forsyth's and Zisserman's work and Nayar's work), and rough matte surfaces (Nayar's and Oren's work). These models have been used for recovering the shape of real objects in laboratory conditions. At the same time, good physical models of color and polarization have been developed and used in various visual tasks (see Healey's, Shafer's, and Wolff's work for example).

The main advantage of the models developed using the empirical approach is that they require a small number of parameters which can be easily manipulated. By contrast, it is difficult to construct abstractions of accurate but complex physical models for use in high-level vision tasks.

5.3 Model Complexity

Whether based on physical analogies or actual physical processes, most physics-based models are defined by fine-grained discrete structures whose behavior is governed by complex differential equations. This makes it difficult to use these models in more abstract visual tasks such as object recognition. More precisely, the large number of parameters and the complexity of the relationships between those parameters makes it difficult to compare two object models.

Model complexity is a less critical issue in applications in which human intervention is acceptable in order to aid in building object descriptions. Object models for rendering, virtual reality, etc. fall in that class of applications. This is the approach favored by, for example, Terzopoulos for the use of complex models. Although recognition is the natural application for most object representation research, it may not be so for physics-based models.

6 Panel 4: Summary and Recommendations

The purpose of the fourth panel was to summarize and synthesize the findings and discussions of the workshop. The panelists were: Tom Binford, Katsushi Ikeuchi, Jitendra Malik, Joe Mundy, Jean Ponce (moderator) and Linda Shapiro.

6.1 General Comments

It seems clear that there is no universal object representation for all aspects of computer vision. Since most of the discussions, and a large percentage of the presentations, during the workshop focussed on object recognition, it seemed reasonable to use this final panel to synthesize the ideas about representation for recognition that have emerged during the workshop. Accordingly, this is also the focus of the remaining sections of this report.

At the same time, we should keep in mind that there are many other areas of computer vision, with possibly different representation needs, for example, digital libraries, inspection, medical imaging, robot navigation, video annotation and virtual reality. Some of these domains are relatively new, and it may be a bit premature to say that some prominent representation schemes have already emerged. This obviously does not mean that representation research in these areas should not be done.

6.2 Mature Research Areas and New Trends

Certain areas of research in object representation for recognition are mature; the corresponding problems may not be fully solved yet, and their partial solutions may not have been fully integrated in working systems, but the basic technology is available. A good example is 3D pose estimation, a very important problem in CAD-based vision and robotics: most object representations, including discrete point sets, polyhedra, superquadrics, algebraic surfaces, solids of revolution, various types of generalized cylinders and deformable surfaces, support pose estimation from range and/or silhouette data, either in closed form or through various forms of linear and non-linear optimization. Formal comparisons of existing techniques are beginning to appear, and research in this area may have matured enough to move from inventing and evaluating new representations to actually integrating them and the corresponding algorithms into working systems.

It is perhaps interesting to note that there now seems to be a consensus -reflected by the workshop discussions- about the fact that both two- and three-dimensional data are important for three-dimensional computer vision: depending on the application, one may be more appropriate than the other, but the difficulty of object recognition is in the problem itself more than in the dimensionality of the data.

At the other end of the spectrum, other research areas are not as mature but are undergoing fast and exciting progress. For example, while most of the work in object recognition relies on geometry as its main reasoning engine, recent research has demonstrated how intensity and color information could be used to guide the recognition process, using new spline-based representations of object appearance. Another example is artificial intelligence: with the recent emphasis on formal, rigorous approaches in both the artificial intelligence and computer vision communities, it may be time for AI to play a renewed role as we progress toward higher-level reasoning. However, it may again be premature to try and assess what constitutes a good representation in these new areas.

Somewhere in the middle ground, there are certain ideas that are now emerging, and whose promise and problems we can probably assess in the near future. This is important because model- based vision must go beyond pose estimation and into actual object recognition: for model databases containing thousands of objects, we cannot afford to try every model, estimate its pose, then verify its presence in the image using the estimated pose. We must also tackle the difficult problems of extracting the relevant information from images (segmentation), automatically constructing the object models, indexing in sub-linear time the model database and eventually integrating the corresponding modules into working end-to-end recognition systems.

In the remainder of this section, as in the actual panel, we focus on the following areas: segmentation, invariants and quasi-invariants, parts and system issues.

6.3 Segmentation

Most of the research discussed during the workshop relies heavily on strong primitive shape assumptions (objects being described by say, algebraic surfaces or superquadrics). Geometric representations that rely on weak shape assumptions (e.g., representations based on the medial axis, on the relationship between surface parabolic lines and contour inflections, or on epipolar surface parameterization) may not provide constraints as powerful as the representations relying on strong ones (e.g., assuming that the observed objects are piecewise solids of revolution), but they are a priori more general.

Between the two extremes, certain representation schemes rely on strong but generic shape assumptions (e.g., non-parametric generalized cylinders, as intended in their original definition), as first identified in Panel 1. While segmentation remains one of the major unsolved problems in computer vision, it may be a domain where generic object representation can play a fundamental role; more precisely, general assumptions on the shape and nature of the objects observed may be used to quickly verify segmentation hypotheses. Various systems now include elements of this approach, and they perform reasonably well for moderately complex images with little texture. Generic shape models could be augmented by statistical models of texture for example.

6.4 Invariants and Quasi-Invariants

Affine and projective invariants have been used quite successfully for recognizing various classes of objects (including discrete planar point sets and general two-dimensional shapes, bi-laterally symmetric objects, solids of revolution and various types of generalized cylinders) from a single image. So far however, there has not been comparable success for general three-dimensional shapes. Of course, what is a "general" shape is an ill-posed question, but certainly arbitrary discrete point sets, polyhedra, or solids bounded by piecewise-smooth surfaces seem to qualify as rather (maybe too) general. In fact, discrete point sets in general position do not yield monocular projective invariants, and this fact has been the source of some objection to the use of invariants in object recognition.

There are at least two possible answers to that objection. The first one, based on a concept introduced by Tom Binford several years ago, is to replace invariants by quasi-invariants, defined informally by properties of the image features that remain approximately true over a large set of viewpoints. This is an extremely attractive idea and various types of quasi-invariants have been found for objects such as pairs of line segments or certain classes of generalized cylinders. However, most quasi-invariant properties have been established empirically only for individual instances of object classes (for example, for a particular instance of a straight homogeneous generalized cylinder, or for a particular pair of line segments). It is important to assess the power of these quasi-invariants in a more analytical manner, for classes of objects rather than instances. It is also important to give sharper, operational definitions of quasi-invariants that everyone agrees upon and can use to develop new recognition algorithms.

The second answer is to model objects as patchworks of primitive shapes that do admit invariants. This brings us to the more general notion of representation by parts.

6.5 Parts

As seen in the previous sections, part-based representations may be the key to the general applicability of invariant or quasi-invariant representations to object recognition. More generally, from the discussions in the second panel emerged a clear need for a part-whole object representation: the shape, structure, or functionality of a complex object simply cannot be captured by a single primitive instance, whatever class the primitive may belong to.

Of course, deformable surface models have been very successfully used to model complex shapes, compute their pose from image data, or track their deformation in sequences of images, but they have not yet proven to be as useful for actual object recognition.

On the other hand, as demonstrated long ago by Marr and Nishihara, we easily recognize animals from stick figures, and the rough shape of a person is well captured by a hierarchy of cylinders (and probably, for that matter, by a hierarchy of any simple shape primitives). There is another pragmatic, combinatorial reason for using part-whole object descriptions: because of their hierarchical nature they afford divide-and-conquer indexing mechanisms with sub-linear performance.

There are two major approaches to defining parts: as the pieces of an object which are well approximated by some primitive shape (e.g., as proposed by Pentland, a superquadric surface), or as the pieces of an object which are separated by some primitive discontinuities (e.g., as proposed by Hoffmann and Richards, by a concavity). Of course, these two views are not incompatible, e.g., as proposed by Binford, a part may be defined as something which is approximately a generalized cylinder whose cross-section undergoes smooth deformations, so that cross-section area discontinuities delimit separate parts (i.e., two connected co-axial cylinders with different radii will yield two parts). Very different approaches are also possible, e.g., Mundy's notion of parts as repeated, or at least similar, sub-structures.

Whatever view is taken, we need to be careful to give a rigorous operational meaning to our definitions. For example, what does it mean to be "well approximated"? Certainly any smooth object can be locally approximated rather well by a polyhedral mesh, a superquadric surface, a polynomial patch etc., but parts should probably capture more global information (for example the relationship between surface elements). If parts are generalized cylinders, how do we find them before we find the discontinuities separating them? Should we use surface-based or volumetric notions of parts? How do parts manifest themselves in images? It seems clear that much work is required to clarify these issues.

Another important point is to re-iterate that, except in very special situations, (parts of) real objects will not be exact instances of simple primitive shapes; for example, a person can be approximated rather well by a simple hierarchy of cylinders, but the limbs or the head are not cylinders. Nor are they superquadrics, algebraic surfaces, solids of revolution, etc. In particular, if we are to use simple primitive shape descriptions as building blocks for shape representation, we must be careful that the "good" (invariant or other) properties of these primitives still hold reasonably well for shapes that are reasonably well approximated by the primitives. This is, in a sense, reminiscent of quasi-invariants: a good property should remain approximately true over a range of shapes "centered" about the prototypical one. Of course, this has to be quantified and, we need again precise, operational definitions as well as analytical results.

6.6 System Issues

As noted earlier, it is time to actually construct end-to-end recognition systems integrating the mature part of object recognition technology. It is also time to plan for future integration of leading-edge research. This will be difficult in a purely academic context because of the short- term, periodic structure of the prevalent single investigator cum graduate students paradigm. Although the lifetime of industrial products is a priori longer, a focus on short-term benefits may also hamper the long-term development of vision technology in a purely industrial context.

This suggests multi-university and university/industry collaborations, which would give system- and idea-building a longer lifetime.

To some extent, the ARPA Image Understanding Environment is a successful example of this type of interaction Similar initiatives are taking place in the computational geometry community for example. All these efforts are oriented at developing basic software libraries (of course they include a major representation component for the design of common data structures). It seems that parallel system-level efforts with longer-term perspectives would stand a real chance of fulfilling the promise of object recognition technology.

Finally, for system initiatives to really succeed, it appears necessary to assess the limitations of current representation/recognition schemes and in general to evaluate the merit of these schemes according to some objective criteria. The next section is an essay by David Forsyth that explores these limitations.

7 What Current Recognition Systems Cannot Do

The conclusions of the panel reported above show clearly that the issues of recognizing objects and modeling them cannot be separated. More precisely, the requirements for building effective object models are dictated by the performance of the available recognition systems. Therefore, understanding the needs and limitations of current recognition systems is critical in understanding which object models should be used. In order to address this need, this section, added after the workshop but motivated by the panels, is an attempt at drawing an inventory of the limitations of the current recognition.

The state of the art in object recognition is best understood by assessing the kinds of systems that can be built with existing technology, because this view most clearly identifies and evaluates the difficulties with what is known. Components of object recognition systems tend to have limited relevance on their own. Furthermore, redundancies appear to be widespread in the whole process of recognition so that it is difficult to assess the significance of the difficulties that particular approaches encounter, if the approach is viewed outside a wider context. It is hard to assess quantitatively what represents good performance, but there is a reasonably widespread sense in the community that "good" systems have large model bases, can operate on "complex" pictures and offer insights that can be used to generate better approaches.

Although the last 10 years have seen significant advances in the capacities of object recognition systems, very real problems remain. Typical systems of ten years ago could manage model bases of one to two objects in complex scenes. These systems used members of a family of object recognition algorithms that match objects to model bases by different kinds of searches over correspondences between object points or lines, and image points or lines. Such algorithms, including *interpretation trees, alignment, Hough transforms and pose clustering,* cannot construct abstractions of object properties, and so are ineffective in managing large model bases.

More recent systems can handle larger model bases - though typically, of more constrained objects - with methods which often have some hope of scaling. It is widely believed that there are three main threads in object recognition - *representation, control* and *matching* - though the relative significance of these ideas is disputed. Recent matching strategies involve constructing some form of object representation based on image information, and matching that representation to the representations constructed from the model base.

The representations used are expected to abstract what is distinctive and reliable in image measurements so that it can be used in matching. Research on building representations has concentrated almost exclusively on geometrical properties. The geometry of the projection of an object in an image depends strongly on viewpoint, and strategies for overcoming this dependency are essential; there are two main strategies, yielding either *object-centered* or *view-centered* representations. Ideally, object -centered representations are constructed in such a way that the representation appears in the same frame and form for distinct views of an object, so that matching is referred to, and occurs in, a fixed frame. This has the advantage that differences between a representation recovered from an image and the representation of the target occur because the two have distinct geometries. Typically, constructing such representations requires both inferring 3D structure from image information, and inferring an appropriate coordinate frame for the structure.

Present object-centered representations can represent only a small class of objects, for two reasons. First, they must infer 3D structure from image information, and for single images this is possible only for object classes that satisfy strong geometric constraints. Second, even if depth data is available from other sources, the process of abstraction is usually interpreted as fitting globally constrained surfaces and then using the parameters of the fitted surface as a representation. However, the constraints that result from the limited class of surfaces handled are reflected as constraints on image outlines, so that segmentation can be simplified. As a result, systems that use object-centered representations can generally recognize rather simple objects in quite complex scenes. It is not known whether larger classes of primitives can be satisfactorily extracted in an object-centered fashion, or whether a more general representation process than fitting exists.

It is a commonplace that complex objects can be represented as a conjunction of simple parts. To date, part-based representations have been ineffective, because it is not known *how parts should be defined and so how objects should be decomposed* and it is not known how a part based representation should be matched. There are a number of current definitions of parts, including an operational definition (the parts are what a part-detector finds), a view based definition (parts are defined purely by local image properties) and a geometric definition (parts are defined by 3D events). It is not known *whether canonical or stable decompositions are accessible from image information alone.* Matching part-based representations remains mysterious. If a particular part distinctively identifies a particular object, then recognition should proceed by using it as a salient feature, which is a relatively well understood process; however, in many cases, the parts to which an object decomposes are relatively nondescript - for example, a set of fingers into cylindrical joints - and it is the relationships between the parts that must be extracted and represented. It is not known *what kinds of relationships between parts can be extracted from images, nor how these relationships should be used to constrain recognition.*

In contrast, view-centered representations effectively match image information to an expanded model base, which consists of every significant view of the objects originally in the model base; the meaning attached to the term significant varies among investigators. This approach has the advantage that inferring 3D structure is unnecessary, but poses significant matching problems. View-centered representations generally lead to systems with defects different from those resulting from object-centered representations. Typically, existing systems find segmentation extremely hard, so that large numbers of objects can be recognized, but only in extremely simple circumstances, with little or no clutter present. A particular difficulty is that these systems cannot employ constraints derived from the model base to segment objects. Furthermore, a satisfactory definition of a significant change of appearance does not exist - the topological considerations used in aspect graphs lead to too large a set of cases to be tractable - and a series of operational definitions, some fairly complex, have had to suffice. It is not known *how pure view- centered representations can be used for abstraction, classification or partial recognition in any useful sense.* Matching view-centered representations requires one or another form of search, usually over a rather arbitrary metric on image differences.

Controlling the flow of activities in object recognition is not well understood. It is clear that as systems become larger, and scenes more complex, a better understanding of the relative worth of activities in a system and of the interactions between recognition hypotheses, will become important. At present, it is known that recognition hypotheses imply camera properties; these implications must be consistent, and this constraint has been used to infer camera properties and the relative pose of objects. Other forms of consistency include consistency in light-source position (which has been used in processing satellite images) and in occlusion where, given an object hypothesis, verification and grouping for other hypotheses in the region it occludes are affected. *The full extent of interaction between recognition hypotheses* is not known.

It is widely believed that some image events, such as parallelism or collinearity of line segments, are unlikely to be accidental and so are strong local grouping cues or strong cues to the presence of an object. This view is oversimplified, as it does not account for the context provided by either the model base or what is already known about the scene; for example, parallelism or collinearity are irrelevant if the model base contains only spheres. Furthermore, this context makes the widespread belief that grouping should occur only on the basis of local relationships between image features increasingly implausible. Image groups that reach higher levels of processing should satisfy stronger constraints, that are more typically global, for example, the constraint that they could possibly have resulted from something that lies in the model base. It is not known *how to formulate the constraint and the context that a model base places on grouping* nor *how to formulate the usefulness of a given activity in the context given by the model base and what is known about the scene.*

Acknowledgments This work was sponsored jointly by NSF and ARPA under NSF Grant IRI- 9407040, and by the CUNY Graduate School and University Center. The views and conclusions contained in this document are those of the authors and should not be interpreted as representing the official policies, either expressed or implied of NSF, ARPA or the US government.

8 Appendix

8.1 List of Participants

Paul Besl (General Motors)	Gérard Medioni (USC)
Tom Binford (Stanford)	Dimitri Metaxas (U. Penn)
Terry Boult (Lehigh Univ.)	Joe Mundy (General Electric)
Octavia Camps (Penn State)	Chahab Nastar (INRIA)
David Cooper (Brown)	Shree Nayar (Columbia)
David Forsyth (U.C. Berkeley)	Alex Pentland (MIT)
Patrick Gros (LIFIA-INRIA)	Sylvain Petitjean (CRIN-INRIA)
Ari Gross (CUNY)	Jean Ponce (U. Illinois)
Martial Hebert (CMU)	Stan Sclaroff (MIT)
Dan Huttenlocher (Cornell)	Linda Shapiro (U. Washington)
Katsushi Ikeuchi (CMU)	George Stockman (Michigan State)
Visa Koivunen (U. Penn)	Gabriel Taubin (IBM)
David Kriegman (Yale)	Demetri Terzopoulos (U. Toronto)
Jitendra Malik (U.C. Berkeley)	Richard Weiss (U. Mass.)

8.2 Workshop Panels and Panel Chairs

Panel 1: Appearance-Based	Panel 2: Primitive-Based
David Forsyth	Paul Besl
Patrick Gros	David Cooper
Dan Huttenlocher	Martial Hebert (moderator)
Shree Nayar	David Kriegman
Jean Ponce (moderator)	Gérard Medioni
George Stockman	Sylvain Petitjean
Richard Weiss	Gabriel Taubin

Panel 3: Physics-Based	Panel 4: Recommendations
Terry Boult (moderator)	Tom Binford
Octavia Camps	Katsuchi Ikeuchi
Ari Gross	Jitendra Malik
Dimitri Metaxas	Joe Mundy
Chahab Nastar	Jean Ponce (moderator)
Alex Pentland	Linda Shapiro
Demetri Terzopoulos	

Object Recognition: The Search for Representation

J.L. Mundy

General Electric
Corporate Research Center
Schenectady, NY
mundy@crd.ge.com

Abstract. Object recognition in visual scenes by computer has proven to be more difficult than anyone would have thought three decades ago, at the beginning of the research program to achieve this goal. A central issue for further progress is the design and effective implementation of an object representation which captures all of the requirements for description and perceptual organization. In this paper, the major issues surrounding the development of such a representation are established and placed in the setting of relatively recent discoveries in the philosophy of recognition and object classification. From this viewpoint, definitions for *representation*, *recognition*, *identification* and *classification* are established and related to standard approaches to object recognition in visual scenes.

The use of biological models and introspection as a source of design ideas for representation is discussed. It is argued that the most profitable source of ideas will emerge from an engineering approach, based on principles from geometric reasoning, photogrammetry. and signal processing. The role of context in object recognition is outlined with emphasis on its use throughout all of stages of recognition. The paper concludes with a description of a object recognition system, called MORSE, which embodies many of the principles derived from these philosophical considerations.

The trouble is that these ideas have lots of cross-connections. My explanations rarely go in neat, straight lines from start to end. I wish I could have lined them up so that you could climb straight to the top, by mental stair steps, one by one. Instead they're tied in tangled webs.

Marvin Minksy in Society of Mind

1 Introduction

1.1 Motivation

In spite of the best efforts of two generations of computer vision researchers[1], we still do not have a practical, robust object recognition system which can organize and label the objects in a 3D scene, based on the observations taken from a single intensity view. Even adding more views and the use of dynamic image sequences does not seem to provide a clear avenue to the construction of a successful recognition system for real world scenes. We should be asking ourselves: Why is the problem of object recognition so hard?

It has been hoped that a magic representation will be discovered, likely based on our understanding of the human visual system, which can solve all of the difficult issues. Indeed, much of the intervening research has promoted one representation or another, with claims that robust recognition is a consequence of imposing the representation on the image and scene data.

In my view, it is time to step back a bit from the battleground of representations and take another look at the foundations of our approach. It seems that the difficulty lies in our conception of the problem. We need to redefine what recognition means and how it relates to effective computations which can be made on intensity data as well as higher level representations of geometry and function.

In this quest, I decided to review the major philosophical discoveries concerning the nature of recognition. This plan is supported by Ayer[2] who states:

> It is the philosopher's business to give a correct definition of material things in terms of sensations.

The focus in this paper is on the major philosophical ideas concerning the identity and classification of objects. The plan is to develop a philosophy of machine perception which will clarify the goals of recognition and establish the principles by which a representation for recognition can be defined.

The most difficult aspect of this program is establishing a basis for object classes according to the nature of sensory information. Many of the great philosophers have struggled with this exact question and we should be aware of their conclusions and the consequence for computer recognition systems.

1.2 Representation and Duality

Representation is a central issue in all fields of artificial intelligence. Is is now realised that the classic struggle over the mind-body problem can be laied to rest since formal algorithms can provide a basis for intelligent activities without referring so some non-material mind force. A representation is in the link between

[1] Perhaps the earliest substantial effort on the recognition of 3D scenes was carried out by Roberts[1] more than thirty years ago. His work focussed on polyhedral objects in uncluttered scenes.

mind and the mechanical underpinnings of thought. This point considered by Dennett[6] to be one of the major achievements of artificial intelligence research.

Dennett sees formal representation as a key breakthrough in avoiding the duality of mind and body which has dominated most of the history of philosophical thought concerning intelligence and perception. At a high level, we can represent recognition by categories which seem to be conceptual and associated only with the mind, say linguistic concepts. But these concepts are in turn decomposed into ever more formal and mechanical representations until finally it becomes obvious how to implement the required data structure or process in a direct link to sense data. There is no remaining need for a vital force or *mind.*

1.3 Key Questions

The following is a list of key questions we consider in the sequel.

How should the act of recognition be defined so that a computer can do it effectively? I will show that class definitions are arbitrary and we can define categories, and classification methods, to satisfy the limitations of image processing and geometric computation. Recognition is something different, it involves both figure/ground isolation and the essence of an individual object. Here is where the physical world imposes complexity beyond the scope of existing representations.

How does a specific object instance generate an abstraction which generalizes the instance? The process of abstraction from an individual is the really strange aspect of recognition. In some way, an object induces a large set of similar instances. Plato thought this happens because there exists a prior ideal form that the object instance participates in [2].

Does an abstract class define a procedure for recognition? The existence of abstractions of sensory experience is obviously of great advantage for summarizing and organizing data. Is there any reason to expect that the categories used for human purposes can be of benefit to the design of computer algorithms for recognition? Perhaps most of the interesting stages of recognition are pre-cognitive and therefore prior to our abstractions.

Is recognition contingent? How well do we have to account for observables to have valid recognition? It has been demonstrated by Hume[3] that all empirical conclusions are *contingent.* That is, a theory about some physical phenomena can be upset at any time when a counter-example or inconsistency arises. Therefore any philosophy of recognition has to define the principles for an adequate interpretation.

[2] μετεχειν - To partake of or to share in.

Does an assembly of parts have identity beyond the parts and their relations? The whole issue of the decomposition of a structure and the role of the parts in recognition has been debated since Plato's time. Is the idea of part-whole decomposition of any real significance to computer vision recognition algorithms?

What is context and what is its role in recognition? It can be easily demonstrated that a piece of image is completely uninterpretable outside of a larger context. Context is hierarchical in that a simple feature can have many meanings which are induced from structures formed from such features. A composite structures can itself be part of a more complex structure, and so on. Ultimately, the evolving structure becomes distinctive enough so that a single interpretation remains. Our investigation must include a clarification of the role of context and its interaction at each stage of recognition.

How should we treat different representations (models) for the same image observables? Some attempts have been made to represent organic shapes by mathematical data models such as superquadrics, algebraic surfaces or their image outlines by snakes of various types. With sufficient degrees of freedom, it is possible to achieve any degree of accuracy in representing a single instance of such objects. In what way are these representations useful for recognition? How does the infinite space of such representations describe the underlying individual and the class to which it belongs?

2 Approaches to Representation

2.1 What is Representation?

Before considering the problem of recognition itself, I will lay out the definition of *representation* itself. After all, if we are searching for something, it is important to be able to define what it is. A similar quest is underway in the AI community as a whole. The article by Randall Davis *et al* is particularly illuminating[5]. Davis lists five rôles for a knowledge representation:

1. A surrogate or a substitute for reality. A formal model which behaves the same way reality does.
2. A conceptual framework. A way of thinking about the world. Perhaps just as important, a statement about what shall be ignored.
3. A structure to support reasoning about the world. The rules of logical inference enabled by the representation.
4. An efficient organization of the data encompassed by the representation. A control structure for computations using the representation.
5. A medium of man-machine communication. Terms of the representation should be understandable by humans.

The last four rôles are central to our problem of defining a representation for recognition. We want the representation to organize computation efficiently and to provide a medium for explanation and description. However the first rôle is not completely applicable, since we are not necessarily trying to represent how objects actually *are* in the world. Instead, we should be representing the world so as to enable machine recognition.

Rôle four emphasizes that representation must include a specification of *process* as well as declarative elements such as geometry and geometric constraints. For example, feature grouping can be viewed as a process where partially formed groups *actively* seek additional elements. The management of hypotheses is clearly dominated by search and refinement control mechanisms. This point is obvious, but must be emphasized since many recent proposals for representations for recognition do not make clear how the construction or interpretation of a representation is to be processed and controlled.

Let us now examine the various choices we have for motivating the design of a representation for recognition.

3 Designing a Representation

Dennett[6] classifies approaches to constructing a system for recognition in terms of the motivation underlying the design of system:

1. Biological - How do humans do it.
2. Introspective - How we think we do it.
3. Kludge[3]. - Works well, but not according to formal principles. Constructed by repairing failures and tuning performance.
4. Engineering - An efficient, simple implementation based on formal principles.

In my view, there is little proven value in motivations 1 and 2, as I shall now argue.

3.1 Biological Models

The great attractiveness of the biological motivation extends beyond scientific and engineering motivations – almost a kind of mysticism. The usual claim is that the best vision system around is man: Just look under the skull for the wiring diagram. The claim is that we can discover the biological principles underlying visual processing and then simulate the processes.

The traditional AI approaches, based on the use of logic and mathematics to carry out symbolic processing is under severe intellectual attack by the rapidly growing community of biologically-motived artificial intelligence[9]. Their primary attack is that formal AI methods, including computer vision, have not worked in spite of years of generous funding. Instead, the way forward is to simulate biological principles, such as neural circuits, natural selection (genetic

[3] My term, not Dennett's. He uses the word *performance*

algorithms) and artificial life. One approach even involves constructing circuits which include live neurons!

In my view such enthusiasm is very premature, if ever ultimately sound. Successes in modeling simple reflex networks will not soon have much to say about high level perceptual functions, such as recognition. For example, an extensive and innovative approach to neural modeling has been carried out by Edelman[10] who introduces the additional idea of Darwinian competition for effective organizations of neural circuits. His successful simulation of these ideas, Darwin III, is able to learn effective reactions to sensory stimulus. However, in speculating about a simulation which could achieve scene analysis capability, Edelman says:

> It may be possible to construct such artifacts someday, but right now it is so unlikely as to be unworthy of too much reflection.

There may eventually be some value in observations made concerning visual processing and associated verbal behavior in the presence of brain injuries[7]. Also complex global patterns of neural activity have been revealed by PET[8] imaging of the brain while the subject carries out various perceptual tasks. These experimental results demonstrate a modular structure to the perceptual process, though it seems unlikely that the complex associations revealed by these experiments can be of direct benefit to our immediate goal. The results can be explained by a multitude of theories which bear little or no predictive ability. Worse, there is no real basis for a design which can extend down to deterministic and formal calculations.

Suppose that the biology-based computer vision path is followed. Such studies will by necessity use the scientific method to validate models and their success in predicting new phenomena. Science proceeds by imposing mathematical and logically sound models onto the empirical phenomena. The final success would be to explain visual processing in terms of organic and biological mechanisms. But such a biological interpretation is needlessly constraining and not likely to be the best approach for designing computer implementations.

For example, chess programs have achieved nearly championship status by using approaches very unlike human chess strategy. Similarly, the most successful geometry theorem proving programs are based on algebraic inference mechanisms, like the Groebner basis[11] which do not bear any relation to human geometric theorem proving. The definition of a circle can be captured in a few lines of algebra, but is it likely that the concept can be learned or effectively represented by neural models?

3.2 Introspection

What about introspection? This approach seems very appealing since we can imagine extracting a useful representation by understanding of our own visual processing. Certainly much has been written and discussed about our subjective processes in visual scene analysis. However there is little evidence to support the claim that such introspection is actually connected to visual processes themselves. In discussing the description of robotic assembly of block structures, John McCarthy[12] points out:

> The difficulty in expressing such facts is indicated by the limitations of English in expressing human visual knowledge. We can describe regular geometric shapes precisely in English (fortified by mathematics), but the information we use for recognizing another person's face cannot be ordinarily transmitted in words.

Dennett[13] gives a good example of the poor extensibility of introspection. He describes the experiments of Shepard and Metzler[14] where subjects are asked to mentally compare two objects which might be rotated copies of a single object. The subjects reported that they "visually" rotated the objects in their mind to see if they could align. A typical case is illustrated in figure 1. The experiments

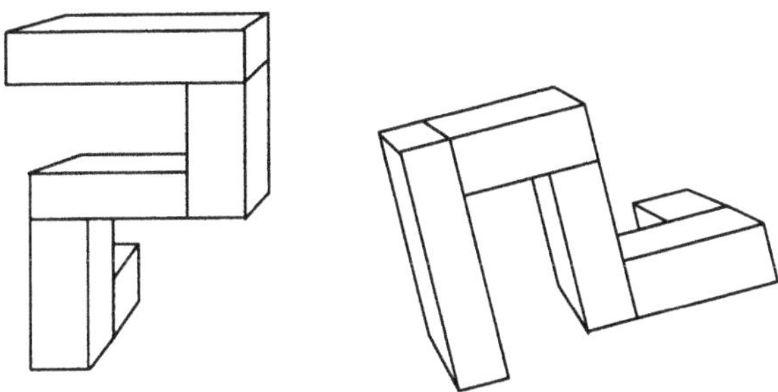

Fig. 1. A sample of the visual alignment task.

showed that as more rotation is required to align the objects, the response time of the subject increases in rough proportion to the required rotation.

Dennet then illustrates that similar visualization tasks don't seem to have such an introspective quality. Consider the task shown in figure 2. Here the goal is to report whether or not the cross can be seen through the square aperture. Subjects reported that they tried to mentally *rotate* the structure but that didn't work. Dennett concludes that the task is harder and we don't have a conscious mechanism for doing it which can be articulated. There are many examples of the failure of introspection to account for 3D perception. Consider figure 3, which shows two identical copies of a cube. We can easily see that they are equivalent, but there is no conscious process of analysis we can report. The success of applying introspection at lower levels of visual processing is even more remote. We are hard pressed to articulate how we organize the boundary information in an image into a coherent and segmented object. For example, consider the standard problem of perceptual grouping. There is little we can say about the conscious mental process we use to discover the bicycle in David Lowe's famous example of grouping shown in figure 4[15].

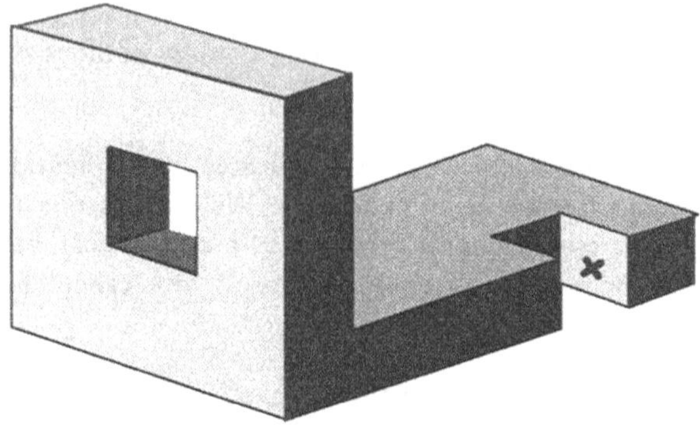

Fig. 2. The introspective visual task is to tell whether or not the cross can be seen through the aperture.

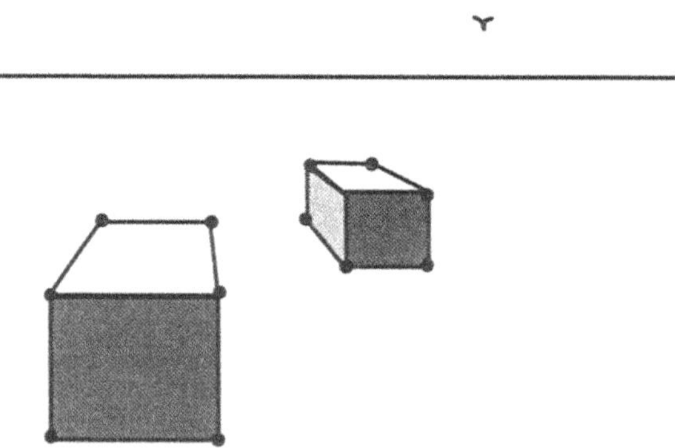

Fig. 3. Two identical polyhedra, under perspective projection with the horizon line shown.

Fig. 4. An example of perceptual organization given by David Lowe

3.3 Kludge vs Engineered

It is easy to dismiss the kludge approach, even though many, even practical, vision systems have been designed by trial and error leading to quite respectable performance. The main issue is that a technical discipline cannot be firmly established on anecdotal experience concerning specific vision applications. Indeed, the rather dismal experience of machine vision companies during the 1980's demonstrates the difficulty of extending a system developed for one application to another. For example, the successful implementation of a vision system for inspecting machine parts bears little relation to the algorithms required for lumber inspection.

Eventually, it is necessary to extract general principles which can be used to construct new systems to satisfy new requirements and to predict performance. This information is the sort which can be taught in courses and transcends current technology limitations. The goal is to define the act of recognition in reasonably formal terms and integrate this formalization with the representation of data which can be reliably and *predictably* extracted from visual images.

I hold that the system problem has been largely ignored in the development of representations for recognition. It is necessary to have an extensive software environment to handle the complexity of real scenes and to provide efficient processing of geometric relations among thousands of elements. Clearly, we need to transcend the amount of software development which can be achieved in three years by a graduate student. Such an integrated system which can support collaboration among multiple research groups will not come into existence without a formal set of agreed principles. Is visual recognition an easier experimental and computational problem than high-energy physics? If not, then we should be thinking in terms of clearly defined engineering principles and large collaborations.

4 What is Recognition?

There is often considerable confusion about various terms which often appear to be synonyms for recognition, or closely related to recognition. Recognition is obviously based on prior experience[4]. The implication is that recognition is intimately associated with memory, a storage mechanism for past sensory experiences. But: What is the difference between recognition and classification, or between detection and perception? A careful study of the philosophical understanding of recognition has been carried out by Sayre[16] and he provides formal definitions for such terms which will prove very useful later.

Sayre defines *classification* as determining a property which all members of a class share, i.e. an *invariant* of the class[5]. This definition is clearly not equivalent to *recognition* or even a satisfactory formal explanation of part of the act of recognition. Some differences:

[4] Note that the word can be emphasized as RE-cognize, i.e., cognize again.

[5] I should point out that Sayre introduces this term and is not a result of bias by the author.

- People can recognize objects without actually being able to define the class invariants. As we shall see later, Wittgenstein shows that such invariants do not even exist for many categories!
- People can learn new classes which require new invariants not articulated before seeing the new object.
- Classification can be carried out independent of any perceptual process. Sayre gives the example of classifying eggs by attempting to pass them through hoops of increasing diameters.

The key conclusion is that recognition leads to new information, classification merely restructures the information at hand.

Sayre defines *recognition* as:

$$\forall O\ \forall x\ \text{Recognize}(O, x) \equiv \text{Perceive}(O, x) \wedge \text{Identify}(O, x)$$

Where O is some observer and x is some object. This introduces two new terms which we need to define.

Identification Sayre establishes the meaning of *identify* as establishing the link between an object and the description for the object. For example, we can ask: Who killed the butler? When the murderer is identified then we can answer the question. In this case, the description is available but the object is initially unknown. Conversely, we can ask: What did that man do? – and deduce that he killed someone by observing the blood on his shirt.

It might seem that *identify* is very close in meaning to *classify*, however, there is one important difference. In our definition, we do not consider classification to be error-prone. By merely attaching a unique class attribute to an object, the job of classification is done, albeit poorly. Identification can be wrong, since we insist that the correct description should be associated with an identified object.

Perception Perception corresponds to the act of figure/ground isolation. An object is extracted from the background and grouped into an individual entity. For example, it would be correct to say that a person *looked* at the picture and *perceived* the vase. Looking is a process and perception is an achievement which results from looking. Note also that perception does not require having seen an object before.

Taking these three concepts together, visual recognition is an achievement, rather than a process, which identifies an organization of image features with a description. The description can be reduced to a short-hand label using classification. I will adopt this definition in the following discussion. However, the concepts of description and classification certainly need further development.

As an interesting aside, Sayre also defines the term *detection* as:

> Simulating a visual event so that all distinguishable manifestations of the event are produced.

If we can simulate the outlines of objects in a scene with adequate fidelity, then each object has been *detected.* Detection, or what in modern computer vision is called *verification*, is a powerful means for achieving recognition[17]. In this approach, a geometric model is used to predict image features, such as edgel chains or line segments. To the extent that the projected model aligns well with image features, an instance of the model is declared to be present in the scene.

The main objection to detection (verification) as a complete solution to recognition is that an object model is necessary to simulate the visual appearance of an object. In the case of a large library of models, a large combinatorial space of hypothesize and test states must be searched to find the best verification. We seek a representation which provides guidance at *all* stages of visual processing and hypothesis formation. Also verification does not appear to provide a fruitful avenue to understanding the problems of class abstraction and perceptual organization.

5 A Review of the Philosophy of Class

Following the definition of Sayre for recognition, the purpose of a representation for recognition must support both the organization of features into a perceptual segmentation of the scene, but identify the individual entities. The latter problem of identification leads us to inquire about the basis for object classes. In this regard, it will prove useful to review the evolution of ideas about descriptions and categories.

5.1 Plato and Aristotle

Plato introduced the concept of *forms* which are inate and a priori ideas that structure reality. Forms are independent of reality and are dimly perceived by the human mind. We owe whatever scientific truths we can acquire to our limited awareness of forms. It is interesting that Plato's thought evolved through a number of relationships between an object and its form. Two of his later interpretations are very similar to modern thought.

μετεχειν To participate. An object participates in an ideal form. The form establishes a framework for the existence of an object according to a specific class.

παραδειγμα To serve as an example or prototype. In this view, an object instance is a prototype for the entire class.

Both of these concepts presume that there is an ideal form to which reality conforms. Humans perceive and identify classes because they exist prior to reality. In Plato's view, reality is a poor imitation of the ideal forms which is made even worse by the defects in our sensory mechanisms.

Aristotle was largely in agreement with these ideas but differed with Plato on the nature of reality. In Aristotle's view a form actually exists in the world and exists in addition to the particular instances of a class defined by the form. An object has both substance and form. This leads to the modern notion of entity and attribute, or subject and predicate. Aristotle's terms for subject and predicate are:

'υποκειμενος Literally to lie hidden beneath. Referring to substance.

κατηγορουμενος To signify or indicate. Referring to an attribute.

Attributes may change, substance does not, and substance is the ultimate form of reality. The debate between Plato and Aristotle is about universals. Are universals only in our mind or are they reality itself? Aristotle defined a set of fundamental categories, some of which still dominate our thinking about classes[18]. Aristotle's categories are:

1. Substance: What a thing is, its essence Primary substance is an individual, Secondary substance is a species the individual belongs to. For example, John the individual is primary, John the man is secondary.
2. Quantity: How much or how large Quantity can be either discrete or continuous. Numbers are discrete. Geometry is continuous. Time is continuous. Quantities can always be partitioned. Quantities never have contraries.
3. Quality: What sort or type. Aristotle distinguished quantity and quality in terms of comparison. We say that two quantities are equal or unequal, greater or smaller. We say that two qualities are like or unlike. Aristotle gives an example of the qualities whiteness and blackness.
4. Relation: How is an object related to other objects.
5. Where: Position in space.
6. When: Position in time.
7. State: How Circumstanced.
8. Action: What is in process.
9. Affection: How passive or what suffering.

One interesting point is how do we decide if qualities are alike? It appears necessary to convert quality into quantity so that computer algorithms can carry out the decision. One approach to establish the quality of a property is classification by intervals or distributions on measurements of the properties. This approach leads to descriptions like those employed in qualitative physics or fuzzy logic[19]. For our purposes, it might be said that a quality is simply a poorly estimated quantity. This issue becomes central for functional classification where there is often no precise set of attribute values corresponding to successful functional performance. For example, what is a good shape for a vase?

In the dialog, Parmenides[4], Plato shows that abstractions like *part* and *whole* can be used in so many guises that no meaningful conclusions can be drawn. The following is an account of how Parmenides, a philosophical rival, shows how something is a part and a whole at the same time:

> When he (Parmenides) wanted to show that I was many he would say that I have a right and left side ... for I cannot deny that I partake of multitude; when on the other hand, he wants to show that I am one, he will say that we who are here assembled are seven, and I am one (of the seven)..

This problem does not give us much difficulty in the modern development of representations. It is easy to define a hierarchy of part-whole relations where an entity is part of one structure but itself is composed of parts.

However, I do have a real concern as to the utility of part-whole relations in recognition. Part decomposition is essential for designing and manufacturing. Also many natural objects can be usefully described and differentiated by a reduction into component structures. In the case of recognition, it is not clear that parts and their relations can provide a useful set of constraints with indexing power. Consider an airplane which has wings and a body and a rudder, ...etc. However, a wing is just a rough quadrilateral with little saliency, that is, such quadrilaterals can be part of a large number of objects. For many objects, visual decomposition into parts is only feasible after recognition has occurred. This issue is taken up again in the section on context[6]

5.2 Leibnitz

After Aristotle, there was much discussion about the nature of substance and attributes. Suppose we change one attribute or part of an object at a time. Is it the case that we still have the same the object? Mach gives the example of his desk which may be stained with an ink spot or have one of the legs replaced. But what if we replace all of the legs and the top? This line of thinking leads to the horrifying thought that an object can exist independent of its attributes or parts. The key issue is: How does a substance differ from its attributes? There are two possible schemes for establishing the identity of a substance:

a) There is a special attribute called *essence* which gives a substance its identity.
b) An individual is defined by the totality of its attributes, what Leibnitz calls a monad.

A corollary of b) is that if two substances have all attributes in common, then they are the same substance. In choosing b) Leibnitz was driven to the logical but unpleasant conclusion that relations between substances cannot exist, since for two substances to be related, they would have to share attributes which violates the unity of the monad.

In more modern terms, say in object-oriented programming, it is impossible to locally establish the equality of two class instances if the class contains pointers to other classes. Equality depends on the state of the associated classes which may in turn be related to other classes. The attributes, or slots, of a class must eventually be run down across the pointer chain to atomic terms such as numbers or strings. Finally, equality depends on the entire network of instances

[6] Section 6.

and pointers, so one might as well treat the network as a single isolated entity and rule out equality for individual instances.

5.3 Hobbes and Locke

Eventually, it became clear that part of the apparent confusion and contradictions with common sense in understanding substance and the classification of substance is caused by language itself. We can manipulate words by logic, but that does not guarantee meaning. For example, the term *round quadrilateral* is inherently meaningless. Hobbes insisted on the following chain:

Event $\rightarrow$ Sensation $\rightarrow$ Thought $\rightarrow$ Concept

In this view everything originates in experience. There are no innate ideas. The notion of quality or essence are those attributes in common with all instances. Common attributes are an abstraction of individual experience.

From a linguistic point of view we hold the concept of an *individual* for some classes but not others. For example we have individuals for the classes, car, tree, vase, but not for mass terms such as, lump of clay, snow drift, pool of water. The whole idea of defining an individual is troublesome. By Locke, the problem of individuals can stated in terms of new definitions. In the case of *qualities* the real and nominal essence of a substance cannot be distinguished. By nominal we mean the individual, i.e. a unique name. By real we mean the actual underlying essence which identifies an object with a class. However for *kinds* (mass terms) the individual essence can be defined by formal concepts, such as snow, grass, gold, i.e., in the case of gold, the individual essence is molecular weight.

This definition of individual essence does not provide any solution for individuals made out of *mass stuff*. For example, how do we define an individual *gold nugget*: Its weight or volume? This problem persists today in the attempt to describe individual natural objects where material or surface properties are more important to our idea of the object than its segmentation into a distinct shape.

5.4 Hume

These discoveries concerning the limitations of language and ultimately the limits of human reasoning itself lead to a rapid rise of the empirical point of view which is dominant today. This quote from Hume in 'An Inquiry Concerning Human Understanding' could characterize the paper review process of a conference like CVPR:

> If we take in our hand any volume; of divinity or school metaphysics, for instance; Let us ask, Does it contain any abstract reasoning concerning quantity or number? No. Does it contain experimental reasoning concerning matter of fact and existence? No. Commit it then to the flames. For it can contain nothing but sophistry and illusion.

In other words, the mind and its reasoning about metaphysics is no longer considered a source of truth, only opinion which is outside of the scientific method.

Hume also pointed out that establishing causality requires connecting events over time. Therefore any conclusion about causality is contingent. That is, the apparent cause-effect relation can cease to hold at any time. This conclusion can also be used to endanger the notion of objectivity. The essence of an independent individual object must endure over time and therefore is also contingent.

For us this means that a recognition act is always contingent. The best we can do is to find an interpretation of a scene which has the least ambiguity and contradiction with observed image events. We may eventually revise our interpretation, for example in the famous old-woman/young-woman reversal illusion, but there is no *true* interpretation.

5.5 Kant and Frege

Kant showed that human reason is limited to the manipulation of symbols. He asserted that there are other sources of true statements which can established through the senses. Kant identified two kinds of truth:

a Analytic - True by definition and manipulable by reason.
b Synthetic - True by virtue of the nature of reality, but not established by experiment, instead by intuition.

The categories by which we structure reality are synthetic. However Kant maintained that both sources of truth are true *a priori*. That is, they represent the fundamental reality of the universe, as revealed through human thought and sensation. The laws of arithmetic are an example of analytic truth, which Kant held to be true by observations of reality, i.e., synthetic. He claimed that we know that $1 + 1 = 2$ because we observe the gathering of one object next to another to make two objects.

Frege rejected Kant's notion of *a priori* synthetic truth. Frege showed that the laws of arithmetic can be derived by logic alone. Amazingly, all of mathematics can be shown to be just a tautological consequence of logical principles and a few definitions about number and equality. Mathematics is inherent in our linguistic framework. A quote from Eddington[20] captures the idea:

> We see our own footprints in the sand.

Our explanations of the world in terms of mathematics is an imposition of our language on reality.

5.6 Wittgenstein

Conversely, language is shaped by reality. Wittgenstein[21] shows that words *mean* in how they are used and nothing more. *Description* is made up of elements which are named objects and named attributes and relations. In particular, the descriptions we are interested in for object recognition are constituted by names for objects and the object attributes which can be readily measured from image data. Such descriptions can be generalized by specifying relations between attributes and removing attributes which are not relevant to recognition.

However, attribute subsets are not a complete mechanism for recognition descriptions. Wittgenstein showed that the idea of basing classes on common attributes cannot account for many of the linguistic categories which are in common use. In his view, a class is a network of similarity relations, i.e. a network and not a central hub of common properties. For example, the class *game*. There is a wide spectrum of activities we associate with a single class symbol, but they are similar in *different* ways, .e.g, number of players, the use of tokens, scoring, physical activities, ...etc.

Language can be quite tricky, particularly when we come to the notion of an individual or class. Wittgenstein gives an example from the use of the standard meter. There is one object which we cannot say whether or not it is one meter long, the standard meter in Paris. This paradox follows because the definition of *meter* means a length approximately equal to the Paris meter stick. Any object is always equal to itself, and so it is not sensible to say an example used to define a descriptive element is or is not the element.

Another example points out the limitation our language entails. We feel quite satisfied claiming the principle of exclusion:

Two things cannot occupy the same place at the same time.

However consider the crossing patterns in figure 5. This figure shows two eight connected digital lines crossing at point **a**. Clearly, a point on both the black and white line exists at the same point. While this contradiction can be eliminated by eliminating point **a** from both lines, the general point is germane to our problem of generating descriptions for object recognition. We are limited by our language in the development of class descriptions which might be effective for recognition algorithms. The exclusion principle leads us to search for a unique

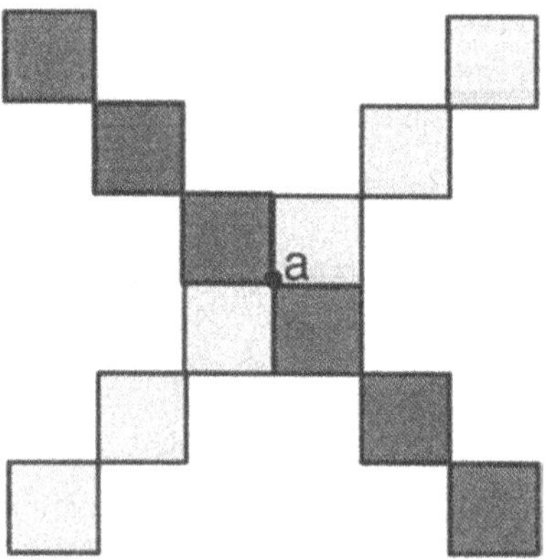

Fig. 5. An example of two colors at the same point.

and dominant representation which may not be justified. For example, there are an infinite number of representations for the square region shown in figure 6. That is, we can have two triangles and two rectangles in the same place at the same time. We could invoke some idea of simplicity in description, but that is only one criterion which is in no way fundamental or *a priori*. This observation

Fig. 6. Multiple representations for a square.

also offers proof that a composite object is more than its primitive parts and the relations between. Such a description is not unique. We can define new primitives which are various combinations of the original set. The new structure is describing the same object and therefore there must be something beyond any individual description. It is more reasonable to view the square region as *all* of its possible descriptions – an equivalence class of descriptions. If we do wish to single out a particular description it should be based on the basis of performance in visual processing, and not some mathematical or aesthetic motivation.

5.7 Wantanabe

Even more startling is the observation by Wantanabe[22] that similarity of attributes between two objects is not a valid basis for classification! The ugly duckling theorem:

> Insofar as we use a finite set of predicates that are capable of distinguishing any two objects considered, the number of predicates shared by any two such objects is constant, independent of the choice of two objects.

The theorem can be established by counting arguments on the lattice defined by boolean combinations of object properties. We can find as many attributes different about two objects as the same. For example consider a duck and a black swan (the origin of the name of the theorem). Both have beaks, one is white the other black, both have webbed feet, one is larger than the other, ... etc.

In general then, we have to single out some properties as more important than others in order to define a class in terms of predicates on attributes. Alternatively, we have to provide a set of examples, class prototypes, which define the importance of various attributes in establishing the classification space. In either case, there is nothing in nature which *a priori* establishes classes, it is our imposition ultimately arising from our evolutionary success in using these classes to carry our our functions.

5.8 The Modern View

To summarize the modern view of categories and classes:

- Only particular objects exist.
- A class corresponding to a general concept, consists of particular objects which resemble each other.

In some way, an individual object *organizes* the more elementary features composing it as well as defines a class of similar objects. These classes correspond to our linguistic instincts which we impose on nature. The correspondence is particularly aligned in the case of man-made objects which involve the shaping of matter according to our *functional* [7] and aesthetic ideas. For example, if we are going to build a system to recognize wrenches, the description of the wrench should be centered on the span of the nut contact surfaces, as shown in figure 7. Building a description on the overall shape of the wrench, might be successful in distinguishing a few wrenches, but fails to capture the fundamental defining characteristic. But how does an individual instance of a new object define a

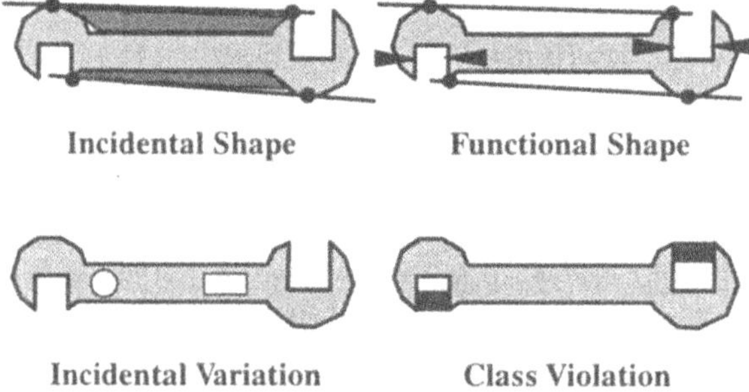

Fig. 7. Various feature representations of a wrench. The fundamental definition should be based on the jaw opening. Incidental artifacts and modifications of the wrench body are irrelevant. The function, and thus the identity, of the wrench can be destroyed by closing off the opening.

category? I argue that one important mechanism is through the elimination of descriptive variables which we agree do not generally affect the class of an individual object. Suppose we have a cube. If we move the cube has its class changed? Most would say not. If we rotate the cube has its class changed? Again,

[7] The idea of a functional base for visual categories has been successful in launching a new promising approach to recognition, e.g. Bowyer and Stark[23]

most would say not. This line of reasoning leads to the following principle of class formation, which I call *meta-invariance*:

> A class is defined by removing or attaching attributes from an object instance which are not considered to be significant for the purpose of its classification.

Classification by the meta-invariance principle follows utilitarian principles since an object description is much more useful if a wide range of sensing environments and object states are covered by the same description. In the case of object recognition, it follows that the description derived from a single image instance is best generalized by removing attributes which affect its image appearance but not its class. Some examples of properties which do not determine the class of an isolated object, but do affect its appearance:

- Illumination and shadow
- 3D position and orientation
- Internal camera calibration parameters.

The elimination of these properties leads to a a 3D projective geometric description of shape and surface texture since these are the attributes which remain fully invariant to these three observation variables.

The principle of attribute perturbation is more general than just an invariant geometric description. It can be extended to functional and aesthetic descriptions as well. For example, can a wrench still be used if we cut off part of the handle? Yes. What if we drill holes in the handle? Still yes. Only if we grind the opening of the nut-gripping surfaces to a non-standard size or remove too much of the handle to enable mechanical advantage, is the definition of the wrench ruined.

It may seem that carrying the meta-invariance principle to its extreme will lead to no description at all, e.g., the Cheshire cat, eventually all we are left with is its smile. Indeed, the more irrelevant effects there are to eliminate, the more complex the object must become to support the extraction of the invariant relations. For example, in 3D projective space, all quadric surfaces are transformationally equivalent to the sphere. A cubic surface is the minimum degree for which a non-trivial projective invariant description can be achieved. One way to reduce the minimum complexity is to share the elimination of class-irrelevant variables between objects. This line of reasoning leads to the need for *context*.

6 Context

6.1 Classification is Relative

It is easy to illustrate that context is important to the specification of attributes for recognition. The description of the color *white* is not an absolute sensory impression, but depends on the adjacent hues. A region may look grey in isolation but white against a darker background as illustrated in figure 8. Similar context-dependence for interpretation extends to geometric features. An interesting example concerning the transformational aspects of context is illustrated in figure 9.

Fig. 8. The quality *white* is relative and requires context to be defined.

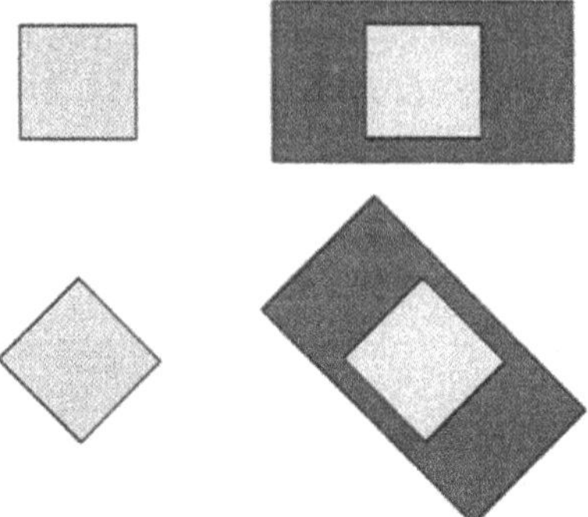

Fig. 9. The interpretation of an isolated square changes to a diamond when it is rotated 45°. If the square is included in an enclosing rectangle its interpretation stays a square.

In both examples, classification depends on the surrounding context. This dependency led Leibnitz (see section 5.2) to propose that individual substances, by definition, do not interact, i.e. no context is required for their definition. He considered that an individual substance should not depend on any other substance for its class definition. Thus the description of an individual must include all of the relations needed to form a complete description. However, as I have just shown in section 5.8, the construction of invariant object descriptions with enough self-contained context to enable recognition requires more structural complexity than is available in many objects we wish to recognize, .e.g., we do want to tell a sphere from a ellipsoid.

The way out of this seeming impasse is to be very definite and explicit about the source and utilization of context. In the design of a representation for recognition, the use of context must be clearly stated and the source of context specified. The formation and management of context becomes a central task of the representation.

6.2 Context-based Vision

Strat and Fischler[26] have defined an innovative new approach to scene analysis called *context-based* vision. Their ideas have been implemented in a system, called CONDOR, for the analysis of outdoor scenes and for natural objects such as trees and rocks. They advocate explicit sources of context which might be defined manually or determined from previous valid object hypotheses in a 3D scene. The key idea is that each decision is made with reference to specific context sources and contingent on the validity of these context assumptions. To present a clearer idea of how context is used in recognition, I will now give two examples at the extremes of contextual assumptions.

Minimal Context Consider an earlier system we built for 3D object recognition. Recognition was achieved for aircraft and buildings from a single 2D intensity aerial image based on a feature called the *vertex-pair*[24]. This system forms object model hypotheses based on clusters in six-dimensional pose parameter space. An individual pose parameter vote is computed from each vertex-pair. The context used in the system is:

- A library of 3D polyhedral object models. It is given that objects outside the library are irrelevant.
- Limitations on object position and orientation associated with an assumption of an aerial view of objects on the ground. I.e., objects are not upside down or under the ground. Objects will not be larger than a certain scale in the image.

The definition of the vertex-pair feature is shown in figure 10. The system verifies

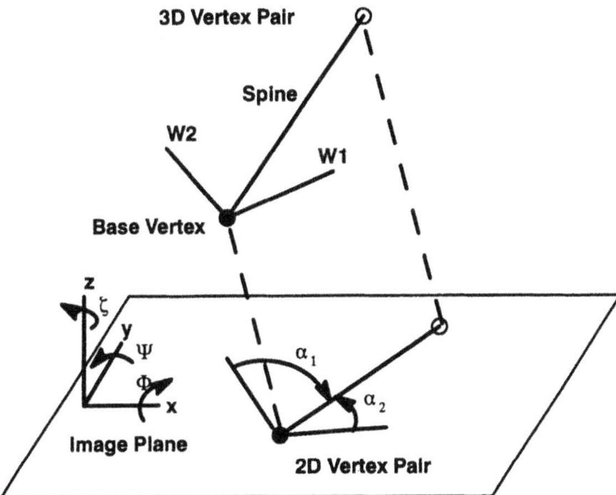

Fig. 10. The vertex-pair feature. Under the assumption of weak-perspective projection, the vertex-pair is sufficient to compute the six pose parameters of the camera.

entire object hypotheses by projecting the 3D model onto the image and collects support for the model edges from fitted images line segments. The recognition algorithm follows the general approach now widely known as *alignment*[25]. In the vertex-pair system, recognition as defined in section 4 is carried out in two parts. The *identification* of the object is carried with the model. The *perception* (feature organization) of the object is achieved by local grouping of line segments to form vertices and clustering in the space of pose parameters.

Extensive Context The context for the airbase scene in figure 10 can be extended by constructing a 3D site model which is assumed to be already aligned with the image. That is, we assume that a camera with known pose with respect to the 3D site coordinate frame is known. An example of a site model for an airfield scene is shown in figure 11. In addition to the building and tarmac areas a set of fixed occupancy detection regions are defined which are aligned with fixed aircraft parking locations. These regions are shown in more detail in figure 12. With this much context, it is feasible to reliably monitor the presence or absence of aircraft in the parking spots, simply from the existence of high contrast edgel patterns within the pre-defined detection regions. The figure shows the presence of edgels in the occupied and empty tarmac regions. The filled regions are easily distinguished from the empty regions by simple histogram measures taken over the range of gradient magnitudes where edgels are detected. The context used

Fig. 11. A site model for the Schenectady airfield.

Fig. 12. An implementation of site model context to provide very restricted recognition interpretations.

here is:

- Known projection of the Earth ground plane onto the image.
- Delineated parking regions on the 3D ground plane which narrowly define the position of aircraft.
- The specification that only aircraft are present in the parking area.
- A reference region in the same image which is known to be empty and representative of the texture of an empty parking area.

In this example, recognition is carried out with *a priori* identification of the object, i.e. we know there is going to be an airplane present, when anything is present. The perception part of recognition is simply based on the existence of edgels. The amount of site constraint might seem to be contextual overkill, albeit a robust solution. In many applications, the environment will be too fluid and the manual effort prohibitive to permit such extensive *a priori* context. We seek a middle ground where context is supported across all representational layers, but is mostly derived from the recognition process.

7 Classes For Recognition

7.1 What Should a Class Do?

I assert that the major purpose of an object's class is to enable recognition of an individual member of the class. This rôle extends beyond the domain of computer vision and can be considered a major function of the categories which help us structure the complex flood of sensory data and sense memory. In order to effectively support recognition, the class description must provide the following functions:

1. **Distinctiveness** - A description must separate instances manifested in image features into the defined classes. It is useless to construct a lexicon of categories which cannot be operationally distinguished. The description should center on those attributes which can be reliably measured from image data and ignore image properties not related to object identity.
2. **Induction** - Our description schemes must support the addition of new instances to existing classes as well as the inductive construction of new classes. Class extension and refinement, in the Wittgenstein sense, should be a consequence of the recognition process itself.
3. **Constraints** A class description should provide sufficent constraints to permit a significant test of its existence in image data. There is no point in defining a class or event the existence of which cannot be decided by computation on the data. The non-existence of an event or class should be detected by the failure of the implied constraint relations.
4. **Context** - The description should be explicit about the context it requires to produce an unambiguous classification. The new contextual constraints produced by the identification of a member of a class should also be explicit.

These requirements can be largely satisfied by basing the class description on geometry and geometric relations. A geometric description is powerful because many 3D geometric relations and properties are preserved under perspective projection. Geometric descriptions can be derived from image intensity discontinuities at occluding boundaries, and some internal surface boundaries, with reasonable invariance to illumination and viewpoint. The main drawback to basing object class descriptions on geometry is that relatively high image resolution is required to obtain accurate geometric measurements[8]

7.2 Symmetry

An ideal example of a description for recognition which meets all of our criteria is given by *symmetry*. Symmetry is a powerful descriptive mechanism which imposes tight constraints on image features under general perspective projection[28] as well as supporting a framework for organizing specific object instances into useful subclasses. This last point is illustrated by several symmetry subclasses:

[8] The smallest object feature of interest should be at least 10 pixels in extent.

- Translational Symmetry - Many objects are formed through repetition and translation, e.g. keys on a keyboard, books on a shelf fence posts ..etc.
- Rotational Symmetry - Surfaces of revolution(continuous rotation) have been extensively studied with encouraging results[29, 30]. Discrete rotational symmetry is inherent in many objects, e.g. petals of a flower and the pattern on a soccer ball. If three-fold or more multiplicity is available, then scaled Euclidean properties of an object can be recovered from a single, uncalibrated perspective view.
- Affine Symmetry - Even greater flexibility in the construction of objects is realized by the use of affine symmetry. The simplest form is translation and simple scaling. Examples are leaves along a branch, such as a fern and two ends of a wrench.
- Bilateral Symmetry - This case is similar to rotational symmetry except the transformation is reflection across a symmetry plane.

It has been argued that symmetry is such a useful description because symmetry is inevitable in nature[31]. Structures are more stable if forces are balanced, i.e. symmetrical. Shape duplication in crystals is a direct consequence of molecular symmetries, which result from stable force balances. Duplication in plant and animal structures follows from genetic templates, i.e. reuse of the same design.

7.3 Assumption and Confirmation

Consider the *butterfly* configuration shown in figure 13. It is assumed that we

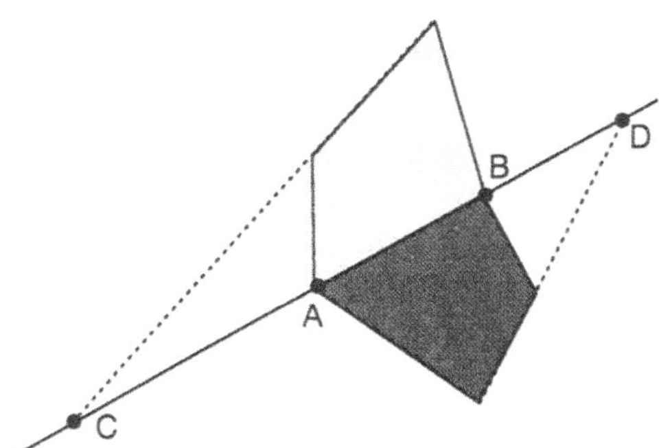

Fig. 13. A set of six points which support direct measurement of a 3D invariant property from any perspective view.

have no information about camera viewpoint or internal calibration parameters. The invariant is constructed by extending the edges of the *wings* of the butterfly until they intersect their line of intersection, as shown in the figure. The four points on the line of intersection define an invariant cross ratio[32] which is the same in the image projection of the butterfly as in 3D space. If the wings are two rectangles, then points C and D merge into a single vanishing point where all lines parallel to the wing edges meet. In this case, the cross ratio is identically zero and conveys no distinguishing power.

In the butterfly example, it is assumed that the wings are coplanar and in return it is possible to measure an invariant cross ratio. But, other than indexing a model library to see if any stored object description has the measured cross ratio value, there is no further constraint available for confirmation of the assumption. From our requirements, it must be possible to confirm an assumption about a geometric class configuration without referring to specific instances.

With a bit more geometric structure, enough geometric relations are available to check the consistency of an assumption. Consider the case of two identical cubes which are related by translation. The lines joining corresponding points on each cube will meet in an epipolar point, ϵ_t as shown in figure 14. Assume

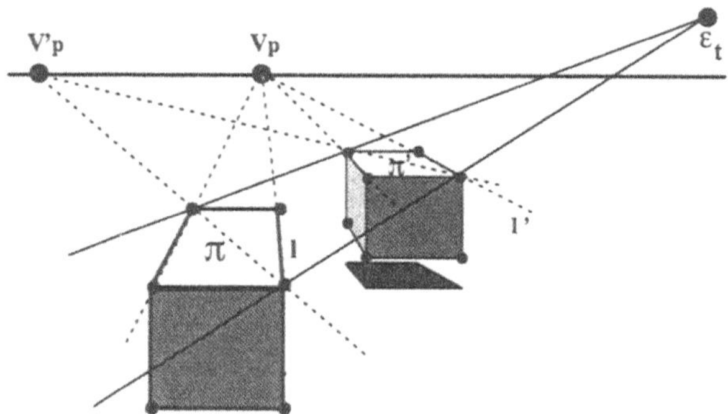

Fig. 14. Two identical cubes, repeated by translation.

that the pairs of correspondences lie on the plane, π and its repeated copy, π'. This assumption can be verified immediately by checking that corresponding lines in π and π' are parallel. This check can be accomplished by choosing two corresponding line pairs from π and π' to define a pair of vanishing points, e.g., V_p for **l** and **l**'. The other vanishing point, V_p' is formed by the lines defined by the epipolar correspondence points in each plane. These two vanishing points define a line at infinity. A projective plane with a distinguished line at infinity becomes an affine plane. Parallel planes have the same line at infinity. Thus, all other corresponding repeated lines in π and π' should be parallel, i.e, intersect on the line at infinity, if the assumption is valid.

The hypothesized translation vanishing point, ϵ_t, permits the direct construction of additional correspondences. From these correspondences, a 3D affine model of the cube, or any structure with translational symmetry can be constructed without knowledge of camera calibration. The 3D affine invariants, e.g. length ratios, provides discrimination power to index to specific model instances.

The repeated structure also supplies context for additional investigation of the scene. The vanishing points of common object edge directions constrains camera pose with respect to the coordinate frame of the object models and leads to a partial calibration of the camera. These camera constraints can then be used to reduce the complexity of structures which are self-confirming. For, example if it is further assumed that objects share a common ground plane, these vanishing points define a horizon line which can be used as a *wild card* in the construction of invariants. For example, a planar invariants on the ground plane requires five lines; if we use the horizon line, then only four lines are required.

7.4 Deformable and Natural Objects

In the development of geometric classes along these lines, there is often some criticism that the expressive power is not sufficient. It is pointed out that deformable or natural objects which cannot be captured by such rigid formal geometric class descriptions. I will illustrate the discussion by taking an example from natural objects such as the peppers shown in figure 15. The data about the shape for an individual pepper can be modeled in many ways, e.g. superquadrics or algebraic surfaces[33]. But such mathematical descriptions do not capture *class* in Wittgenstein's sense. For example, peppers grow in diverse shapes, maintaining a general tendency to form 4 lobes, but this is not even necessary since it depends on a complex history of sun, shade, nutrients etc. There is no very definite

Fig. 15. Several peppers selected from the same plant to illustrate geometric diversity in shape.

constraint on shape except for flutes, and stem and a rough aspect. Therefore the geometric parameters derived from an algebraic fit to the outline or surface of an individual pepper will not convey much about class membership nor will there exist very strict constraints on the model parameter space to confirm or deny the class assumption.

The point of this example is to demonstrate that the language of geometry is able to capture many aspects of object classification of interest to human concept, but not universally so. It seems obvious that if such shapes are to be understood by a computer recognition process that a teleological[9] representation is necessary which simulates the growth process. At present we don't know how to link such models to hard image constraints that can enable recognition. Ultimately the class *pepper* will take on a Wittgenstein family relational structure like that that for the class *game*, i.e., there is no single set of attributes which characterizes all members of the class.

7.5 Man-Made Objects

A key defense of the use of geometric descriptions, like symmetry, is that they capture the essential form of man-made objects which justifies their study and implementation, even though many object classes lie outside of geometric expression, as we now understand geometry. This defense requires a more careful definition of *man-made* since many natural objects, such as crystals, possess complex and perfect geometric symmetries. On the other hand, many objects constructed by man are not readily described in geometric terms at all, for example a pillow or a shoe. The man-made defense should be restated as:

> Geometry provides a conceptual framework which captures many attributes of both natural and man-made object classes. We should demonstrate that object recognition can be carried out reliably on a large library of geometric instances before considering classes for which we have vague or unknown descriptions.

7.6 Closed World

In a similar vein, I advocate that recognition experiments be carried out in a closed world before venturing to an open world where unknown objects can appear[34]. The closed world argument is almost tautological: Since object classes must account for the appearance of instances in image data there can be **no** significant unmodeled effects which would not lead to misclassification. Almost by definition, a computational approach to recognition has no response when a large effect occurs which is not accounted for in the description of objects. If recognition succeeds, it will be only to the extent that the unknown event generates sufficiently small perturbations to the assumptions of the class description. If there are large unknown effects, the only recourse is a statistical model which is likely to be a poor predictor due to unavoidably small sample sizes and uninformative prior distributions.

This closed world framework has been adopted in the analysis of video sequences of football games[10]. In this application, the closed world assumptions

[9] This term has been adopted by the graphics community to indicate geometric models which are defined by physical laws and physical constraints.

[10] I gave a talk on MORSE (see section 8) at the MIT AI lab in the fall of 1993 where I mentioned the closed world idea.

involve the background patterns on the field as well as restrictions on the number and types of uniforms and players[35]. The resulting tracking of player motions is much more reliable, and in some instances would be infeasible, without a closed world.

8 The Representation

The implementation of some of these ideas into a representation is underway in a project called MORSE[11]. A schematic of the representation is shown in figure 16. In broad outline, MORSE follows the conventional approach taken by geometrically oriented recognition systems, however the use of general geometric classes to provide specific control strategies for grouping and scene organization more fully meets the requirements for a representation. Some of the key aspects

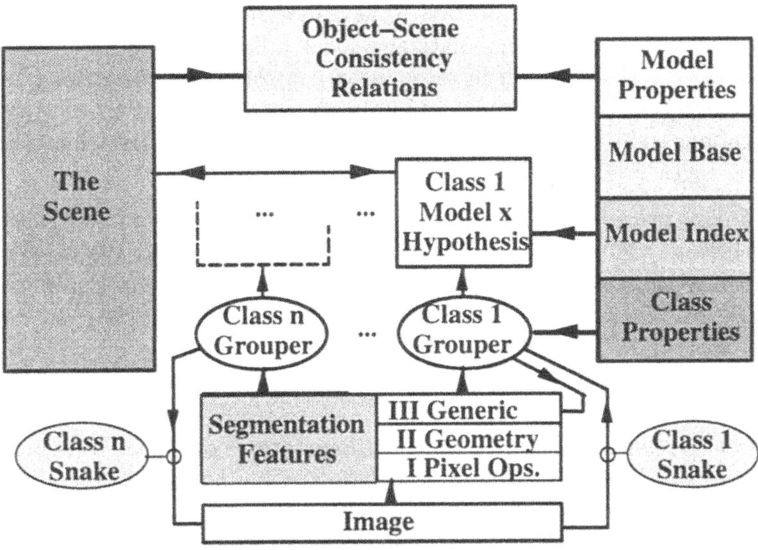

Fig. 16. The architectural structure of the MORSE representation.

of the design are:

- MORSE uses geometric classes such as rotational symmetry, canal surfaces and polyhedra to define a hierarchy of descriptions for recognition. The influence of class constraints extends down to the edge detection and edgel linking stages to control local image event interpretations, such as topology.

[11] Named for the Oxford detective in stories by Colin Dexter. The acronym derives from Multiple Object Recognition by Scene Entailment.

- MORSE exploits geometric consistency constraints to strengthen recognition hypotheses, and to simplify indexing, essentially by piecing together consistent 3D world models. Most current recognition research is focussed on detecting single objects without regard to scene constraints.
- MORSE description processing is phrased in terms of database interactions which provides an effective interface for grouping operations. Finding structures which satisfy class constraints is formulated as queries to a relational network of geometric constraints.
- The MORSE representation includes multi-agent control structures to ensure efficient handling of hypotheses. This control approach allows propagation of context in a uniform manner across all levels of description. A uniform notion of control is also essential for managing the priority of hypothesis processing and testing. A balance must be established between depth of interpretation, which can provide more context, and breath of interpretation which can improved material for hypothesis formation.
- Tight feedback loops of hypothesize and test are important to the efficiency of representation. It is important to have descriptions which provide enough constraints to enable immediate and local validation of a hypothesis. The most substantial problem in recognition control is not uncertainty or noise but ambiguity. At early stages of interpretation it is essential to verify interpretations so as to avoid combinatorial expansion of incorrect hypotheses in later stages.

MORSE will provide a laboratory for exploring the issues of recognition described here as well as provide a vehicle for collaboration. At the moment, the MORSE representation is being developed at four vision labs in Europe and the US.

9 Conclusions

The effort in reviewing the philosophical foundations of object recognition has proven well worth the effort, in my own case, and I hope some of these ideas will prove useful to others who are thinking about the recognition problem. Along the way I have identified some interesting issues which will influence my own research program in the future.

Of particular interest is the problem of inducing a class from a few instances of an object. The problem of what attributes and relations really are critical to the class and what can be ignored is a problem central to progress in recognition. The more difficult problem of acquiring Wittgenstein's *similarity networks* as class descriptions, even for classes which are totally defined by formal geometric descriptions, has barely been touched in computer vision research.

Finally, we cannot readily make use of context which is not directly linked to the terms and attributes of a particular description. Some form of translation is needed to convert linguistic context to image and geometric attributes. This capability was introduced as rôle five of a representation[12]and is essential so

[12] See section2.1.

that both human descriptions and computer-derived context information can be brought into the recognition process. I expect that the way forward here is to establish a symbiotic interface between the recognition algorithm and human interpretation. Much can be learned about linking the two conceptual frameworks by mutual pointing at sample image data and comparing interpretations.

References

1. Roberts, L. G., *Machine Perception of 3D Solids* in Optical and Electro-Optical Information Processing, Tippet, J. T., *et al* eds., MIT Press, 1965.
2. Ayre, A.J., *Language Truth and Logic*, Dover Books, 1935, with a new introduction, 1946.
3. Hume, D., *An Enquiry Concerning Human Understanding*, 1748, in The English Philosophers From Bacon to Mill, Burtt, E. ed., The Modern Library, 1939.
4. Jowett, B., *The Dialogs of Plato*, vol. 3, Scribner, Armstrong and Co., 1874.
5. Davis, R., Shrobe, H. and Szolovits, P., *The AI Magazine*, Spring 1993, p. 17.
6. Dennett, D., *Brainstorms*, MIT Press, 1981
7. Weiskrantz, L., *Thought Without Language*, Oxford:Clarenden, 1988.
8. Raichle, M., *Visualizing the Mind*, Scientific American, April, 1994.
9. Freedman, D., *Brainmakers*, Simon and Schuster, 1994.
10. Edelman, G., *Bright Air, Brilliant Fire*, Basic Books, 1992. p. 194.
11. Kapur, D. and Mundy, J.L., editors *Geometric Reasoning*, MIT Press, 1989.
12. McCarthy, J.,*Epistemological Problems of Artificial Intelligence*, in Readings in Knowledge Representation, Brachman, R. and Levesque, H., eds., Morgan Kaufmann, 1985.
13. Dennett, D., *Consciousness Explained*, Little, Brown and Company, 1991.
14. Shepard, R. N., Metzler, *Mental Rotation of Three-Dimensional Objects*, Science, **171**, pp. 701-703.
15. Lowe, D., *Perceptual Organization and Visual Recognition*, Kluwer Academic Publishers, 1985.
16. Sayre, K. M., *Recognition: A Study in the Philosophy of Artificial Intelligence*, University of Notre Dame Press, 1965.
17. Grimson, W.E.L., *Object Recognition by Computer: The Role of Geometric Constraints*, MIT Press, 1990.
18. Aristotle, *Aristotle: Categories on Interpretation and Prior Analytics*, Loeb Classical Library, Cooke, H., translation, 1938.
19. Hayes, P., *The Second Naive Physics Manifesto*, in Readings in Knowledge Representation, Brachman, R. and Levesque, H., eds., Morgan Kaufmann, 1985.
20. Kline, M., *Mathematics and the Search for Knowledge*, Oxford University Press, 1985.
21. Wittgenstein, L., *Philosophical Investigations*, Ancombe, G., translator, The Macmillan Co., 1953.
22. Wantanabe, S., *Pattern Recognition, Human and Mechanical*, John Wiley and Sons, 1985.
23. Bowyer, K. and Stark, L. *Form and function; a theory of purposive, qualitative 3-D object recognition*, Artificial Intelligence and Computer Vision, Proc. 7'th Israeli Conference, pp 137-146, 1990.

24. Mundy, J.L. and Heller, A., *The Evolution and Testing of a Model-Based Object Recognition System* in *Computer Vision: Advances and Applications*, Kasturi, R. and Jain, R. editors, IEEE Computer Society Press, 1991.
25. Huttenlocher, D. and Ullman, S., *Object Recognition Using Alignment*, Proc. First International Conference on Computer Vision, 1987.
26. Strat, T. and Fischler, M, *CONDOR*, IEEE Transactions on Machine Intelligence and Pattern Analysis, Oct. 1991.
27. Mundy, J. L. and Vrobel, P., *The Role of IU Technology in RADIUS Phase II*, Proc. ARPA Image Understanding Workshop, Morgan Kaufman, 1994.
28. Mundy, J.L. and Zisserman, A., *Repeated Structures: Image Correspondence Constraints and 3D Structure* in *Proc. 2nd Workshop on Applications of Geometric Invariants in Computer Vision*, Lecture Notes on Computer Science, 825, Springer Verlag, 1994.
29. Zerroug, M. and Nevatia, R., *Using Invariance and Quasi-invariance for the Segmentation and Recovery of Curved Objects* in *Proc. 2nd Workshop on Applications of Geometric Invariants in Computer Vision*, Lecture Notes on Computer Science, 825, Springer Verlag, 1994.
30. Liu J., Mundy J.L., Forsyth D.A., Zisserman A. and Rothwell C.A., *Efficient Recognition of Rotationally Symmetric Surfaces and Straight Homogeneous Generalized Cylinders*, Proc. Conference on Computer Vision and Pattern Recognition, 1993.
31. Arnheim, R. *Art and Visual Perception*, University of California Press, 1971.
32. Mundy, J.L. and Zisserman, A. editors, *Geometrical Invariance in Computer Vision*, MIT Press, 1992.
33. Sullivan, S., Sandford, L. and Ponce, J., *On Useing Geometric Distance Fits to Estimate 3D Object Shape, Pose and Deformation from Range, CT and Video Images*, Proc. Computer Vision and Pattern Recognition Conference, 1993.
34. Mundy, J.L., Forsyth, D., Zisserman, A. and Rothwell, C., *MORSE: Multiple Object Recognition by Scene Entailment*, Draft Report, GE Research Report, Feb. 1994.
35. Intille, S. and Bobick, A., *Tracking Using a Local Closed-World Assumption: Tracking in the Football Domain*, MIT Media Lab Report 296, August, 1994.

Appearance-Based 3D Object Recognition

Linda G. Shapiro[1,2] and Mauro S. Costa[1] *

[1] Department of Electrical Engineering, FT-10
[2] Department of Computer Science & Engineering, FR-35
University of Washington
Seattle WA 98195
U.S.A.

Abstract. General, three-dimensional object recognition is still an unsolved problem. A major handicap in many systems is the use of standard point and straight-line-segment features for recognition. We believe that general object recognition can only be accomplished by utilizing the appropriate sensors for each object class and the appropriate features that can be reliably extracted using those sensors. We also believe that the analysis of complex scenes will require an active system. In this paper we define a new representation called an *appearanced-based model* and discuss its use for hypothesize-and-test object recognition in an active environment.

1 Introduction

Three-dimensional object recognition has seen a great deal of activity in the last decade, as has been pointed out in recent surveys ([2], [4], [11], [25], [30]). Most systems fall into three main categories: 1) systems that use intensity data alone ([1], [5], [8], [7], [20], [23], [24], [27], [31]), 2) systems that use range data alone ([3], [6], [13], [14] , [15], [22], [16], [21], [28]), and 3) systems that use both range and intensity (sometimes including color) data ([17], [18], [29]).

In intensity-image-based systems, points and straight line segments are still the most commonly-used features. In fact, the recent popularity of the alignment method [19] has led to a significant number of systems that blindly match triples of points or line segments from the image to similar triples from the model, using little or no contextual information. These algorithms all make the assumptions that a) the points or line segments are reasonably reliable features of the class of objects to be recognized or located and b) the pose of the object can be uniquely determined from a small set of these features. These assumptions are only true for polyhedral objects or those with a number of sharp, straight edges. They fall apart for most curved-surface objects. Figure 1 illustrates the kind of object where points and line segments make good features, while Figure 2 illustrates a simple object with both curved and planar surfaces where they are not very useful.

* This research was supported by the National Science Foundation under grant number IRI-9023977, by the Boeing Commercial Airplane Group, and by the Washington Technology Center.

In range-image-based systems, primitive surfaces (usually planar or quadric) are the most common features, but 3D line segments and points are also used. Because surfaces from range data are more reliable than surface regions from grayscale, a number of systems use the properties of and relationships among surfaces in matching algorithms. This type of approach has worked well for simple objects with a small number of simple surfaces. Systems that work with more complex, free-form surfaces generally look for interest points and perform point matching. Again the reliable detection of feature points is crucial to success.

A few systems have been developed that make use of more than one kind of image data. The use of color or grayscale intensity images registered with range images has been successfully demonstrated in several domains. We believe that this approach is in the right direction and should be extended further to use as many types of data and extractable features as needed to best recognize and locate objects in a given domain. In this paper we formalize this approach and describe a system we are building that will make use of it.

Fig. 1. Image of a polyhedral object whose junctions and line segments make useful features for recognition and pose estimation.

Fig. 2. Image of a nonpolyhedral object for which line segments and junctions alone are virtually useless as recognition features.

2 Appearance-Based Models

In most real applications, the type of object is restricted to a particular, well-defined class. Different object classes produce images with very different kinds of features, from polyhedra that produce clear, sharp edges to medical images where edges are useless features. The features that should be used to recognize and locate an object depend on 1) the available sensor(s) for the recognition task and 2) the features that are reliably detectable with those sensors. Features can be predicted by applying graphics software to CAD models ([8], [16]), but artificial images are not yet good enough to use this method alone [26]. Instead, the artificial images must be augmented by a set of real images from the real sensor(s) to be used. We will assume in this paper that a set of suitable training images has been obtained.

An *appearance-based model* of an object is a description of the object in terms of features that are detectable in images of the object. A feature is *detectable* if there is a computer program that can extract the feature from an image of the object, through some well-defined procedure. The definitions of these procedures are still part of the "art" of image processing. Appearance-based models can be *full-object models* including all the features that appear in any view of an object, or they can be *view-class* models in which an object is represented by a small set of characteristic views, each having its own distinct feature set. We will use view class models of industrial parts made of metal with planar, cylindrical, and threaded surfaces as illustrations in this work.

Let C be the class of objects to be recognized, and let T be the set of feature types to be used in the recognition task. Features may be 2D or 3D, depending on the sensors used to detect them. Each type of feature has a 3D source, an explanation for the appearance of the feature in an image. 3D image features, in general, correspond to 3D object features. For example, 3D surfaces detected in range images correspond to object surfaces and 3D edges correspond to object edges or boundaries between objects. Some 2D features correspond directly to 3D features of the object. Straight and curved 2D segments, for instance, can correspond directly to 3D straight and curved edges. 2D ellipses can correspond directly to 3D circles or ellipses. Other 2D features come about due to a mixture of factors including the geometry of the object, the viewpoint, and the lighting. Limb edges are viewpoint dependent and have no corresponding 3D edge at all on the object. Highlight regions are caused by the shape of the object and the lighting; they are also highly viewpoint-dependent.

Let $S_{V,M}$ be a set of training images for view class V of object model M. The images are all taken from a connected volume of the viewing sphere, which is assumed to be centered at the origin of the object and to be of a fixed radius equal to the maximum possible viewing distance. If the object has symmetries, then the connected volume can be replaced by the union of several connected volumes. Each image $I \in S_{V,M}$ is processed to yield a set of features F_I. A feature f_n^I from image I is *equivalent* to another feature f_m^J from image J if they have the same type and are judged (interactively by human or automatically by computer program) to have come from the same 3D source. The set of features that represent the view class is the set $F_{V,M}$ of equivalence classes of the union of the feature sets. The feature types we are investigating for use in our system are: coaxial circular arcs (two-cluster, three-cluster, and multi-cluster), ellipses, triples of line segments (U-shaped and Z-shaped), junctions (V-junction, T-junction, Y-junction, and Arrow), parallel line segments (close and far apart).

Figure 3 shows some of the features that were extracted from the image in Figure 2. Figure 3(a) shows the raw edge image obtained by applying a Canny edge operator to the image of Figure 2. Figures 3(b) and 3(c) show the straight line segments and circular arcs, which were extracted by the Object Recognition Toolkit (ORT) package [12]. Figure 3(d) shows the line features identified by our system. Line segments 5 and 7 form a far-apart, parallel pair, while line segments 8 and 9 form a V-junction as do line segments 5 and 6. Figure 3(e) shows the single, cluster of three coaxial arcs detected by our system, and Figure 3(f) shows the single ellipse detected.

The image of Figure 2 belongs to View Class 1 of Model 1 of our model database. Of the five samples we have analyzed so far of that view class, the three coaxial arcs and the ellipse were detected in all five; the V-junctions were detected in three; the far-apart, parallel lines were detected in three; another pair of parallel lines was detected in two; and a U-shaped triple of line segments was detected in two.

Features alone are not always sufficient in a recognition task. The properties of the features and the relationships among them can often be helpful in object recognition. A *structural description* $D_{V,M}$ of a view class is a triple

$$D_{V,M} = (F_{V,M}, P_{V,M}, R_{V,M})$$

where $P_{V,M}$ is a representation of the properties of the features, and $R_{V,M}$ is a representation of the relationships among the features. If properties and relationships do not exist or are not useful for the particular recognition task, the sets are merely empty.

Our current system works with pairs of gray-tone images of a multiple-object scene. The two images are taken from the same viewpoint, but with two different lightings, one with the light source at the left and one with the light source at the right. The combination of the two images allows the elimination of shadows and some highlights. We have also developed a method that utilizes polarizers to detect shadow regions and highlight regions [9].

We will use the subscript v to denote a viewpoint and, to simplify notation, include the positions of the camera *and* the two light sources in v. We denote an image pair at viewpoint v by $I_v = (I_{Lv}, I_{Rv})$ where I_{Lv} is the image with the light source at the left, I_{Rv} is the image with the light source at the right, and the subscript v can be dropped when the viewpoint is understood. The image of Figure 2 was taken with the light source at the right. The combined processing of this image and the corresponding image from the same viewpoint with the light source at the left produced the features shown in Figure 3.

3 Scene Analysis Using Appearance-Based Models

We wish to use appearance-based models to analyze multiple-object scenes (see Figure 4) whose objects are all expected to come from a known object class on which the system has been trained. Suppose that we have a database *DB* of models. In our system, each model would be a structural description $D_{V,M}$ of a particular view class V of a particular object M. The structural descriptions would all have been derived from a large set of training image (pairs) via our feature extraction process which uses the ORT package and some home-grown procedures. We envision the following scene analysis paradigm.

1. Acquire a new image pair I_v at current viewpoint v.

2. Process I_v to produce structural description $D_v = (F_v, P_v, R_v)$.

3. Select an element E_v (or a small subset of elements) of D_v and use it to index into the database *DB*. The result is a list L_v of *hypotheses*. Each hypothesis H is of the form

   $$H = (V, M, C, S)$$

   where V and M refer to view class V of object M. C defines the *correspondence(s)* between F_v (the features of the image pair I_v) and $F_{V,M}$ (the features of view class V

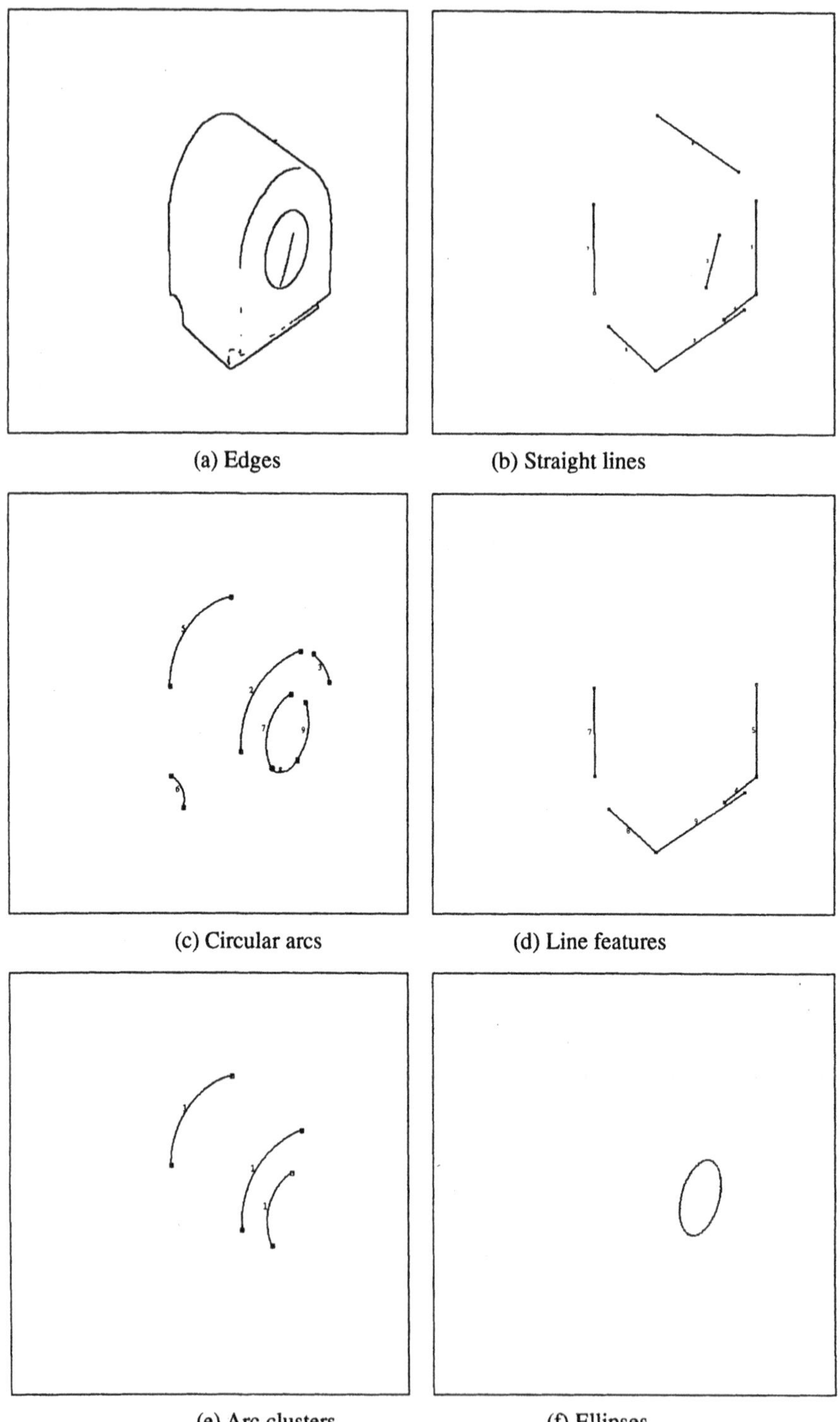

(a) Edges

(b) Straight lines

(c) Circular arcs

(d) Line features

(e) Arc clusters

(f) Ellipses

Fig. 3. A set of features extracted from the training image of Figure 2.

of object M). S is a rating of the strength of the hypothesis. The list may be returned all at once if computational resources are adequate or may be returned incrementally, as needed, if not. If all elements of D_v have been tried and have failed, this step fails and the system backtracks.

4. If none of the hypotheses is strong enough, go back to the previous step to try another image element. Otherwise, for each hypothesis H (in order of strength), determine if the current image I_v and its structural description D_v are sufficient to verify H and compute pose. If so, succeed. If not,

5. Determine a set of actions A_H that will help verify or disprove hypothesis H. In our system, each action involves moving the camera (or the light sources) and leads to a new viewpoint and a new image pair. Furthermore, since the hypothesis H specified a model and view class, the new image pair is constrained to come from the same model and a related view class. This reduces the database to be searched to a fraction of the original database. It also limits the region of the new image pair where the object to be verified ought to appear. The system can thus begin again with reduced database DB', new image pair $I_{v'}$, and a constrained search region to be processed.

4 Some Initial Experiments: Relational Indexing

We created a database of appearance-based object models for a set of mechanical parts that have both flat and curved surfaces, holes, and threads. Each object is represented by a small set of view classes for the major characteristic views of the object. Each view class model consists of a set of features that have been detected in one or more training images of the view class. We performed a large set of experiments to determine the features and their frequencies from a training set of real images. We currently have 7 models, 4 view classes of each, and 5 training image pairs of each view class (280 images in all). We have compiled the distinct, detectable features for each view class of five of theses models and their associated frequencies in single-object images, and we have begun using these features in recognition experiments.

We have developed a new matching technique, called *relational indexing* to work within our general image analysis framework described in Section 4. The idea is to utilize the structural description $D_{V,M}$ and represent each model-view as a relational graph of the features $F_{V,M}$. As in our previous work on hashing [10], the method has two phases: an off-line preprocessing phase, and an on-line matching phase. In the off-line processing, for all $D_{V,M}$, small relational subgraphs of size n are encoded and used as indices to access a hash table. The bin corresponding to a particular encoded subgraph will store the information about which model-views gave rise to that particular index. This is done for all possible subgraphs of size n and for all the models in the database. In the on-line phase, a relational graph is constructed using the features and relations

Fig. 4. A sample multi-object pair. The top image is illuminated from the right, and the bottom image is illuminated from the left.

detected in the scene. All the subgraphs of size n are encoded and used to index the hash table. Each subgraph index votes for all the model-view classes stored in its corresponding bin. After all possible subgraphs have been used to index the table, the model-views with sufficiently high votes are taken as possible hypotheses.

The hypotheses produced in this manner will rule out some of the model-views, but a number of hypotheses will be generated that index incorrect model-views. This is an expected by-product of the procedure and merely means that a verification phase is essential. The information stored in the hash table is actually more than just the model-view that gave rise to a particular subgraph index. It also contains information about which specific features (and their attributes) are part of the subgraph index. This information is necessary for hypothesis verification.

In the experiments we have conducted so far we have been using subgraphs of size two. This means that each index, in the general case, is made up of a combination of two features and two relations. In our implementation, each feature type and each relation are represented by distinct labels (numbers). Therefore, each subgraph index is uniquely represented by a 4-tuple of numbers (f_1, f_2, r_1, r_2). We chose to implement our hash table as a two dimensional array. The two features f_1 and f_2 are used to index the x dimension, and the two relations r_1 and r_2 are used to index the y dimension. The size of the hash table is determined by the number of possible combinations of two feature types at a time (x dimension) and of two relations at a time (y dimension). By utilizing this scheme, the coding of the subgraph into indices that can be directly used to access the table becomes trivial, and the additional source of errors in hashing techniques, namely the quantization of indices, is avoided. It is obvious that this proposed scheme is only viable if the number of feature types and relations is relatively small, which so far it is. If the size of the subgraphs increase, it may be necessary to devise an alternative hashing scheme.

Although our goal is to handle complex multi-object scenes, we decided to perform the initial tests on images of single objects. We used some of the training images for the first group of tests. Notice that since the relational models derived make use of all five training samples for each model-view, the use of one of the five training samples does not necessarily imply the use of "perfect data". Each training sample contains only a subset of the features and relations of its model-view, so it cannot perfectly match the model-view. Typical results of the hypotheses generation process using the relational indexing technique is given in Table 1. This table shows the results of matching sample 1 of view 1 of model 1 against the database of models. The count shows how many votes a particular model-view obtained. These results are encouraging since the number of possible model-views to try has been reduced and the correct one (model-view 1-1) is the strongest candidate for verification.

For purposes of illustration, we have taken the model-view with the highest votes in the experiment of Table 1 through pose computation (without verification). The result of the transformed model features overlaid on the scene is shown in Figure 5.

Table 1. Typical results of the hypotheses generation process.

Model-View	Count
1-1	5
1-2	0
1-3	1
1-4	1
2-1	2
2-2	2
2-3	0
2-4	2
3-1	3
3-2	4
3-3	4
3-4	1
4-1	4
4-2	0
4-3	0
4-4	2
5-1	1
5-2	0
5-3	1
5-4	3

5 Discussion

There are many different domains in which object recognition can be applied. We have used our current work on recognition of metal industrial parts to illustrate the concept of appearance-based models. In this work, the sensor is a CCD camera; the sensed data consists of pairs of gray-tone images taken from the same viewpoint with light sources at two different positions; and the features are constructed from line and curve segments. If we were using range data, then the set of features would be completely different. If we had both range and intensity data, both sets of features could be used. If the objects had many different colors and textures, then color and texture features would be added. The human visual system *is* often able to recognize objects from incomplete data, such as line drawings or silhouettes. But when high-quality data of several different types is available, it makes recognition simpler and potentially faster.

In any real machine vision application, there is a fixed domain and often a fixed set of objects. The challenge in any recognition/localization task is to determine the set of sensors to use and features to extract that make the task as simple as possible. In the past, we have often concentrated on the recognition algorithms and put little or no effort

Fig. 5. Image of a test scene overlaid with the features of hypothesized model match.

into the training phase. It is now recognized that training on images that truly represent the task domain is crucial to the success of the system. It is also crucial that the system be thoroughly tested on a different set of images, as has been done in statistical pattern recognition systems.

Object recognition research has come a long way in the past few years. There are successful systems that can rapidly recognize polyhedral objects from single gray-tone images, including images of cluttered scenes. There are also successful systems that use range and sometimes color data to recognize objects with simple curved surfaces. Specialized techniques have been proposed that work well with surfaces of revolution. There is, as of yet, no general theory on how to recognize any class of objects. Such a theory must solve several problems: 1) how to determine the best sensor(s) for the task, 2) how to determine a suitable set of images for training the system, 3) how to automatically or semiautomatically determine the best features for this class of objects, 4) what statistics to compile for the selected set of features, 5) how to use the statistics of the features in a matching algorithm, and 6) how to test the system thoroughly enough for use in a real application. There is still a lot of research to be done.

References

1. J. Ben-Arie and A. Z. Meiri. 3D Object Recognition by Optimal Matching Search of Multinary Relations. *Computer Vision Graphics and Image Processing*, 37:345–361, 1987.

2. J. P. Besl and R. C. Jain. Three-Dimensional Object Recognition. *ACM Computing Surveys*, 17(1):75-154, 1985
3. R. C. Bolles and P. Horaud. 3DPO: A Three-Dimensional Part Orientation System. *International Journal of Robotics Research*, 5(3):3-26, 1986.
4. J. P. Brady and N. Nandhakumar and J. K. Aggarwal. Recent Progress in Object Recognition From Range Data. *Image and Vision Computing*, 7(4):295–307, 1989.
5. J. B. Burns and E. M Riseman. Matching Complex Images to Multiple 3D Objects Using View Description Networks. In *Proc. of the IEEE CVPR*, pp. 328-334, 1992.
6. C. H. Chen and A. C. Kak. A Robot Vision System for Recognizing 3D Objects in Lower Order Polynomial Time. *IEEE Transactions on Systems Man and Cybernetics*, 19(6):1535–1563, 1989.
7. C. H. Chen and P. Mulgaonkar. Automatic Vision Programming. *CVGIP: Image Understanding*, 55(2):170–183, 1992.
8. O. I. Camps, L. G. Shapiro, and R. M. Haralick. Image Prediction for Computer Vision. In *Three-dimensional Object Recognition Systems*, A. Jain and P. Flynn (eds). Elsevier Science Publishers BV, 1993.
9. M. S. Costa and L. G. Shapiro. Design of an Active Object Recognition System. *ISL Technical Report*, University Of Washington, July 1993.
10. M. S. Costa, R. M. Haralick and L. G. Shapiro. Optimal Affine Invariant Point Matching. In *Proceedings of 10th ICPR*, volume 1, pp. 233–236, 1990.
11. R. T. Chin and C. R.Dyer. Model-Based Recognition in Robot Vision. *ACM Computing Surveys*, 18(1):67-108, 1986.
12. A. Etemadi. Robust segmentation of edge data. In *Proceedings of the IEE Image Processing Conference*, 1992.
13. T. J. Fan and G. Medioni and R. Nevatia. Recognizing 3D Objects Using Surface Descriptions. *IEEE Transactions on Pattern Analysis and Machine Intelligence*, 11(11):1140–1157, 1989.
14. O. D. Faugeras and M. Hebert. The representation, Recognition, and Locating of 3D Objects. *International Journal of Robotics Research*, 5(3):27-52, 1986.
15. P. J. Flynn and A. K. Jain. BONSAI: 3D Object Recognition Using Constrained Search. *IEEE Transactions on Pattern Analysis and Machine Intelligence*, 13(10):1066–1075, 1991.
16. K. D. Gremban and K. Ikeuchi. Appearance-Based Vision and the Automatic Generation of Object Recognition Programs. In *Three-dimensional Object Recognition Systems*, A. Jain and P. Flynn (eds). Elsevier Science Publishers BV, 1993.
17. L. Grewe and A. Kak. Interactive learning of multiple attribute hash table for fast 3d object recognition. In *Proceedings of the Second CAD-Based Vision Workshop*, pages 17–27, February 1994.
18. I. Higuchi, H. Delingette, M. Hebert, and K. Ikeuchi. Merging multiple views using a spherical representation. In *Proceedings of the Second CAD-Based Vision Workshop*, pages 124–131, February 1994.
19. D. P. Huttenlocher and S. Ullman. Recognizing solid objects by alignment with an image. *International Journal of Computer Vision*, 5(2):195–212, 1990.
20. D. P. Huttenlocher. Three-Dimensional Recognition of Solid Objects from a Two-Dimensional Image. Ph.D. Dissertation, Cambridge, MIT, 1988.
21. K. Ikeuchi and T. Kanade. Towards Automatic Generation of Object Recognition Programs. *CVGIP: Image Understanding*, 76(8):1016–1035, 1988.
22. A. K. Jain and R. Hoffman. Evidence-Based Recognition of 3-D Objects. *IEEE Transactions on Pattern Analysis and Machine Intelligence*, 10(6):783–802, 1988.
23. D. G. Lowe. Three-Dimensional Object Recognition from single Two-Dimensional Images. *Artificial Intelligence*, 31:355–395,1987.

24. H. Murase and S. K. Nayar. Visual Learning of Object Models from Appearance. *International Journal of Computer Vision*, in press. Also Tech. Rep. CUCS-054-92.
25. A. R. Pope. Model-Based Object Recognition – A Survey of Recent Research. *Technical Report 94-04*, University of British Columbia, January 1994.
26. K. Pulli. TRIBORS: A Triplet-Based Object Recognition System. *Technical Report 95-01-01*, Department of Computer Science and Engineering, University of Washington, January 1995.
27. M. Seibert and A. M. Waxman. Adaptive 3D Object Recognition From Multiple Views. *IEEE Transactions on Pattern Analysis and Machine Intelligence*, 14(2):107–124, 1992.
28. F. Stein and G. Medioni. Structural Indexing: Efficient Three Dimensional Object Recognition. *IEEE Transactions on Pattern Analysis and Machine Intelligence*, 14(2):125–145, 1992.
29. T. M. Strat and M. A. Fishler. Contex-Based Vision: recognizing Objects Using Information from Both 2D and 3D Imagery. *IEEE Transactions on Pattern Analysis and Machine Intelligence*, 13(10):1050–1065, 1991.
30. P. Suetens and P. Fua and A. J.Hanson. Computational Strategies for Object Recognition. *ACM Computing Surveys*, 24(1):5-61, 1992.
31. S. Zhang and G. D. Sullivan and K. D. Baker. The automatic Construction of a View-Independent Relational Model for 3D Object Recognition. *IEEE Transactions on Pattern Analysis and Machine Intelligence*, 15(6):531–544, 1993.

Using quasi-invariants for automatic model building and object recognition: an overview.

Patrick GROS

LIFIA - INRIA Rhône Alpes
46, avenue Félix Viallet, 38031 Grenoble Cedex 1, FRANCE

Abstract. We address the problem of automatic model building for further recognition of objects. Our initial data are a set of images of an object. In a first stage, these images are put into correspondence using quasi-invariants, epipolar geometry and an approximation of the apparent motion by an homography. The different aspects of the objects may thus be computed and each aspect gives raise to a partial model of the object. In a second stage, these models and indexed in a data base which is used for recognition. This work is based on the idea that aspect graphs may (should?) be learned from examples rather than computed from CAD models, and that a planar representation associated with geometric quasi-invariants is a relevant tool for object recognition.

1 Introduction

This paper takes place in the frame of object recognition. This problem may be stated as follows:

- a system contains some "knowledge" about a collection of objects;
- when a new image is given to it, the system can determine if some of those objects are present in this image.

Usually the knowledge contained by the system is a set of what is called models, one for each objet, organized in a data base. When designing such a system, one has thus to decide which information is necessary to recognize an object, how this information can be obtained, i.e. how the models have to be computed.

The applications of such a system in a robotic environment are numerous: recognizing objects allows a robot arm to grasp them, a mobile robot to keep away from them when moving or to recognize its position according to high level markers. Furthermore, recognition is a bridge between low level environment description in terms of free space and shapes, and a high level description in terms of objects, rooms and ways.

The current approaches to this modeling problem may be classified according to two criterions: the kind of data used to construct the model and the dimension of the model constructed. Data may be 2D or 3D, man made or obtained from a sensor. The model may be 2D or 3D. Such a classification is presented by Flynn et al. [FJ91] and is used here to compare the different systems.

3D man made data: they usually come from a CAD system. The data are made of a description of the object in terms of its geometrical and mechanical properties. The problem is thus to infer the object visual aspects from these data. The model building step using CAD data has been intensively studied, creating a new field of vision called CAD-based vision [Bha87].

2D man made data: another way of using CAD data is to compute the 2D aspects of the modeled object [KD79, PPK92]. Each aspect is topologically different from the others and they are organized in a graph called aspect graph according to their associated viewpoint. The model of the object thus consists of the set of all its aspects. Even simple objects may have several tens of different aspects.

3D sensed data: they concern mostly two fields of vision: medical imagery using 3D volumetric sensors and robotic applications using 3D range sensors. In the first case the sensor gives a complete 3D image, while it gives only a depth map from a given viewpoint in the second case. Surveys of these techniques are given by Besl [Bes88] and Nitzan [Nit88].

2D sensed data: these data are usually images of the object to be modeled, taken from different viewpoints. Modeling and recognition systems using such data are very numerous. They differ in the kind of information they extract from the images, and in the dimension of the model (2D or 3D). Connell and Brady [CB87] use intensity data, Arbogast [AM91] uses occlusion contours, Mohr et al. [MVQ93] use points, Rothwell et al.[RZFM92] use numerical invariants associated with some configurations of points, lines and curves, Weiss uses differential invariants associated with algebraic curves [Wei92].

Our approach falls in the last category. It relies on the idea that aspect graphs should be learned from examples rather than computed from CAD models, and that a planar representation associated with geometric quasi-invariants is a relevant tool for object recognition.

The input consists of a large set of images. These images represent the object to be modeled and are taken from different viewpoints. The aim of the method is to find out which of these images represent the same aspect of the object. Such images belong to the same view of the object, and all these "characteristic" views form the object model.

Our method relies upon the matching of images one with another: two images represent the same object aspect if they contain approximately the same features and the same relationship between them. Thus we try to compare the contents of the different images. As the viewpoint changes between the different images, the location of the features within the images also changes and we try to estimate this motion in order to find a correspondence between the features of each image.

Our method models an object directly from what can be seen of this object in images. In this it differs from the methods based on CAD data. With these methods, the main problem is to infer visual information from geometrical properties. This inference is usually not satisfactory and is a weakness of the method. Furthermore, the use of aspect graphs adds another problem: the number of theoretical aspects of an object is much greater than the number of its visual aspects. Theoretical aspects very often differ only in unsignificant details. The complexity of these methods is a real obstacle. Bowyer gives a complete criticism of these methods [Bow91]. On the contrary, our method has a pragmatic notion of aspect. The different aspects are separated according to their visual dissimilarities, and not to their topological differences.

With respect to the methods using 3D models computed from 2D sensed data, our method avoids the reconstruction and projection stages. The reconstruction consists of computing the 3D shape of an object from 2D information. The projection is the opposite operation, i.e. computation a 2D visual aspect of an object from its 3D model. These two stages are complex and sensitive to noise.

Our method is thus more natural: the data used for modeling are 2D sensed data, so are the images to be recognized. The built models stay as close as possible to this kind of data.

In this paper we focus on the two main stages of the method. The algorithm to find the correspondence between two images is described in section 2; section 3 explains how to go from these correspondences to the model, and particularly shows the learning ability of this process.

2 Matching sets of 2D features

2.1 The matching algorithm

The aim of matching is to find which segments of each image are the projections of the same edge of the 3D object. The output is a correspondence between the features (here the segments) of each image.

Matching is a prior stage to many algorithms and usually relies on one of the two following assumptions:

1. first assumption: the motion of the camera between the two viewpoints (or that of the object if the camera is supposed motionless) is approximately known and the location of one feature in an image may be deduced from the location of the corresponding feature in the second image. Foe example, this is assumed by the systems based on correlation techniques [Ana89, Fua90]. Another important case of systems using this assumption is that of tracking. The motion is supposed to be very small or very regular and the location of the features within an image of a sequence may be predicted from the knowledge of the previous images of the sequence [CS90, DF90].
2. second assumption: some of the features or group of features remain qualitatively similar. In this case, matching is based on the search of particular features configurations: small graphs of segments [SH92], the whole graph of all the segments [HHVN90], symmetric features [HSV90].

The first methods are quite limited by their assumption: the motion has to be approximatively known. In many cases, especially those when the camera is not calibrated, the motion is not known at all, even if its kind (pure rotation or translation...) is known. This is also the case if the images are taken with different cameras. The second methods are sensitive to noise. In the case of the use of small graphs of segments, either these graphs are too big and their configuration is never perfectly conserved, or they are too small and are no longer discriminant.

Our method is based on the following idea: matching would not be a problem if corresponding features were in the same place in the two images to be matched.

This difference of position is called "apparent motion". This motion is not a classical geometric transformation because two different features in one image can correspond to only one feature in the second one. On the other hand, in many cases, it can be approximated by a transformation like a similarity or an affine transformation. Our method consists in computing such an approximation. It does not assume that the camera motion is known or that it is very small. It also works with noisy images or occluded objects.

The different stages of our matching method are the following:

1. We have two images containing line segments approximating contour curves. We assume that segments' apparent motion between the two images is a similarity (resp. an affine transformation). We associate numerical invariants to some feature configurations: angle and length ratio defined by every pair of segments having an extremity in common (resp. affine coordinates associated with every set of four connected segment vertices).
2. Invariants and their corresponding segments and vertices are matched according to the invariants' value. As there is some noise in the images, equality is tested up to a noise threshold, in consequence of what all matches are not right.
3. To eliminate wrong matches, a Hough transform technique is used, in order to evaluate the parameters of the approximation of the apparent motion. As a matter of fact, the right matches define the same approximation and the computation of this motion allows to recognize them. When two invariants are matched, there is enough geometrical information to compute the transformation. In our case, when two configurations are matched, it is possible to compute the parameters of the similarity (resp. the affine transformation) which transform one of the two configurations into the second one. Such a computation is done for all the matches done at stage 2, whether they are right or wrong. In this way, each match gives a point in the transformation parameter space.
4. The points corresponding to wrong matches are distributed almost uniformly in the parameter space. This is because they are not correlated. On the contrary, the points corresponding to right matches define all the same real transformation parameters up to a noise factor. Thus they give many points in a small region of the space. This "accumulation point" may be found easily and define the best estimate of apparent motion. Every match which gives a transformation far the best estimate is eliminated.
5. The match between the individual segments are deduced easily from the matches of segment pairs.

2.2 Matching improvement

The algorithm described in the previous section allows to make a first match without a priori informations. The result can contain errors and miss many possible matches. This is due to the apparent motion approximation which is too restrictive for some image regions.

Once this first match is made, two finest tools may be used to evaluate the validity of already computed matches and to find new ones. The first one is an approximation of the apparent motion by an homography, the second one is epipolar geometry.

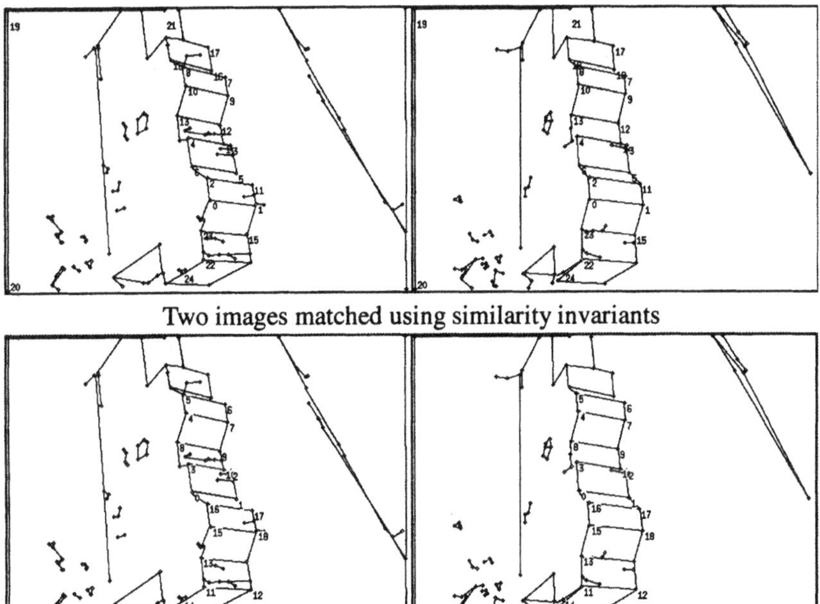

Two images matched using similarity invariants

Two images matched using affine invariants

Fig. 1. Experimental matching results

Computing both of them using a least median square method allows to detect wrong matches. On the other hand, they give a precise evaluation of the localization of the correspondent of every point, and allow to make new matches.

2.3 Experimental results

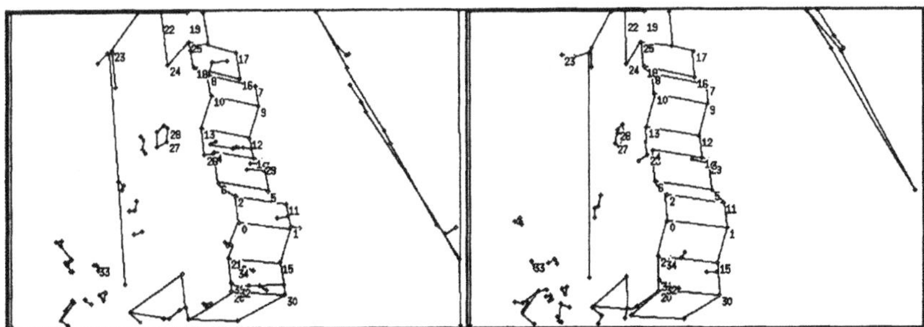

Fig. 2. Matching improvement results.

In this paragraph, we provide some results that show that the algorithm is robust, even when the apparent motion is far from an exact similarity or affine transformation.

Figure 1 shows the match obtained with two images. The corresponding vertices have the same number in the two images. The two upper images were matched using similarity invariants, while we used affine invariants in the two lower images. The first image contains 132 vertices, the second one 105. With the similarity invariants, we obtained 24 correspondences and 25 with the affine invariants. In both cases, all the correspondences are right.

Figure 2 shows the result of the improvement process on the first images shown on Figure 1.

After this first matching stage, it is now possible to compute the epipolar geometry of the two images, or an approximation of the apparent motion by a homography. This information may then be used to detect any eventual wrong match, and to find other correspondences.

3 From matching to modeling

The matching algorithm is the central point of our method. As a matter of fact, it allows to find the characteristic and robust features of one object represented in two noisy images. A model of this object to be used in a recognition system has to contain this information: the robust and characteristic features that will appear in every image where the same aspect of the object is visible.

3.1 Matching more than 2 images

A first problem is that two images may not be sufficient to recover the whole characteristic structure of one aspect of the object if they are very noisy. In the case where we have three or more images, we use the following stages:

1. the images are matched two by two;
2. as incoherences may occur, we compute a global match;
3. the features that appear at least in 60 percents of the images are put in the model; the position of the modeled feature is the average of the corresponding features in the images, after correction of the apparent motion by an homography.

This algorithm assumes that the three images represent approximately the same aspect of the object. The most in this algorithm is the computation of a global match from partial matches. To do that, we represent the different matches as a graph; the vertices of this graph are the features of each image, and the edges represent the correspondences. We then consider the different connected components. They usually contain to many features, and we cut them up in more strongly connected parts until they reach an acceptable size. Two vertices a and b are strongly connected if they are connected and if there exist some other vertices c_i which are connected with both a and b. The degree of connectedness is then given by the number of the existing vertices c_i. A component is of acceptable size if it contains at most one feature of each image and at least $0.8n$ features where n is the number of images.

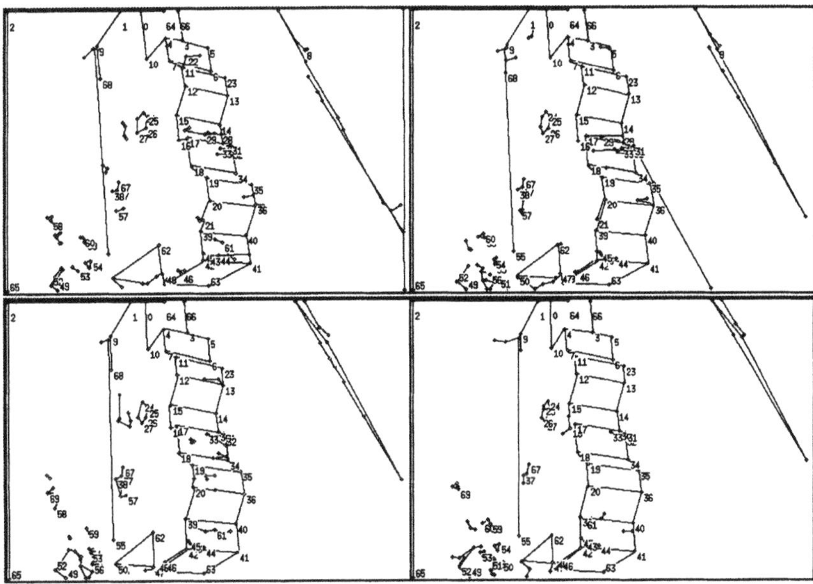

Fig. 3. 4 images from a 10 image sequence globally matched.

3.2 Computing aspect models

When the different images available do not represent the same aspect of the object, the different aspects have to be separated first. Otherwise, the global match algorithm will fail to find features present in more than 60% of the images.

In this case, we add two steps to the previous algorithm:

1. the images are matched two by two and the distance between two images is defined as the percentage of matched features;
2. all the distances are put in a distance matrix and an agglomerative clustering algorithm is used to gather the images according to the aspect of the object they represent;
3. a global match is computed for all the groups found at the previous step;
4. one model is deduced from each group; in this case the model of an object is composed of a set of partial models.

A difficult step in this algorithm is the automatic computation of the number of aspects. This is a general difficulty in all clustering methods.

3.3 Towards recognition

As this part of the work is still under development, this paragraph only gives an overview of the subject. For each object we want to recognize, we learn its model from a set of images, taken from various viewpoints. All the models all gathered in a model base.

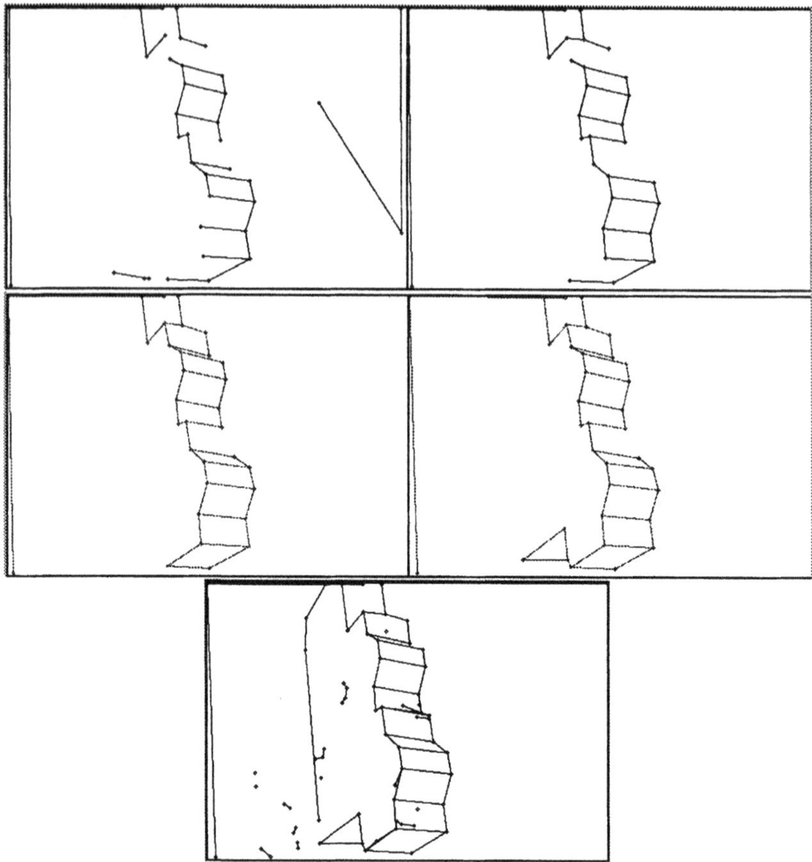

Fig. 4. Models of an object computed from respectively 2, 4, 6, 8 and 10 images of it.

The recognition problem may then be solved as a correspondence problem between a model and a new image. The matching algorithm presented in section 2 may be adapted with this aim in view.

For a kind of transformation (affine or similarity), the invariants of all models are put in a single table. Some invariants are also computed from the image and compared with the first ones. When a correspondence is found, that gives a vote to a model. This vote is expressed as a point in a transformation parameter space. When all the possible correspondences are made, we count the coherent votes of each model, following the method presented in section 2.1.

This allows to predict the models which have the greatest probability to appear in the image. This has to be followed by a verification stage. The features corresponding to the model are removed from the image, and the process may be repeated to find another object.

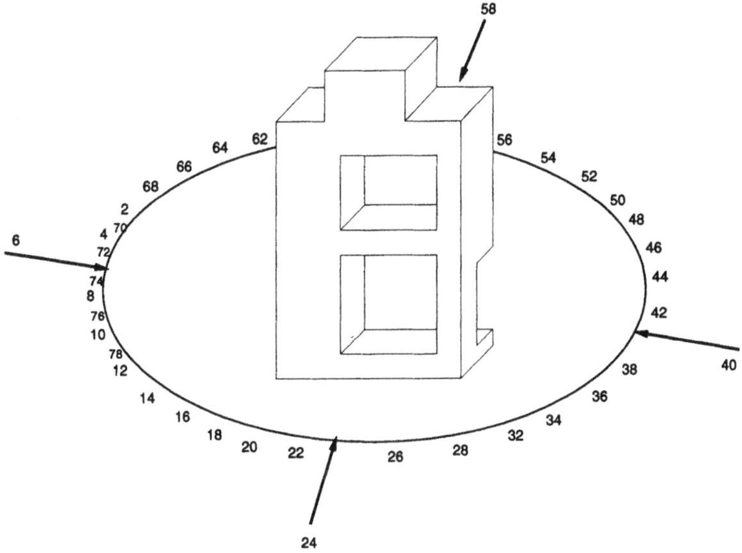

Fig. 5. 80 views of an object.

3.4 Experimental results

To obtain the following results, we used a sequence of ten images of the same object. These images were matched two by two, and a global matched was computed using the presented algorithm. The result is shown on Figure 3: the points labeled by a same number are in correspondence.

After that, we computed different models. Figure 4 show the models computed from 2, 4, 6, 8 and 10 images. The models are clearly more and more complete. In the last model, some texture even appear has stable.

The case where different aspects of a same object are present in the images is presented on Figure 5 and 6. The first one shows an object: 80 images were taken turning around it. The second Figure shows two images and their features for each one of the seven computed aspects.

The object is almost symmetric with respect to a vertical plane. Some images taken from opposite view points are then considered as defining the same aspect.

4 Conclusion

In this paper are presented a algorithm for image matching, in the case where images contain line segments, and a method to construct a model from a set of images of an object.

Even if the learning ability of this last method is clear according the provided results, the use of such models in a recognition system causes other difficulties, like fast indexing, that have still to be studied.

The main contributions of this work are:

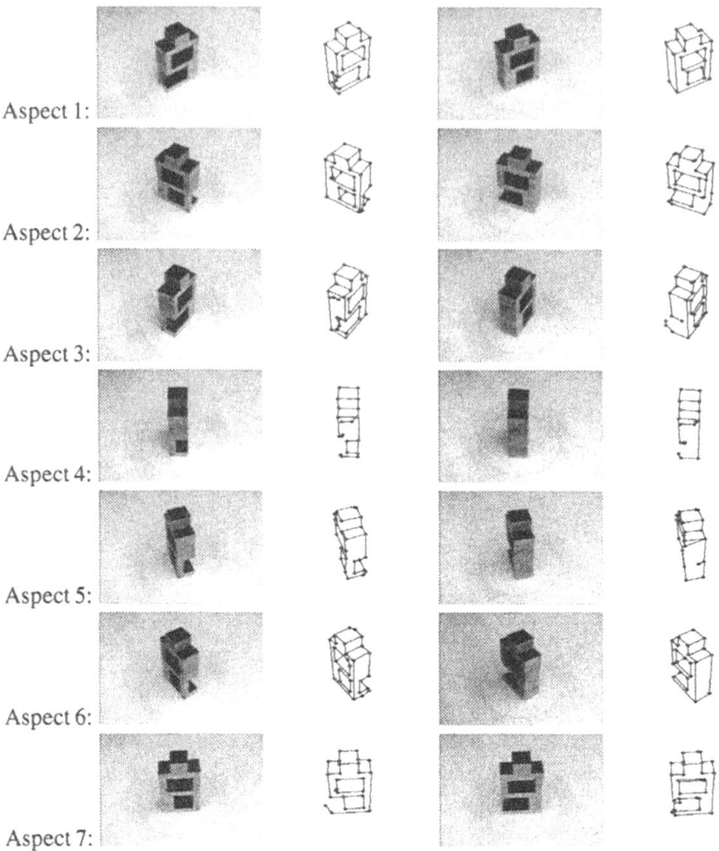

Fig. 6. Two images of each model, with their features.

- the use of local quasi invariants as a robust and discriminant feature in images; they have proven to be more usable than topological structures like subgraphs or even than exact invariants which are sensitive to noise and difficult to compute for non trivial objects;
- a new matching algorithm which works for images containing segments, even if the motion of the camera is unknown; this geometric method is a real alternative to the often used correlation and relaxation techniques;
- a method for "modeling from examples", which can compute a model for the main aspects of an object without computing the aspect graph of the object from a CAD model.

Acknowledgments. This work has been sponsored by European Esprit project No 6769 (the SECOND Project). The author also acknowledges Edmond Boyer and Olivier Bournez for their participation to the project, and Françoise Veillon for her fruitful comments.

References

[AM91] E. Arbogast and R. Mohr. 3D structures inference from images sequences. *International Journal of Pattern Recognition and Artificial Intelligence*, 5(5):749, 1991.

[Ana89] P. Anandan. A computational framework and an algorithm for the measurement of visual motion. *International Journal of Computer Vision*, 2:283–310, 1989.

[Bes88] P.J. Besl. Active optical range imaging sensors. Springer-Verlag, New York, USA, 1988.

[Bha87] B. Bhanu. Guest editor's introduction. *Computer (Special Issue on CAD-Based Robot Vision)*, August 1987.

[Bow91] K. Bowyer. Why aspect graphs are not (yet) practical for computer vision. In *Proceedings of the* IEEE *workshop on Direction on automated* CAD*-based Vision, Maui, Hawaii, USA*, pages 97–104, 1991.

[CB87] J.H. Connell and M. Brady. Generating and generalizing models of visual objects. *Artificial Intelligence*, 31:159–183, 1987.

[CS90] J.L. Crowley and P. Stelmazyk. Measurement and integration of 3D structures by tracking edges lines. In O. Faugeras, editor, *Proceedings of the 1st European Conference on Computer Vision, Antibes, France*, pages 269–280. Springer-Verlag, April 1990.

[DF90] R. Deriche and O. Faugeras. Tracking line segments. In *Proceedings of the 1st European Conference on Computer Vision, Antibes, France*, pages 259–267. Springer-Verlag, April 1990.

[FJ91] P.J. Flynn and A.K. Jain. CAD-based computer vision: from CAD models to relational graphs. IEEE *Transactions on PAMI*, 13(2):114–132, February 1991.

[Fua90] P. Fua. A parallel stereo algorithm that produces dense depth maps and preserves image features. *Machine Vision and Applications*, 1990.

[HHVN90] L. Hérault, R. Horaud, F. Veillon, and J.J. Niez. Symbolic image matching by simulated annealing. In *Proceedings of the British Machine Vision Conference, Oxford, England*, pages 319–324, September 1990.

[HSV90] R. Horaud, T. Skordas, and F. Veillon. Finding geometric and relationnal structures in an image. In *Proceedings of the 1st European Conference on Computer Vision, Antibes, France*, pages 374–384. Springer-Verlag, April 1990.

[KD79] J. Koenderink and A.V. Doorn. The internal representation of solid shape with respect to vision. *Biological Cybernetics*, 32:211–216, 1979.

[MVQ93] R. Mohr, F. Veillon, and L. Quan. Relative 3D reconstruction using multiple uncalibrated images. In *Proceedings of the Conference on Computer Vision and Pattern Recognition, New York, USA*, pages 543–548, June 1993.

[Nit88] D. Nitzan. Three-dimensional vision structure for robot applications. IEEE *Transactions on PAMI*, 10(3):291–309, 1988.

[PPK92] S. Petitjean, J. Ponce, and D.J. Kriegman. Computing exact aspect graphs of curved objects: algebraic surfaces. *International Journal of Computer Vision*, 9(3):231–255, 1992.

[RZFM92] C.A. Rothwell, A. Zisserman, D.A. Forsyth, and J.L. Mundy. Fast recognition using algebraic invariants. In J.L. Mundy and A. Zisserman, editors, *Geometric Invariance in Computer Vision*, chapter 20, pages 398–407. MIT Press, 1992.

[SH92] H. Sossa and R. Horaud. Model indexing: the graph-hashing approach. In *Proceedings of the Conference on Computer Vision and Pattern Recognition, Urbana-Champaign, Illinois, USA*, June 1992.

[Wei92] I. Weiss. Noise-resistant projective and affine invariants. In *Proceedings of the Conference on Computer Vision and Pattern Recognition, Urbana-Champaign, Illinois, USA*, pages 115–121, June 1992.

Object Representation for Recognition-by-Alignment

George Stockman

Computer Science Department, Michigan State University,
E. Lansing, Michigan 48824

Abstract. We present an approach general enough to apply to recognition of complex rigid 3D objects from either a single intensity image or a single range image. Within the general paradigm of recognition by alignment, we address (1) definition and detection of primitives, (2) indexing to model hypotheses, (3) constructing view sphere models from sensed data, and (4) aligning model and sensed features for verification. The overall paradigm is not new, but rather fits within theory already espoused by Lowe and Ullman and many others: our position is therefore both a synthesis of and endorsement of much other work toward recognition of rigid free-form objects.

1 Introduction

We advocate recognition by alignment for rigid, but otherwise general objects and report some of the results of five years on work in the area. The advantages and disadvantages of the general paradigm are discussed.

One important lesson learned from artificial intelligence (AI) and computer science in general is that problem representation is a very important part of problem solution. Because the problems in the field of computer vision (CV) are so varied, it seems to follow that no one representation and no single process will serve all purposes. Accepting this, one must then look for a catalog of representations and processes, each of which is applicable to certain tasks and domains. Having stated this position in [9] and mentioning it here, we proceed within just the alignment paradigm.

We assume a domain of rigid objects and will primarily exploit shape information in recognition. Change of albedo (*surface marks*) observable in an intensity image will also be used. Our object domain is admittedly ill-defined; we would like our programs to incorporate as few assumptions as possible. Generally we follow the paradigm proposed by Lowe [17]: incomplete sets of image features are used to index into model space in order to activate top-down processes which attempt to verify the presence of an object in a certain pose. The task of model verification, although fairly well-defined, is complicated because objects are 3D and images are 2D. Also, scene objects may be partly occluded. Object models need not be 3D, however. In Section 2 below, aspect models are described which allow simplification of matching by representing the object features in terms easily related to the observed image. Section 3 describes work done in creating aspect models from intensity images and using these models to match an object observed in a single intensity image. Section 4 explores how range images might be used in order to improve the capability of matching.

2 Recognition using image-based aspects

There has been a great deal of CV work within the recogniton-by-alignment paradigm. Often, alignment is achieved by the ability to extract salient features from the observed data and make correspondences between those features and model features (e.g. [20, 14, 11]). Only one global model containing the aggregation of labeled geometric features is needed. Another significant advantage of this approach is that an alignment transformation can be computed in closed form from a small set of feature correspondences. Disadvantages are that (a) a general method must possess a large set of these special feature definitions and (b) the method may not be applicable at all to objects that do not have enough useful salient features.

Another promising approach is through fitting 3D parametric models to 2D or 3D sensed data [1, 21]. The main advantage of this approach is that a global view-independent model of roughly a few dozen parameters is obtainable via fitting the model to 3D sensed data. Disadvantages include a) model fitting often assumes very good segmentation; b) the models may not be sensitive enough to handle slight differences between objects or instances of the same kind of object; c) complicated objects need to be modeled by composing parts, which mixes a segmentation problem with the alignment problem. Here, we advocate the view-based approach, where 3D objects are modeled by a collection of 2D views of the object, each view being fairly close in geometry and topology to a sensed image [3, 10, 12, 18]. The advantages are that learning can be used in a straightforward manner to build models and that matching can be done with a representation that is close to the sensed data. The disadvantages are that the representation is verbose and full of seams and that the indexing scheme must not only produce candidate objects but also particular views of that object.

2.1 Primitive features

There is good agreement among both CV researchers and psychologists that edge segments are fundamental units for low level representations. Edges are generally present in images and if context allows us to bound scale, many good edge detectors exist. A great deal of 3D shape information is available from the observed silhouette; edges internal to the silhouette refine the overall shape, differentiate similar objects, and differentiate object pose. Moreover, because of their precise locality, edges are critical for alignment. Therefore, edge segments and structures immediately constructed from them are the primitives used in our representation. The models proposed in Sections 3 and 4 are thus both image-based and edge-based. Being edge-based rather than surface-based, we lose the capability of generating complete images, which some other approaches can do. Although an edge representation is sparse, we do represent some local surface information. In the representation described in Section 3, local surface approximation is stored for each smooth rim point seen in a set of intensity images. Section 4 describes how contrasting surface information is encoded along with each edge primitive extracted from a range image.

2.2 Primitive Vocabulary and Indexing

We need a finite vocabulary of symbolic primitives to be used to index into a small set of candidate model aspects based only on the representation extracted bottom-up from an observed image. A good indexing strategy would operate in $O(lg\ N)$ runtime for a model base of N models, but some would even argue for $O(1)$ runtime. The indexing scheme may use (a) the primitive symbols, (b) numerical attributes of the symbols, and (c) relationships or combinations of (a) and (b).

To support indexing, the primitives and indexing scheme must have the following properties.

1. We must be able to extract the primtive regardless of significant variation in translation, rotation, and scale. (Since we are view-based, this variation can occur in 2D, not 3D.)
2. The numerical attributes of the indexing primitive must be invariant as above.
3. The indexing scheme must account for occlusion: it must work when characteristic primitives are missing or added due to an interposing object.

For intensity images, we propose segmentation of object silhouettes based on an intermediate codon representation [13, 19]. Our initial experience with these previously proposed techniques has been very encouraging. The parts resulting from the segmentation according to curvature extrema do have the properties prescribed above. Ideas and methods of indexing have been reported by Breuel [4] and Califano and Mohan [5].

In range images, surface information is usually available in the neighborhood of an "edge"; in fact, "edges" are extracted using the contrasting surface properties. In Section 4, we describe some experiments using primitives which we call "wings" that encode adjacent surface information along with an edge segment. Such primitives should have more discriminating power due to the symbolic shape attributes available on either side of the edge segment. Details of our work with such a vocabulary are given in Section 4.

3 Modeling and recognition using intensity images

We have used the curvature method of Basri and Ullman to construct aspects of objects which represent not only the local 3D structure at the rim but also the 3D location of creases and marks. A model aspect can be used to predict all significant object contours in an intensity image taken from a viewpoint within that aspect. A global X-Y-Z coordinate system is assumed fixed to the object: our convention is to assign the Y-axis to a canonical vertical direction on the object; for example, for a car the Y-axis would be perpendicular to the ground plane made by the four wheels. A model aspect is made from 5 images taken by rotating the viewpoint up,down,left and right of a central viewpoint [2]. Typically, we have used rotations of 20 degrees, but the size of the model aspect is dependent on the complexity of the object and the precision of image modeling and pose computation needed for the recognition task. Model aspects need not *partition* the viewsphere and, depending on the task, model aspects may or may not cover the entire viewsphere. What is important is that a recognition system can index from an image to candidate model aspects or that a motion tracking system can transition cleanly from one aspect to another as the object moves. To monitor cars at an intersection or entrance, perhaps only 9 such model aspects are needed, each taken from an oblique downward viewpoint. (9 x 40 degrees give a full circle.) For objects on a conveyor we could use 9 aspects for each stable state, or we could cover the entire viewsphere with about 40 aspects and handle any object pose. To inspect an object in a nominal position, perhaps only one aspect is needed.

For each model aspect, we store the three rotational parameters that relate the local coordinate frame of the aspect to the global coordinate frame for the entire object model. The 4 rotation angles (up,down,left,right) used to create the 4 viewpoints are stored in order to define the limits of the aspect. We store each crease or mark point (x_i, y_i, z_i). For each silhouette point, we store the center and radius of curvature which generates the local neighborhood of that point for the small rotations within the aspect. Given pose parameters ω, the image point for each model silhouette or internal edge point can be computed by applying the parameters of rotation, scaling, translation, and projection. (We use weak perspective projection.) Because of the construction process, we do not need any hidden surface or line removal when using this model aspect to generate an image edgemap. Besides the geometric information needed to predict image contours, the object model must also contain the neighborhood relations among the model aspects stored and a feature table which can be used for indexing whenever the object becomes *live* in a domain.

3.1 Steps of recognition

The steps of recognition proceed as follows.

- object extraction
  The object or jumble of objects must be extracted from the background: this is the difficult figure-ground problem in general. For indoor scenes, we have controlled the background to allow for simple separation methods. For outdoor scenes we use image subtraction to extract blobs corresponding to moving objects. We have only done minor work on this step.
- image processing
  The first step yields the boundary of the objects. Edges internal to the boundary are obtained using the Canny operator: small segments are discarded. We have only done minor work on this step.
- indexing to model aspects
  The boundary is segmented into codon representation and then the codons are segmented into parts [19]. Parts may be convex or concave. For each part, several features are computed which are invariant to scaling, translation, and rotation. Also, for each part, at least three points are available to determine a model-to-object alignment for just this part. The set of parts, together with their invariant features are used to index into the model aspect database. The indexing scheme allows for added or missing parts due to occlusion, imperfect object extraction, or imperfect feature extraction. Work with 2D objects has been very successful; testing on the approximating silhouettes of 3D objects is still in progress [8].

- aligning model aspects to images
  Using feature points from parts extracted in the previous step, approximate translation, scale, and rotation about the view axis (Z-axis) is available. A crude alignment is thus available. (Major advantage of an aspect model.) The best alignment is defined in terms of the mean squared distance between certain image edges and corresponding projected model edges. The Levenberg-Marquardt algorithm is used to search for the six pose parameters $\omega = (s, t_x, t_y, \theta, \phi, \rho)$ that produce a satisfactory alignment. Heuristics are used to (a) identify corresponding edge segments in different areas of the image, (b) discard badly matching edge segments, and (c) balance the weight of silhouette versus internal edge matching errors. The first two heuristics are designed to counter the effects of occlusion and imperfect feature detection, while the third is designed for more precise pose parameters. A fourth heuristic is used to synthesize error vectors from corresponding edge segments in order to refine pose parameters in the next iteration. Some results of our alignment work are given below.

3.2 Results with intensity images

In Figure 1 below, four experimental objects are shown with the final alignment of model edgemaps. In one test of the viability of the model aspects, for each input image we initialized the model pose ω at regularly spaced points on the viewsphere within the aspect and studied the convergence of the alignment algorithm. Table I shows the results. Note that translation and scale are not varied systematically in Table I; they were set arbitrarily by hand. Other tests have shown that any translation within the image frame and scale variations between 0.8 and 1.2 are routinely handled by the algorithm. The column $\sharp$ *(Error < Et)* in Table I records the number of times that the alignment algorithm terminated with an alignment error less than the tolerance of *Et* =1.0 pixels. The column $\sharp|\omega - \bar{\omega}|$ gives the number of the computed pose vectors ω that were within a small box around the ground truth parameters $\bar{\omega}$. Rotation is defined in terms of axis direction cosines **N** and angle θ. For each test object a high percentage of the starting viewpoints converged to an alignment within the error tolerance *Et*. Moreover, there was no significant decline in performance for the cases where part of the object was occluded. The average number of iterations (not shown in the table) ranged from 9 for Ellipsoid1 to 19 for the Squirrel.

4 Modeling and recognition using range images

For range images, we proposed primitives which we called *object wings* [6] to represent meaningful object structure useful for both indexing and alignment. Wings are, however, small enough so that some of them are likely to survive significant occlusion and feature detection error. We defined a *simple wing* as a triple consisting of a pair of 2 1/2-D surface patches separated by a 2D contour segment.

`contour symbols` `surface symbols`

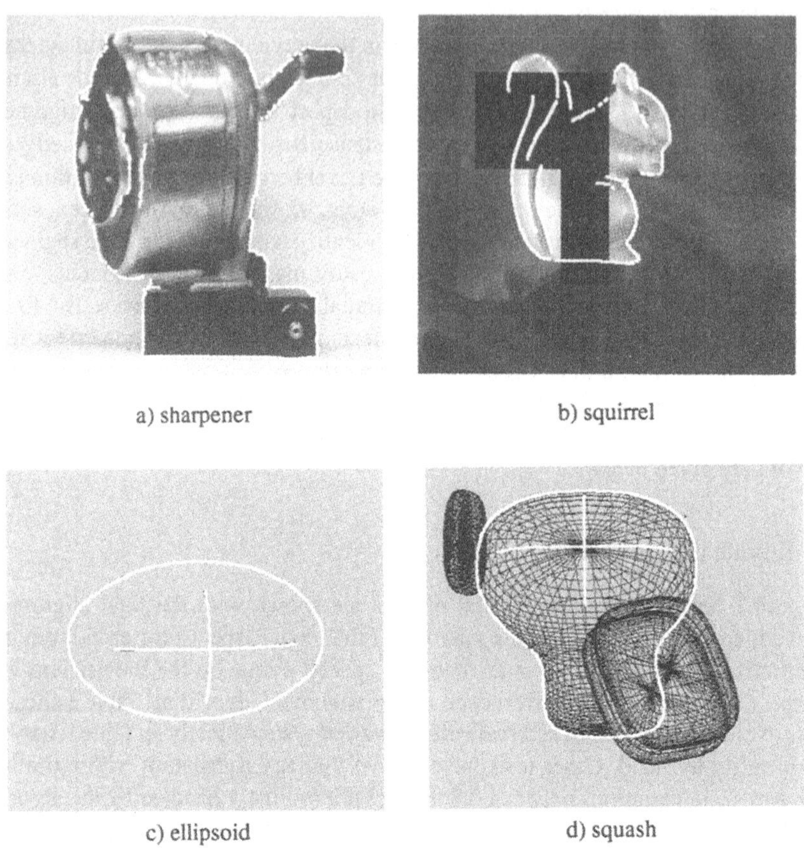

Fig. 1. Model fitting of various objects (from Chen94). (a) pencil sharpener; (b) partly occluded squirrel; (c) a superquadric model (ellipsoid); (d) a partly occluded superquadric model ("squash").

```
!    silhouette/jump    NIL  unknown/background
<    occluding/blade    +    convex surface
<<   occluding/limb     -    concave surface
+    convex crease      S    saddle
-    concave crease
```

From combinatorical considerations only, there could be $4^2 5 = 80$ wing primitives, but only about half are actually possible. Figure 2shows the wing representation for 3 distinct 3D objects which yield essentially identical contour segments. Clearly, the additional information in the wing primitive would not only be necessary for efficient indexing, but it would also be convenient for lower level grouping processes used in

TABLE I

Results of Alignment Trials

(Error Tol. Et = 1.0 pixels; Max iterations = 100)
(Eb: $|\theta - \bar{\theta}| \leq 3$, $\|n - \bar{n}\| \leq 0.1$,
$|t_x - \bar{t_x}| \leq 2$, $|t_y - \bar{t_y}| \leq 2$, and $|s - \bar{s}| \leq 0.01$.)

Object	Using Silhouette and Internal Edges		
	♯ Trials	♯ (Error < Et)	$\sharp\|\omega - \bar{\omega}\| \in Eb$
Pencil Sharpener	20	20	20
Squirrel	240	239	233
Ellipsoid1	240	236	240
Ellipsoid2	240	235	240
occluded Squash	240	240	216
occluded Squirrel	240	237	226

bottom-up extraction of an image representation.

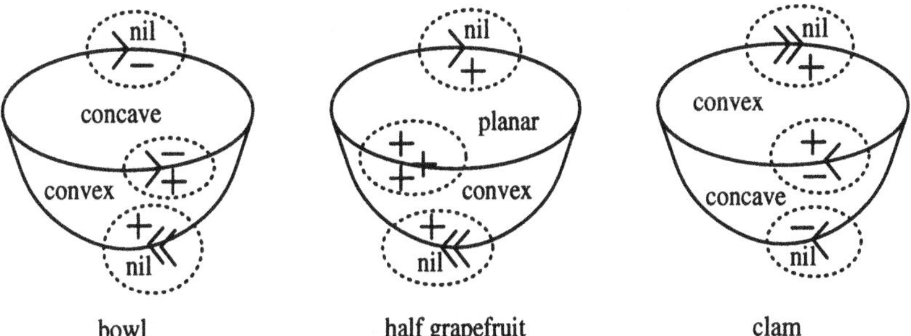

Fig. 2. Wings from 3 different objects with similar image contours

4.1 Segmentation and indexing using wings

Work with polyhedral and origami objects confirmed the utility of the wing-based approach [16]. Larger object parts could be derived from multiple wings and occluding parts could be identified using a bottom-up grouping algorithm. The representation computed from the grouping process was called a *Labelled Line Drawing Graph* (LLDG); a representation simliar to that of Huffman and Clowes, but with geometrical attributes recorded. These results were unexciting since many other methods can handle polyhedra. Chen performed experiments [6] with sculptured objects and found that segmentation was difficult and that model aspects could be quite ambiguous. Improved range scanners which could produce fine silhouettes in the image would allow the part segmentation work described in Section 3 to be directly extended to range images. The 3D surface tags along the contour segments could augment the 2D invariants in any indexing scheme. We need to return to explore this in future work.

4.2 Using specific surface hypotheses

Lee explored using specialized fitting to extract wings from objects with quadric surface patches, as should be permissible with many industrial objects [16]. He used both 2D contour points and neighboring 3D surface points to fit small windows of a fused range and intensity image and found that the contour points had a significant focusing effect on the surface model fitting. Others had shown how such data could be usefully combined, but Lee applied the concept to parts of parts and showed that only a dozen silhouette and a dozen surface points could yield a good estimate of part shape. Table II is one example of many similar ones we obtained for the results of trying several focused fit hypotheses windows of fused range and intensity data. In this case, cylindrical data is in the window and competing hypotheses are {*cylinder, cone, sphere, plane*} . Only 15 range samples were used in fitting, and, when contour data was used ("Fused" column) only 15 contour samples were added. Table II shows that with low noise level, surface data alone can be used to make a fairly reliable surface classification, but under higher levels of noise, use of associated contour samples is needed for acceptable classification.

TABLE II
FITTING ALL MODELS TO CYLINDRICAL SURFACE PATCH

True surface form: CYLINDRICAL Number of trials: 1000 Number of points sampled per trial: 30									
		Surface-Only				Fused			
		con	cyl	sph	pln	con	cyl	sph	pln
$\sigma = 0.01$	*bad fits*	530	58	882	736	824	61	913	571
	χ^2	5.38	0.30	6.76	5.63	5.11	0.21	6.10	5.21
	iterations	23.72	6.80	8.53	3.61	29.00	6.82	4.55	3.41
	Best χ^2	58	905	6	13	4	933	0	40
	% misclassified	7.8%				4.5%			
$\sigma = 0.05$	*bad fits*	661	728	996	857	859	233	942	637
	χ^2	7.17	6.53	10.09	8.85	6.09	4.39	7.94	6.87
	iterations	21.16	5.43	4.00	3.63	29.10	6.51	4.67	3.45
	Best χ^2	276	235	1	87	45	706	0	96
	% misclassified	60.8%				16.6%			

Lee's goal of achieving a valid LLDG for such curved objects is sketched in Figure 3. Derivation of a valid LLDG as shown in the bottom block of the figure is not only unnecessary but probably impossible in general. Within the context of the paradigm presented in Sections 2 and 3 any representation intermediate between the middle and bottom block of Figure 3 should be adequate to support the indexing stage of recognition-by-alignment.

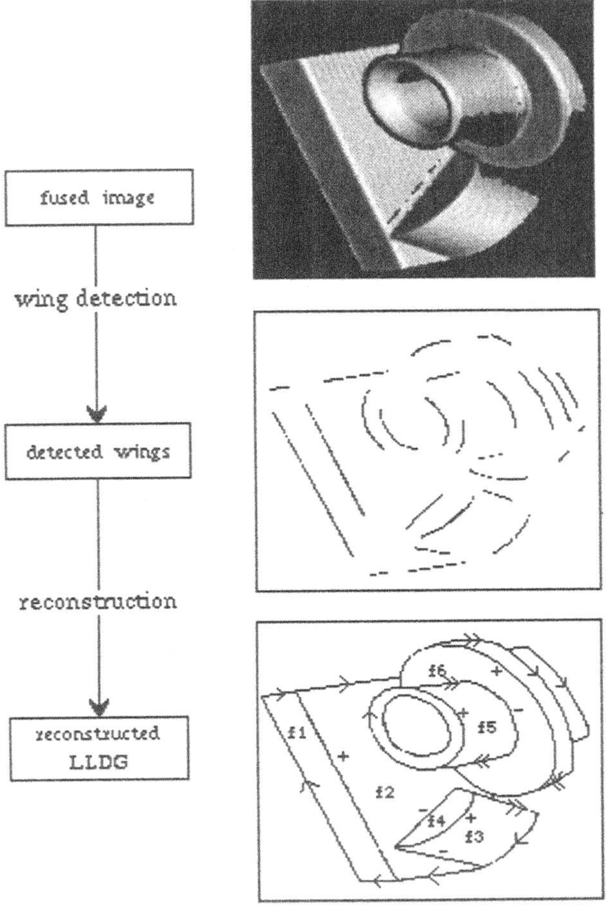

Fig. 3. Extracting wing features from a range image and using them for richer bottom-up description.

5 Discussion

This paper tried to establish these points: (1) object representation is critical for recognition but is both object domain and task specific; (2) recognition-by-alignment is a paradigm that might be practical for all rigid objects; (3) a promising implementation within this paradigm exists, which uses image-based object models, codon-based segmentation for indexing, and an optimization technique with heuristics to achieve alignment of model aspect to image contours.

A large number of research efforts can be interpreted as fitting within the pardigm presented; space limitations prevented adequate citation of all important contributions. In our own work, we have experimented with alternative implementations of the various stages in the alignment paradigm. Much of our recent efforts have gone into model-building and the alignment algorithm. Current work is focused on indexing and we are concentrating on intensity images rather than range images. There is still much to do.

Acknowledgements

Many ideas in the paper were developed jointly with Sei-Wang Chen, Greg Lee, and Jin-Long Chen [6, 16, 7]. Any implementations reported in Section III are due to Jin-Long Chen while those reported in Section IV are due to Sei-Wang Chen and Greg Lee. This work was supported in part by NSF Grant CDA-8806599 and by Texas Instruments.

References

1. R. Bajcsy and F. Solina: Three-Dimensional Object representation Revisited. Proceedings 1st Int. Conf. on Computer Vision, (1987) 231–240
2. R. Basri and S. Ullman: The Alignment of Objects with Smooth Surfaces. Proceedings 2nd Int. Conf. on Computer Vision, (1988) 482–488
3. K. Bowyer, D. Eggert, J. Stewman, and L. Stark: Developing the Aspect Graph Representation for Use in Image Understanding. In Proc. DARPA Image Understanding Workshop, Palo Alto,(1989) 831–849
4. T. Breuel: Adaptive Model-based Indexing. In Proc. 1989 DARPA Image Understanding Workshop, (1989) 805–814
5. A. Califano and R. Mohan: Multidimensional Indexing for Recognizing Visual Shapes. IEEE T-PAMI, Vol 16., No. 4, (1994) 373–392
6. S.-W. Chen: Wing Representation for Rigid 3D Objects. In Proc. 10th IAPR, Atlantic City, NJ, (June 1990) 398–402
7. J.-L. Chen, G. Stockman, and K. Rao: Recovering and Tracking Pose of Curved 3D Objects from 2D Images. Proc. IEEE Conf. on Computer Vision and Pattern Recognition, New York, NY, (June 1993) 233–239
8. J.-L. Chen and G. Stockman: Indexing to model aspects using invariant contour features. Computer Science Dept. Tech. Rep., Michigan State Univ., (Nov 1994)
9. J. Ponce, D. Forsyth, L. Shapiro, R. Bajcsy, D. Metaxas, M. Hebert, K. Ikeuchi, S. Sclaroff, A. Pentland, T. Binford, A. Kak, and G. Stockman: Object Representation for Object Recognition. Proc. IEEE Conf. on Computer Vision and Pattern Recognition, Seattle, WA, (June 1994) 147–152
10. S. Edelman and D. Weinshall: A Self-organizing Multiple View Representation of 3D Objects: Biological Cybernetics, Vol.64, (1991) 209–219
11. P. J. Flynn and A. K. Jain: BONSAI: 3D object recognition using constrained search: IEEE-T-PAMI, Vol. 13, No. 10, (Oct 1991) 1066–1075
12. C. Goad: Special Purpose Automatic Programming for 3D Model-based Vision. In Proc. DARPA Image Understanding Workshop, Arlington, VA, (1983)
13. D. D. Hoffman and W. A. Richards: Representing smooth plane curves for recognition. In Proc. AAAI, Pittsburg,PA (1982) 5–8
14. D. Huttenlocher and S. Ullman: Recognizing solid objects using alignment: Proc. DARPA Image Understanding Workshop, (April 1988) 1114–1122
15. K. Higuchi, H. Delingette, M. Hebert, and K. Ikeuchi: Merging Multiple Views Using a Spherical Representation. In Proc. IEEE 2nd CAD-Based Vision Workshop, Champion, PA (Feb 1994) 124–131

16. G. C. Lee: Scene Representation from Fused Imagery. PhD Dissertation, Michigan State Univ., (August 1992)
17. D. G. Lowe: Three-dimensional Object Recognition from Single Two-dimensional Images. Artificial Intelligence, Vol.31 No.3, (1987) 355–395
18. A. Pentland, B. Moghaddam, and T. Starner: View-Based and Modular Eigenspaces for Face Recognition. Proceedings IEEE Conf. on Computer Vision and Pattern Recognition, Seattle, WA, (June 1994) 84–91
19. P. L. Rosin: Multiscale Representation and Matching of Curves Using Codons. CVGIP: Graphical Models and Image Proc., Vol. 55, No. 4, (1993) 286–310
20. G. Stockman: Object Recognition and Localization via Pose Clustering. CVGIP, Vol. 40, (1987) 361–387
21. G. Taubin and F. Cukierman and S. Sullivan and J. Ponce and D. J. Kriegman: Parameterized Families of Polynomials for Bounded Algebraic Curve and Surface Fitting. IEEE T-PAMI, vol.16 no.3, (Mar 1994) 287-303
22. S. Ullman: An Approach to Object Recognition: Aligning Pictorial Descriptions. Technical Report AI Memo No.931, M.I.T. AI Laboratory, (1986)

Distinctive representations for the recognition of curved surfaces using outlines and markings

David Forsyth[1], Andrew Zisserman[2] and Jitendra Malik[1]

[1] Computer Science Division, U.C. Berkeley, Berkeley CA 94720, USA
[2] Robotics Research Group, Oxford University, Oxford, UK

Abstract. Recognising 3D objects from single images presents a range of significant problems, mostly to do with the nature and distinctiveness of the representations that can be recovered. In such special cases as polyhedra, surfaces of revolution, general cones, canal surfaces and algebraic surfaces, the geometry can be recovered with varying degrees of success and of ambiguity. We discuss these volumetric primitives, comparing their utility to that of surface primitives.
For a model based recognition system, representation is not simply concerned with particular geometric primitives, but the entire recognition process. In our view, representations should be motivated by the way quantities that are measurable in an image influence decisions throughout the recognition process.
When some geometric information is available, its potential distinctiveness can often be substantially enhanced by constructing representations that capture surface markings in an appropriate frame on the surface itself.
Keywords: Object Recognition, Computer Vision, Invariant Theory, Surface Representation, Surface Markings.

1 Introduction

The fundamental question in discussing representations is "What is the representation for?". Representations, such as depth maps and Bezier patches, that may be appropriate for tasks such as geometric modelling, graphics, navigation, or grasping may not be appropriate for model based recognition. In this paper we explore representations for object recognition, with a particular emphasis on representations for curved, 3D objects that can be extracted from a single image.

For recognition systems to be successful and useful, they will have to have large modelbases, containing a wide variety of objects. For systems with large modelbases, searching over object-image correspondences to estimate pose and then verify, is impractically expensive. Existing recognition schemes where recognition is phrased in this way include *pose-clustering* [35], *alignment* [16, 17], and *interpretation trees* [15]. Fortunately, *object recognition is not pose recovery*; representations can be defined which yield *viewpoint-invariant* descriptions from images. Such invariant descriptors directly identify models, without first computing pose, avoiding searching the model base. As a result, defining and managing appropriate object representations is the central question in building object recognition systems.

If the representation is too impoverished then it will not admit object-independent constructions that take image data alone and yield object identity, as recent papers have shown [4, 5, 22]. A number of the pose based schemes cited above are forced to search their model libary because the motivating notion of objects, as semi-coherent clouds of points or line-segments, is too parsimonious: clouds of points are a poor representation for most objects, because representing an object as a cloud of points wastes the available and potentially rich structure of outline and of markings.

It is convenient to distinguish between two basic types of representation:

- *Volumetric primitives*, which are drawn from globally constrained geometries such as generalized cylinders and algebraic surfaces, and can represent, for example, vases, cans and boxes. A great deal is known about extracting invariant representations of volumetric primitives.
- *Surface primitives*, which consist of patches of surface, defined by their curvatures for example, and subject to little or no constraint - they represent shirts, Henry Moore sculptures, and trees. Very little is known about what information should be extracted or what is available.

Typically, representations should (and occasionally do) include information both about the geometry of objects, and about the markings that lie on the objects. Geometry and markings are, to some extent, interdependent; measurements of markings are unreliable unless referred to some coordinate system on a curved surface itself (to avoid the effects of foreshortening), and markings generate image clutter that obstructs grouping. Admitting markings is essential, because it expands the scope of any representational scheme - it is hard to distinguish between soft-drink cans without exploiting marking information.

In what follows, we use the following terms:

- **markings:** patterns of reflectance changes on a surface - for example, the patterns on the cover of a book.
- **texture:** markings that are more structured or statistical patterns of reflectance changes - for example, the patterns on a wooden desktop.
- **outline:** the points in an image plane where the surface is tangent to the ray through the focal point and the image point.
- **contour generator:** the points on the surface that project down to the outline.

2 What is known

2.1 Indexing

A system that must handle large numbers of models requires *indexing*, where a process that is wholly or largely model-independent, is used to compute representations that are largely or wholly unaffected by the position and intrinsic parameters of the camera, and that differ from object to object. These descriptions, often known as *indexing functions*, have the same value for any view of a given object, and so can be used to index into a model base without search (for example, [5, 7, 18, 28, 29, 31, 34, 39, 41]). Note that schemes where the model must be known to compute an index (e.g. [40]) do not escape searching the modelbase, and cannot be seen as solving the indexing problem in a useful way.

Invariant representations are a natural goal, and have a long tradition in vision. The evolution of technique in representation in the last decade or so can partly be charted by the extent to which the representation is invariant: the tangent angle vs. arc length representation [1] is invariant under *similarity* transformations, which correspond to presenting an object in a fronto-parallel plane - a very restricted viewing geometry; later, Lamdan *et al.* introduced *affine* invariant representations, corresponding to parallel projection with unknown intrinsic parameters [18]; and then *projectively* invariant representations [7] covered the most general transformation between a planar object and image, with no restrictions on pose or intrinsic parameters. Indexing using invariants yields an attractively simple architecture: in a typical system that works for *plane* objects, projective invariants are computed for a range of geometric primitives in the image; if the values of these invariants match the values of the invariants for a known model, we have good evidence that the image features are within a camera transformation of the model features, and that hypothesis can be either combined with other hypotheses, or verified directly. The efficiency of this indexing process means that systems with non-trivial sizes of model base can be constructed[3].

It must be possible to extract indexing functions from images under realistic conditions. For this reason they cannot be too local, as this compels the use of high derivatives which cannot be measured locally, or too global, and thus dependent on all features being present and grouped. There has been a trend, analogous to the invariance of representations described above, of moving away from global descriptions such as moments and Fourier descriptors (both which are hopelessly inadequate when features are missing - either due to occlusion, or to the ever-present problem of "drop out" with feature detectors) through to modern semi-local descriptors (e.g. [37]).

2.2 Volumetric primitives

In the computer vision literature, there is a long tradition in the use of volumetric primitives, generalized cylinders being the strongest current. Knowing that one is looking at a part of a volumetric primitive characterized by a small number of parameters is a very powerful constraint. An analogy with statistical inference is appropriate–if the data can be well modeled as arising from a parametric model of one of the standard types–the right strategy is to use the data to estimate the (small number of) model parameters and then use the model to answer the various questions one might have.

In contexts when the objects are well modeled by volumetric primitives, not much benefit is derived from aspect graphs. Given that there exists a model with a small number of parameters, the multiple views are redundant.

Of course, just as parametric statistics is appropriate only for a subset of real world situations, simple volumetric primitives are not general enough to deal with the wide variety of man-made and natural objects - crumpled newspaper is one good example.

One approach to indexing a 3D object would be to determine a comprehensive reconstruction of its 3D geometry, and then abstract geometric invariants from that representation; in general, this is not possible. In some cases, for example, surfaces

[3] Current systems using indexing functions have model-bases containing of the order of thirty objects.

of revolution, complete reconstructions are definitely unavailable in a single image (some parameters are unresolved). However, for a number of volumetric primitives it is possible to extract index functions, from a single image, which measure some of the object's 3D geometric properties:

- **Polyhedra:** For polyhedra that are position free, and have "many" quadrilateral faces, projective invariants can be recovered from a single image ([30], after [32]); these invariants are measurable because of the rich incidence structure of these polyhedra.
- **Repeated structures:** A single view of an object with an n-fold reflectional symmetry is equivalent to n views of a section of that object; for polyhedral objects and space curves, this has been used to reconstruct the projective geometry of such objects [24, 30].
- **Surfaces of revolution (SOR):** Cross-ratio's of a series of points, defined by outline bitangent lines, on the imaged axis of a SOR, yield projective invariants of the surface [9, 19]. A further construction allows outlines to be transferred from view to view [36], but the information available in the image is insufficient to reconstruct the surface (two further parameters are required).
- **Straight homogeneous generalised cylinders:** It is possible to reconstruct some SHGC's from image information alone [42]; the ambiguity in the reconstruction is not specified, but appears to be at least an elation through the focal point.
- **Algebraic surfaces:** The complete projective geometry of a generic algebraic surface can be recovered from its outline in a single view through a generic focal point; the algorithm given is too complex for practical use [6].
- **Canal surfaces:** A canal surface is a generalised cylinder where the swept curve is a circle of constant radius, with the plane of the circle orthogonal to the axis curve. In the case of a planar axis, the axis curve can be recovered, from the outline alone, up to an affine ambiguity [26]. Consequently, affine invariants of the axis curve can be used as index functions, but the surface can only be recovered up to an affine ambiguity.

2.3 Surface primitives

For a representation to be effective for recognition, it must be an *abstraction* of the surface description produced by throwing away *irrelevant detail.* A pointwise map of principal curvatures is not much more useful than a pointwise map of normals or depths.

A major program of research in trying to understand surfaces at a higher level of abstraction was initiated by Koenderink and Van Doorn, paralleling work in mathematics on singularity theory. Essentially they tried to understand visual events: how does the topology of the outline change for an observer moving around the object? This leads naturally to the notion of aspect graphs. This line of attack has been pursued to its logical conclusion– the development of algorithms and understanding the computational complexity is now complete for both polyhedra[14] and curved objects[27]. The computational complexity has proved to be very high, making the practical use of exact aspect graphs of general curved surfaces unlikely. An even more serious criticism is that the abstraction is the wrong kind of abstraction. The formalism doesn't take into account the *scale* of the different topological changes of the outline. For a human observer, and a computer vision system that has to start from a brightness image, the scale of the feature has a major impact on its detectability and localizability.

While the usefulness of multiple view representations remains debatable, it seems clear that a concern for what one can hope to reliably extract from early vision needs to inform the choice of representation formalism. From a single image, outline, texture and shading are the main cues available. Texture cues can be treated locally [21] so when we are lucky enough to have visually resolvable texture, we can locally get surface normals and curvatures. It is the shading cue that is more difficult to analyze; global interactions due to interreflections [8] make Horn style shape-from-shading theory unusable in general contexts. As of present writing, surface primitives derived from image data have not been successfully applied in recognition, and, as it is unlikely that this position will change in the short term, we do not treat surface primitives further.

2.4 Markings

There have been few attempts to exploit surface markings explicitly in recognition; most such attempts completely exclude geometry [25, 33]. Some of the difficulties usually cited include variations in colour caused by illuminant effects or interreflections, difficulty of segmenting marked objects, and the effects of foreshortening on markings. However, it is easily shown that, if the geometry of an object can be recovered up to a projective ambiguity from a single view, then the markings on that object can be covariantly recovered by mapping image grey-levels back on to the geometry recovered[4]. This observation has been used to produce representations that incorporate markings (in [11]), but no strategies for exploiting the markings were proposed.

Recognition of 2D textures is a well-studied problem with numerous successful applications in remote sensing, inspection etc. Usually it is approached with statistical classification techniques (see e.g. [12] for a review) given a suitable set of texture features. Various features have been used in the literature, including some derived from co-occurrence matrices, Fourier domain features, and those based on convolving the image with multiscale, multi-orientation linear filters.

3 Fundamental notions

Indexing is a concept so basic to recent work on representation that it had to be expounded (above, section 2.1) before the work was discussed. Clearly, one cannot build a large fast recognition system without some form of indexing, and this gives some clues to appropriate areas for future research. However, indexing alone will not answer all the difficulties of utilising a large model base with a wide variety of objects. Some form of *structuring* or organisation is required at every level of the recognition process. This organisation is centred on the notion of object **class**.

[4] This follows because the process of mapping texture back involves intersecting lines through the focal point with the surface's geometry; since forming intersections is covariant (preserved) under projective ambiguities, backprojecting image texture must be covariant under projective ambiguities.

3.1 Class

The collection of known volumetric cases (SOR, algebraic ...) above induces a natural, utilitarian, *geometric* notion of class [43]: At early stages, constraints derived from the particular class are used to guide segmentation and grouping; At later stages viewpoint invariant or *viewpoint stable* descriptors for that class are computed from these groups; Index functions are computed from these descriptions and paired with appropriate model class sub-libraries, and so on. In this way an appropriately structured representational system can be used to control complexity at all levels in the recognition process.

Using model classes in this way can be thought of as a modern version of Shape from X, where X is now a model class. In an update of Binford's use of models [2, 3], we use model class to facilitate three distinct tasks:

1. **Grouping**
   Recent work by Zerroug and Nevatia shows that one can recover volumetric SHGC descriptions from fragments of the outline in the presence of occlusion, clutter and extraneous edges [42]. For a surface of revolution this grouping can be accomplished without requiring a cross-sectional curve in the image [43]. The grouped outline also enables simpler computation of the next two levels.
2. **Invariant Representation**
   Using the outline alone, 3D surfaces can often only be recovered up to a parameterised family of surfaces that could have projected to the outline. In contrast, an invariant representation can be obtained which is sufficient for recognition. For example, the axis cross-ratios obtained from corresponding outline bitangent for SORs are invariant to projective transformations, and can be used as index functions, but the surface is not determined.
3. **Recovery of 3D shape**
   With additional information, e.g. camera calibration, both pose and 3D shape can be determined. For example, for a SOR if a cross-sectional circle is visible and the camera is calibrated, pose is determined up to a one parameter family (distance from the camera), and the shape of the surface recovered from the outline up to an overall scaling.

This notion of class is organised around measurable image cues. There is no point in drawing class distinctions that cannot be measured in images; furthermore, an ideal notion of class involves a notion of *emerging object identity*, where the behaviour of later recognition modules is conditioned by class hypotheses established earlier. For example, given that an image line is likely to have come from a polyhedron rather than from a surface of revolution, it is more productive to organise the grouping strategy around constructing object faces.

3.2 Consistency

Typically, in systems that use indexing, recognition hypotheses are not monolithic; several image groups may index to object substructures. It is then essential to determine which of these hypotheses can be fused into an hypothesis about object identity, requiring a notion of consistency. As a second example, hypotheses about object identity are equivalent to hypotheses about such matters as camera internal parameters and illuminant colour, and some pairs of hypotheses about object identity may be mutually incompatible. By enforcing compatibility between these hypotheses, we may extract information about the camera [10] or the illuminant.

Consistency appears to offer mechanisms by which the geometry of unknown volumetric primitives can be constrained, using reconstructions of the known objects surrounding them. Without a modelbase, image data can constrain an object's Euclidean geometry to at best a four parameter ambiguity[5]. This ambiguity can be constrained further by the use of, say, such cues as occlusion, support, or the approximate size of typical objects.

3.3 Richer descriptors

It is a commonplace that richer descriptors should make recognition easier. The basis for this argument is simple; the number of cells in an indexing table goes up exponentially with the dimension, meaning that, in principle, cramped tables can be made spacious by making another measurement. However, success in constructing such descriptors has been generally poor. Cues such as colour and markings, which are widely recognised as having great potential, have not yet been used effectively. The present generation of volumetric representations can be made richer in three ways: by increasing the geometric detail recovered, by recovering more distinctive geometric measurements, and by associating texture and marking cues with the geometry recovered.

Increasing geometric detail is not a promising strategy; as the history of 3D from 3D matching shows, a significant component of representation involves identifying and disposing of irrelevant information. For volumetric primitives, understanding how to recover more distinctive measurements and understanding which measurements are more stable and effective in indexing represents an important, largely open, problem. Referring markings to a surface coordinate frame (as in figure 1), and measuring descriptors of object patterns in that frame, will provide substantially increased richness in representation.

4 What should be done

Effective notions of class and of indexing are essential for building large modelbased vision systems; class, because it can be used to organise both the modelbase and the segmentation process, and indexing, to restrict search of the modelbase. These areas represent those in which we believe the most concrete progress is possible. Progress in indexing is likely to be most concrete: it remains important to expand and strengthen notions of the volumetric primitives for which indexing functions can be constructed; to incorporate surface markings into the construction of indexing functions; and to generate a reasonable theory of how to choose amongst possible indexing options.

[5] This is the family of projective transformations that fixes the focal point and the cone of rays joining the object to the focal point; as a result, if this transformation is applied to the world, the image is unaffected, which means it is a fundamental ambiguity.

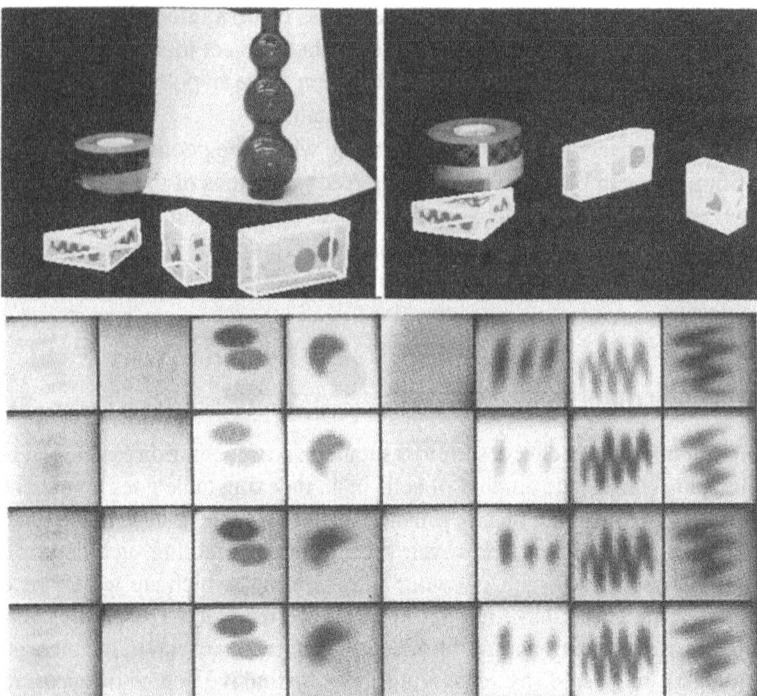

Fig. 1. *The figures on the top show groups of polyhedral objects with polyhedral snakes overlaid by the grouping process. Note the internal boundaries that have been inferred automatically. The markings within the quadrilateral faces are then projectively mapped to a canonical frame. The figure below shows faces in a canonical frame for four views each of eight distinct faces. Note that the resulting sets of markings are fixed (up to the symmetries of the square), meaning that iconic matching of markings is relatively easy. Furthermore, in each case the markings referred to a canonical frame are stable for changes in viewpoint, but largely distinct from face to face. Note the effects of illumination for different views of the same face. Geometrically stable marking data of this form is relatively easily matched using an iconic matcher. Not every object face needs to appear in the modelbase for a face match to be useful; faces are omitted from the modelbase when their markings are not distinctive.*

The extent to which surface markings have been neglected to date is surprising, as they present the best way to expand the size of modelbases possible with currently understood volumetric primitives. We see two approaches to managing the effects of foreshortening in images of markings: constructing pose-invariant features that describe markings; or, using the inferred geometry to refer the image detail back to a surface coordinate frame. The second approach is more general. The neglect of markings is particularly puzzling because the problems presented by markings appear to be concrete and accessible with present knowledge.

To conclude: We have argued in this paper that representation is not simply a matter of whether a particular primitive (super-quadric, SHGC, etc.) is used. Instead, representation in visual recognition is the entire *process* from the early stages of segmentation and grouping, through the extraction of viewpoint-invariant descriptions, hypothesis combination, to finally the recovery of 3D shape. The implementation of this process that we suggest, based on current achievements, is a *geometric* notion of class using volumetric primitives.

Acknowledgements

Supported in part by a grant from United States Air Force Office of Scientific Research AFOSR-91-0361, in part by the National Science Foundation under award no. IRI-92-09729, in part by the European ESPRIT Project 6448 'VIVA', in part by the National Science Foundation under award no. IRI-92-02129, and in part by a National Science Foundation Young Investigator Award with matching funds from GE, Tektronix, Rockwell and Eugene Rikel.

References

1. Asada, H. and Brady, M., "The curvature primal sketch," *PAMI-8*, 1, 2-14, 1986.
2. Binford, T.O. "Inferring Surfaces from Images," *Artificial Intelligence,* Vol. 17, p.205-244, 1981.
3. Binford, T.O., "Survey of model-based image analysis systems," *IJRR*, **1**, 1, 18-63, 1982.
4. Burns, J., Weiss, R. and Riseman, E., "View variation of point set and line segment features," *Proc DARPA IU workshop*, 1990.
5. Clemens, D.T. and Jacobs, D.W. "Model Group Indexing for Recognition," Proceedings CVPR, p.4-9, 1991, and *PAMI*-13, No. 10, p.1007-1017, October 1991.
6. Forsyth, D.A., "Recognising algebraic surfaces by their outlines," *Int. J. Computer Vision*, in press.
7. Forsyth, D.A., Mundy, J.L., Zisserman, A., Heller, A., Coehlo, C. and Rothwell, C.A., "Invariant Descriptors for 3D Recognition and Pose," *IEEE Trans. Patt. Anal. and Mach. Intelligence*, **13**, 10, 1991.
8. Forsyth, D.A., and Zisserman, A., "Reflections on Shading," *Special Issue of I.E.E.E. Pattern Analysis and Machine Intelligence on physical modelling in computer vision*, July, **13**, 7, 671-679, 1991.
9. Forsyth, D.A., Mundy, J.L., Zisserman, A. and Rothwell, C.A., "Recognising rotationally symmetric surfaces from their outlines," *Proc. Second European Conference on Computer Vision*, G. Sandini (ed.), Springer LNCS-x, 1992.
10. Forsyth, D.A., Mundy, J.L., Zisserman, A. and Rothwell, C.A., "Using global consistency to recognise Euclidean objects with an uncalibrated camera," *Proc. CVPR-94*, 1994.
11. Forsyth, D.A. and Rothwell, C.A., "Representations of 3D objects that incorporate surface markings," in *Applications of invariance in computer vision*, J.L. Mundy, A. Zisserman and D.A.Forsyth, (ed.s), Springer LNCS 825, 1994.
12. Fukunaga, K. (1990) Introduction to Statistical Pattern Recognition 2nd Edition, London: Academic Press.
13. Garding, J. and T. Lindeberg, "Direct computation of shape cues by multi-scale retinotopic processing," *Int. J. Computer Vision*, to appear

14. Gigus, Z. and Malik, J., "Computing the aspect graph for line drawings of polyhedral objects," *PAMI-12*, 2, 113-122, 1990.
15. Grimson, W.E.L. and Lozano-Pérez, T. "Localizing Overlapping Parts by Searching the Interpretation Tree," *PAMI*-9, No. 4, p.469-482, July 1987.
16. Huttenlocher, D.P. and Ullman, S. "Object Recognition Using Alignment," Proceedings ICCV1, p.102-111, 1987.
17. Huttenlocher, D.P. and Ullman, S. "Recognizing Solid Objects by Alignment." *IJCV*-5, No. 2, p.255-274, 1990.
18. Lamdan, Y., Schwartz, J.T. and Wolfson, H.J. "Object Recognition by Affine Invariant Matching," Proceedings CVPR, p.335-344, 1988.
19. J. Liu, J.L. Mundy, D.A. Forsyth, A.P. Zisserman and C.A. Rothwell, "Efficient Recognition of rotationally symmetric surfaces and straight homogeneous generalized cylinders," *IEEE conference on Computer Vision and Pattern Recognition '93*, 1993.
20. Malik, J. and Perona, P., "Preattentive texture discrimination with early vision mechanisms," *Journal of Optical Society of America A*, **7** (2), May 1990, pp. 923-932.
21. Malik, J. and Rosenholtz, R., "Recovering surface curvature and orientation from texture distortion; a least squares algorithm and analysis," *Proc. ECCV-94*, Springer Lecture Notes in Computer Science 801, 1994.
22. Moses, Y. and Ullman, S. "Limitations of Non Model-Based Recognition Systems," *Proc. ECCV*, LNCS 588, Springer-Verlag, p.820-828, 1992.
23. Mundy, J.L. and Zisserman, A., "Introduction," in J.L. Mundy and A.P. Zisserman (ed.s) *Geometric Invariance in Computer Vision*, MIT Press, 1992.
24. Mundy, J.L. and Zisserman, A., "Repeated structures: image correspondence constraints and 3D structure recovery," in *Applications of invariance in computer vision*, J.L. Mundy, A. Zisserman and D.A.Forsyth, (ed.s), Springer LNCS 825, 1994.
25. Nayar, S.K. and Bolle, R., "Reflectance Ratio: a photometric invariant for object recognition," *Proc. ICCV-4*, Berlin, 1993.
26. Pillow, N., Utcke, S. and Zisserman, A., 'Viewpoint-Invariant Representation of Generalized Cylinders Using the Symmetry Set', To appear, *Image and Vision Computing*.
27. Ponce, J. and Kriegman, D.J., "Toward 3D curved object recognition from image contours," in J.L. Mundy and A.P. Zisserman (ed.s) *Geometric Invariance in Computer Vision*, MIT Press, 1992.
28. Rothwell, C.A., Zisserman, A.P., Forsyth, D.A. and Mundy, J.L., "Using Projective Invariants for constant time library indexing in model based vision," *Proc. British Machine Vision Conference*, 1991.
29. Rothwell, C.A., Zisserman, A.P, Forsyth, D.A. and Mundy, J.L., "Fast recognition using algebraic invariants," in J.L. Mundy and A.P. Zisserman (ed.s) *Geometric Invariance in Computer Vision*, MIT Press, 1992.
30. Rothwell, C.A., Forsyth, D.A., Zisserman, A. and Mundy, J.L., "Extracting projective structure from single perspective views of 3D point sets," *International Conference on Computer Vision*, Berlin, 573-582, 1993.
31. Stein, F. and Medioni, G., "Structural indexing: efficient 3D object recognition," *PAMI-14*, 125-145, 1992.
32. Sugihara, K., *Machine interpretation of line drawings*, MIT Press, 1986.
33. Swain, M.J. and Ballard, D.H., "Color Indexing," *Int. J. Computer Vision*, **7**, 1, 11-32, 1991.
34. Taubin, G. and Cooper, D.B., "Object recognition based on moment (or algebraic) invariants," in J.L. Mundy and A.P. Zisserman (ed.s) *Geometric Invariance in Computer Vision*, MIT Press, 1992.
35. Thompson, D.W. and Mundy, J.L. "Three-dimensional Model Matching from an Unconstrained Viewpoint," Proceedings ICRA, p.208-220, April 1987.

36. Utcke, S. and Zisserman, A., 'Transfer and Invariants of Surfaces of Revolution', *OUEL Report*, Oxford, 1994.
37. Van Gool, L., Moons, T., Pauwels, E., and Oosterlinck, A., "Semi-differential invariants," in J.L. Mundy and A.P. Zisserman (ed.s) *Geometric Invariance in Computer Vision*, MIT Press, 1992.
38. Voorhees, H. and T.Poggio, "Computing texture boundaries from images," Nature 333, pp. 364-367, 1988
39. Wayner, P.C. "Efficiently Using Invariant Theory for Model-based Matching," Proceedings CVPR, p.473-478, 1991.
40. Weinshall, D., "Model-based invariants for 3-D vision," *Int. J. Computer Vision*, **10**, 27-42, 1993.
41. Weiss, I. "Projective Invariants of Shapes," Proceeding DARPA Image Understanding Workshop, p.1125-1134, April 1988.
42. Zerroug, M. and Nevatia, R., "Volumetric Descriptions from a Single Intensity Image," *Int. J. Computer Vision*, to appear.
43. Zisserman A., Forsyth D., Mundy J., Rothwell C., Liu J. and Pillow N., "3D Object Recognition using Invariance", *Oxford University Engineering Report*, OUEL 2027/94, 1994.

The Epipolar Parametrization*

Richard S. Weiss

Dept. of Computer Science
University of Massachusetts, Amherst, MA 01003-4610

Abstract. The epipolar parametrization arises naturally in the reconstruction of surfaces from profiles with known camera motion. This is a special case of a local parametric representation which is a mesh. One of the advantages of this type of representation is that it is general and applies to any piecewise smooth surface, so it does not restrict the type of surface in any practical way. Local parametric representations can be combined into a global one by computing the transformation in parameter space on the overlap of the patches. This paper also discusses the applicability of this type of model to problems in grasp configuration determination and pose determination.

1 Introduction

When considering any visual task such as pose determination, it is necessary to consider where the object models come from. Different reconstruction algorithms may produce different types of information about a surface. It is important to have a representation that bridges the gap between the modeling and data collection on the one hand and the application on the other. This paper focusses mostly on the model acquisition but also touches on the requirements for some tasks such as pose determination and grasp configuration planning. An assumption made here is that modeling objects will be automated or semi-automated and will directly use images. The modeling process can involve stereo on viewpoint independent features (texture or edges), laser range sensors, or generalized stereo on profiles (reconstruction from profiles with known camera motion).

There are two types of mathematical models for surfaces[2]. One is the *implicit* representation given by an equation $F(X, Y, Z) = 0$, e.g. superquadrics and algebraic surfaces. F is also called an *inside-outside function.* This is convenient as a global representation if the function F has a nice form, e.g. low degree polynomial or superquadric, and fits the surface well. However, if this is not the case, the surface will need to be decomposed into parts such that each part can be represented both simply and with desired accuracy. For pose determination it is not necessary that part decomposition for the 3D model correspond to a part decomposition that can be extracted from an image. The second type is

* The author would like to acknowledge the support of ARPA and TACOM under contract DAAE07-91-C-R035 and NSF under grants IRI-920892 and IRI-9116297.

[2] I am ignoring CSG and volumetric models

a *parametric* surface representation, which is a map from $D \subseteq R^2 \to R^3$, i.e. $(X(u,v), Y(u,v), Z(u,v))$, e.g. generalized cylinders and splines. Superquadrics also have a simple representation as parametric surface patches. The set D together with the map is called a chart or coordinate *patch* because it can be thought of as mapping a u, v-Cartesian coordinate system onto the surface. Parametric representations are more limited in the sense that a surface which might be represented globally by a single simple function, e.g. $X^2 + Y^2 + Z^2 = 1$, may require several parametric patches. Note that any parametric patch representation can be converted to an implicit representation, and in the case of algebraic surfaces this can be done in closed form.

Since 3D data for a surface often comes in the form of discrete points or discrete points that are connected in some way, the concepts of implicit and parametric representations do not apply directly. Sampling a surface as a set of 3D points which are connected by arcs produces a graph on the surface of the object. A graph has some properties of both types of representations. One can think of a graph as being a polyhedron, whose facets together with a point on the inside define the inside-outside function. One can also think of each facet as defining a map from a subset of the plane to the surface, which is a parametric patch. However, for a mesh, which is a graph of degree four, there are two families of polygonal curves that are transversal to each other which also define a parametric patch.

One of the major problems in model acquisition is that of sensor fusion and merging data. This can be handled by meshes and triangulations and requires a representation of uncertainty. A graph-based approach is suitable for representing uncertainty in the data and does not bias the surface estimation. One needs to model uncertainty at the resolution of the data, otherwise fitting data to an implicit (inside-outside) function makes assumptions about the shape of the object and may weight data points inappropriately. For example, when fitting a super-ellipsoid to data there are a number of issues. First, using the evaluation of the inside-outside function is different from using geometric distance. In general, one can approximate geometric distance based on the gradient of the inside-outside function [18, 21], but even then different points will have different covariances which need to be taken into account. However, the most important problem related to representation is that the family of functions itself affects the result, e.g. finding the closest super-ellipsoid to the data may be significantly different from finding the closest polynomial surface of fixed degree. Note that if one actually has some additional constraints on the surface, e.g. one knows that it is a cylinder or surface of revolution, then such models are entirely appropriate and the covariance of each measurement should be used in fitting such a parametrized family of models to the data. Of course, fitting an implicit function model can be done from a graph representation, but it would not be necessary to go through that intermediate step, and the representation of uncertainty would be on the parameter level rather than the data level.

The important issues for representation and modeling are:

- Representing accuracy of the model, which is determined by sensor error and sampling density.
- Determining what parts of the surface have not been sampled, i.e. the boundaries of observation.
- Facility of constructing the model from image data.
- Capturing the information necessary for the application.

These issues are addressed in section 3 in the case of the epipolar representation. The epipolar representation or epipolar parametrization arises naturally in the reconstruction of surfaces from a sequence of profiles where the motion of the observer is known or can be recovered from features visible in the image sequence. This leads to a set of local parametric models on patches that must be combined in order to represent global information. It may happen that only part of the surface is observable, and even then it may require several trajectories by the camera to reconstruct that part. Kutulakos and Dyer [15] have developed strategies for moving the camera or the object so that a maximal subset of the surface is covered by such a set of patches.

2 Epipolar parametrization

Given a smooth surface M and a curve $\mathbf{c}(t)$ of camera centers, we have, for each t, a *critical set* or *contour generator* Σ_t on M consisting of those points $\mathbf{r}$ where the 'visual ray' or viewline from $\mathbf{c}(t)$ to $\mathbf{r}$ is tangent to M (see Fig 1). The *epipolar plane* is the plane spanned by this ray and the tangent to the curve $\mathbf{c}$ of centers. In practice the epipolar plane is computed from a visual ray to one camera center and the baseline connecting the two centers [5, p.170].

On the surface M an *epipolar curve* through $\mathbf{r}$ is defined as one whose tangent is along the visual ray as shown in Fig. 1. As $\mathbf{c}$ moves with time, the visual ray slips along the epipolar curve. Note that this is almost never the intersection of the epipolar plane with the surface, unless the camera moves in a straight line. In general, the critical sets and epipolar curves make a coordinate grid on M: a local parametrization $\mathbf{r}(t, u)$ can be found in which the critical sets are given by t =constant and the epipolar curves by $u =$ constant. In practice one has a discrete set of views and a finite sampling of each profile, which is the image of the critical set in a viewplane or viewsphere. This can result in a mesh where the nodes are the intersections of the two families of curves. It is the 'epipolar parametrization' of M which is used in [1, 2] to reconstruct M from its profiles, although b-splines were used to represent critical sets. In [20] it is shown that reconstruction from the epipolar parametrization is readily transformable into an optimal estimation problem. The epipolar parametrization also has another very interesting property: the viewlines associated with points $\mathbf{r}(t, u)$ and $\mathbf{r}(t + \delta t, u)$ of M will (being lines in space) generally not intersect. However, for the epipolar parametrization, the point at which these lines come *closest* to one another is (as $\delta t \to 0$) on the surface M. So this parametrization is the best one to choose if one uses the pseudo-intersection of viewlines as an approximation to the surface. In addition, the best trajectories for minimizing these distances are planar

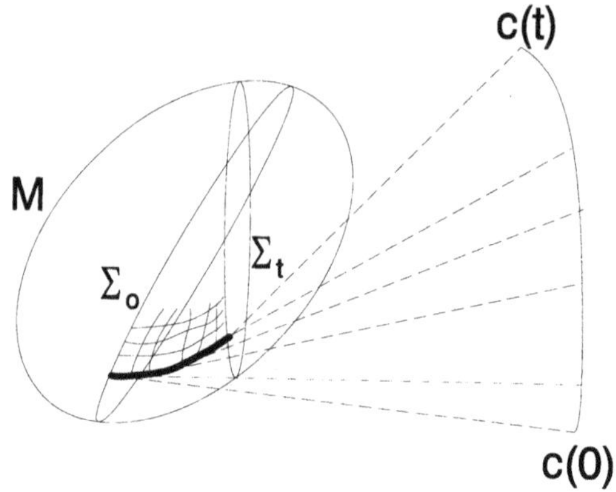

Fig. 1. A surface M and segment of camera path from $\mathbf{c}(0)$ to $\mathbf{c}(t)$. Also shown are the two corresponding critical sets Σ_0, Σ_t, a segment of epipolar curve (drawn heavily), viewlines (dashed) tangent to the epipolar curve, and a local coordinate grid of critical sets and epipolar curves.

trajectories. This leads to strategies that use a sequence of planar trajectories each of which may only produce a model of a small region on the surface. The boundaries of these patches, which depend on the camera trajectories, have been characterized in [10].

There are basically two types of situation which occur at the boundaries of an epipolar patch. The first type is called the *frontier*, where the epipolar plane becomes tangent to M. It is precisely at such points that the epipolar curve becomes singular, and the epipolar parametrization breaks down. To examine the situation at the frontier one can use the 'spatio-temporal surface' as in [9, 10]. The second type of situation occurs in the case of occlusion. Occlusion may happen in two ways. First, the surface normal may turn away from the camera. This event is typically a cusp and is characterized by the fact that the epipolar curve and critical set become tangent. In the image, the profile is seen as ending at a point, and the locus of points on the surface which project to such end points is called the line of cusps. Another type of occlusion occurs at T-junctions. In this case, part of the critical set is obscured by another part of the surface, even though the normal is pointing toward the camera. These points are characterized as the distal points of contact of bitangent viewing rays. Taken together, the frontier and the *natural boundary,* which is the set of points of occlusion, form the boundary of each epipolar patch. A more complete description will appear in [10], and some of the results have appeared in [9].

To summarize those results, the (local) epipolar parametrization of M has a boundary when any of the following occur:

(i) The critical sets form an envelope on M (frontier points).
(ii) The critical set and epipolar curve on M are smooth and tangent to one another. The profile has a singularity which is a cusp or has higher order.

(iii) The viewing ray is tangent to the surface at a point closer to the camera. (T-junction points).
(iv) The critical set on M is singular (having an isolated point or a crossing).

3 The Epipolar parametrization as a representation

One of the issues is how to represent the accuracy of the model in such a way that incorporates both accuracy of the depth estimates and the sampling density of measurements. As noted above, there are criteria in the process of reconstruction from profiles for detecting the boundaries of the reconstruction process. Therefore, it is possible to keep track of parts of the surface that have not been observed. As for the accuracy of depth, the approach of Szeliski and Weiss [20] uses linear smoothing and explicitly models the covariance of the position and curvature at each point in the mesh.

It is clear that local parametric surface patches can be easily produced from reconstruction from profiles. In addition, information about curvature is recovered. One of the important issues is how the parts are combined into a global representation. It is important to do this because combining observation data from different views improves accuracy and is necessary for a consistent model, e.g. the profile segments produced from different patches should agree so that it won't matter which patch is used. This is potentially a difficult problem. If feature points are available on the surface, then they can be used to identify corresponding points in different views. Otherwise, one might try to find the rotation and translation that brings two patches into alignment on maximal subsets. In general this might be difficult, but for the epipolar parametrization there is additonal geometric information available, for example at the frontier, where epipolar curves meet in cusps.

4 Applications: representation of errors

A critical function of vision in navigation, recognition, or enhanced reality systems is to determine the position and orientation of an object relative to the viewer. The pose (i.e. position and orientation) of an object has six parameters: three rotational and three translational. Most pose algorithms take as input an initial "guess" of the approximate object pose and correspondences between a set of projected model features and 2D image features. Chen, Stockman, *et al.* [3] computed pose from profiles using curvature information. One would like to know how closely the profile of the generated view matches the actual view. A graph representation which also stores curvature information is suitable for this algorithm, and accuracy information in the form of a covariance matrix at each node can be used to predict the accuracy of the profiles. A global figure of merit, which could be used to evaluate a model is the maximum over all views of the average distance between the predicted profile and the actual profile.

For grasping, there are a couple of ways to use geometric models. One is finding stable grasp configurations. A stable grasp configuration is given by the null space or force closure equation:

$$\sum_{i=0}^{d} w_i = 0 \tag{1}$$

where d is the number of contacts and w_i is the wrench due to contact i. This expression can be minimized and evaluated at any set of d points for which the normals and friction cones are given. The most time-consuming aspect of this process is finding zeros of this expression. An iterative search is possible and requires a direction for moving the contacts so that the magnitude of (1) is decreased. A local parametric representation such as the epipolar representation, which captures curvature at each node satisfies these requirements.

Another way in which a geometric model can be used is in planning the motion for achieving the grasp. This task requires a prediction of where the manipulator will make contact with the surface if it approaches along a specified ray. From this standpoint, the question that must be answered is where given ray will intersect the surface. This leads again to a representation of accuracy of the surface that is not just a value at each point but a matrix representing the error ellipsoid.

5 Conclusion

This paper shows how a reconstruction algorithm and graph representation fit together to represent the information needed to solve some specific applications. The graph representation is a discrete form of the epipolar parametrization that results from the reconstruction from profiles. It meets the informational requirements of pose determination and stable grasp formation. Issues of speed have not been addressed and may require auxiliary representations. The issues of modeling accuracy and determining the boundary between observed and unobserved parts of the surface are also addressed.

References

1. A.Blake and R.Cipolla, 'Robust estimation of surface curvature from deformations of apparent contours', *Image and Vision Computing* 9 (1991), 107–112.
2. R. Cipolla and A. Blake, 'Surface shape from the deformation of apparent contours', *Internat. J. of Computer Vision* 9 (1992), 83-112.
3. J.-L. Chen, G.C. Stockman, and K. Rao, "Recovering and Tracking Pose of curved 3D Objects from 2D IMages," cvpr93 pp. 233-239.
4. A. Blake and M. Taylor, "Planning planar grasps of smooth contours," *IEEE Proc. Int. Conf. on Robotics and Automation.* IEEE Computer Society Press: Los Alamitos. 1993, pp. II-834-839.
5. O.D. Faugeras, *Three Dimensional Computer Vision*, MIT Press, 1993.
6. O.D. Faugeras. 'On the motion of 3-D curves and its relation to optic flow,' *Proc 1st ECCV*, pp 107–117, Springer Verlag, April 1990.

7. P.J. Giblin, J.E. Rycroft and F.E. Pollick 'Recovery of an unknown axis of rotation from the profiles of a rotating surface', *J.Optical Soc. America* 1994
8. P.J.Giblin and R.S.Weiss, 'Reconstruction of surface from profiles', *Proc. First Internat Conf. on Computer Vision*, London, 1987, Computer Society of the IEEE, pp. 136–144.
9. P.J.Giblin and R.S.Weiss, 'Epipolar fields on surfaces', Proceedings of ECCV, Stockholm 1994, Springer Lecture Notes on Computer Science, Ed. J.-O.Eklundh, 800 (1994) 14-23.
10. P.J.Giblin and R.S.Weiss, 'Epipolar curves on surfaces', to appear in *Image and Vision Computing.*
11. R. Grupen and R. Weiss. "Force Domain Models for Multifingered Grasp Control." In *Proc. Int. Conf. on Robotics and Automation*, IEEE Computer Society Press: Los Alamitos. 1991.
12. K. Higuchi, H. Delingette, M. Hebert, K. Ikeuchi, "Merging Multiple views using a spherical representation." *Proc. Second CAD-Based Vision Workshop.* IEEE Computer Society Press: Los Alamitos. 1994, pp. 124-131.
13. T. Joshi, N. Ahuja, J. Ponce, *Structure and Motion Estimation from Dynamic Silhouttes,* University of Illinois Tech. Rpt. UIUC-BI-AI-RCV-94-01, 1994.
14. K.N.Kutulakos and C.R.Dyer, 'Occluding contour detection using affine invariants and purposive viewpoint control', *IEEE Conf. on Computer Vision and Pattern Recognition*, Seattle, Washington (1994) 323-330.
15. K.N.Kutulakos and C.R.Dyer, 'Global surface reconstruction by purposive control of observation motion', *IEEE Conf. on Computer Vision and Pattern Recognition*, Seattle, Washington (1994), 331-338.
16. J.R.Steenstrom and C.I.Connolly, 'Constructing object models from multiple images', *Int. J. of Computer Vision*, 9 (1992), 185-212.
17. J. Ponce, S. Sullivan, J.-D. Boissonat, and J.-P. Merlet, "On characterizing and computing three- and four-finger force closure grasps of polyhedral objects," *IEEE Proc. Int. Conf. on Robotics and Automation.* IEEE Computer Society Press: Los Alamitos. 1993, pp. II-821-827.
18. S. Sullivan, L. Sandford, and J. Ponce, "On Using Geometric Distance Fits to Estimate 3D Object Shape, Pose, and Deformation from Range, CT, and Video Images," *Proc. IEEE Conference on Computer Vision and Pattern Recognition,* New York, June 1993, pp. 110-115.
19. R.Szeliski, 'Rapid octree construction from image sequences,' *CVGIP: Image Understanding* 58 (1993), 23-32.
20. R. Szeliski and R. Weiss, 'Robust shape recovery from occluding contours using a linear smoother,' *Proc. IEEE Conference on Computer Vision and Pattern Recognition,* New York, June 1993, pp.666-7.
21. G.Taubin, R. Bolle, and D.B. Cooper, "Representing and comparing shapes using shape polynomials," *Proc. IEEE Conference on Computer Vision and Pattern Recognition,* 1989. pp. 510-516.
22. R. Vaillant and O.D. Faugeras, 'Using extremal boundaries for 3-D object modeling,' *IEEE Trans. on Pattern Analysis and Machine Intelligence,* Vol. 14, No. 2, 1992, pp. 157–173.

Using Two-Dimensional Models to Interact with the Three-Dimensional World

Daniel P. Huttenlocher *

Computer Science Department
Cornell University
and
Xerox Palo Alto Research Center

Abstract. This paper describes a method for tracking a moving object in an image, when the camera motion is unknown and other moving objects may be in the image. The method is based on matching two-dimensional geometric structures between successive frames of an image sequence. A bitmap representing the object being tracked at one time frame is matched to features extracted from the image at the next time frame. The transformation mapping the object to the image specifies a new model of the object for the subsequent frame. The approach makes no use of optical flow estimates nor of three-dimensional information. We present examples of the method for tracking moving objects in video sequences and for visual guidance of a mobile robot.

1 Introduction

A difficult problem for dynamic vision systems is tracking non-rigid objects moving in a cluttered environment. This problem is important for a number of applications, including target acquisition, remote surveillance, intelligent teleoperation, and autonomous navigation. The problem of object tracking has received considerable attention in the computer vision research community (e.g., [3, 1, 4, 7, 8, 9, 10, 13, 14]). Our approach to the tracking problem differs from these methods in that it is based on matching two-dimensional geometric models extracted from each image frame. The method makes no use of three-dimensional information nor of an image motion field. It can successfully track an object when there are multiple moving objects and unknown camera motion.

The central observation underlying our method is the fact that the two-dimensional image of an object moving in three-space can be decomposed into two parts

- *a two-dimensional shape change*, corresponding to a different aspect of the object becoming visible or to an actual change of shape in the object, and

* This work was supported in part by National Science Foundation PYI grant IRI-9057928 and matching funds from Xerox Corp., and in part by Air Force contract AFOSR-91-0328.

– *a two-dimensional motion*, corresponding to the motion of the object in the image (this image motion is restricted to a particular class such as translations or similarities).

We represent an object in terms of a sequence of bitmaps, each of which captures the two-dimensional shape of the object in the image plane at a given point in time. The main condition imposed by our method is that the two-dimensional bitmap representing the shape of an object not change greatly between two successive frames of a sequence. Unlike other approaches to tracking, our method imposes no restrictions on the magnitude of the two-dimensional image motion from one frame to the next. Thus the method can track objects with large displacements in the image.

Figure 1 shows the output of the method for six successive frames from a 150 frame image sequence. The left column of each row contains an image frame (320×240 pixels), and the right column contains the image feature bitmap for that frame. The model at each time frame is indicated by a bounding box (in white in the image and in black in the feature bitmap). The model for each frame consists of the subset of the image features contained in the bounding box. The model for each frame is obtained by finding the best match of the model from the previous frame to the current image. This match specifies a new bounding box which is used to define the model. The matching method is described in the following section. The current implementation uses intensity edges as image features, however any other means of deriving binary features from images (or even multiple different sets of binary features) could easily be used.

Figure 2 shows every thirtieth frame of the output for the same sequence as in the previous figure. All 150 frames were used in generating the output, it is just that only every thirtieth frame is shown in the figure (i.e., the model at each time frame was generated using the transformed model from the previous time frame). This figure illustrates the degree to which the two-dimensional model changes across time. The initial model is a nearly head-on view of the bike and is quite small, whereas the final model, 150 frames later, is an oblique view and is much larger. (Note that the initial model only includes the bike and not the biker's torso which is out of the field of view. Subsequent models only track what was contained in the initial model.)

The tracking method operates by comparing the model bitmap at time t to the image bitmap at time $t+1$, in order to find a transformation specifying the best match of the model at time t to the image at the next time frame. The transformation from time t to $t+1$ consists of a translation and scaling, with separate scale factors along each image axis. The scaling allows for the distance between the camera and the object to change, as well as for foreshortening of the object. While we could employ more complicated transformations, such as full affine transformations, we have found in practice that translation and scaling are sufficient for many applications, such as those where image frames are obtained relatively close together in time.

The model at time $t+1$ is formed by selecting a subset of the image bitmap at $t+1$ based on the transformed model from time t. The transformed model from

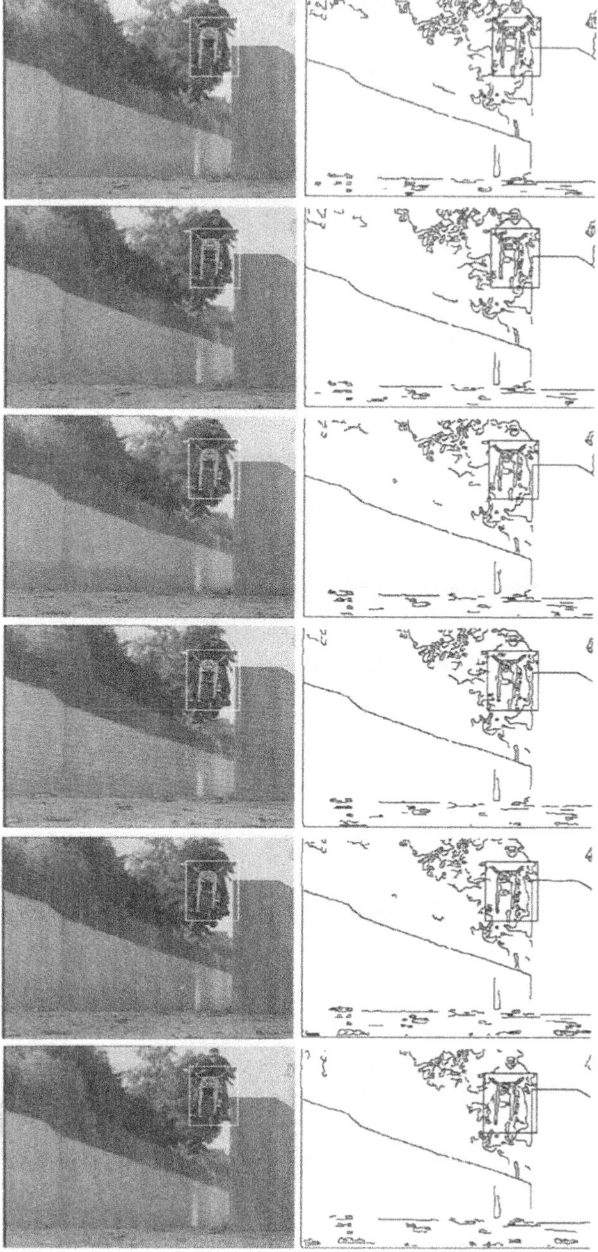

Fig. 1. Six successive frames from a 150 frame motion sequence (see text).

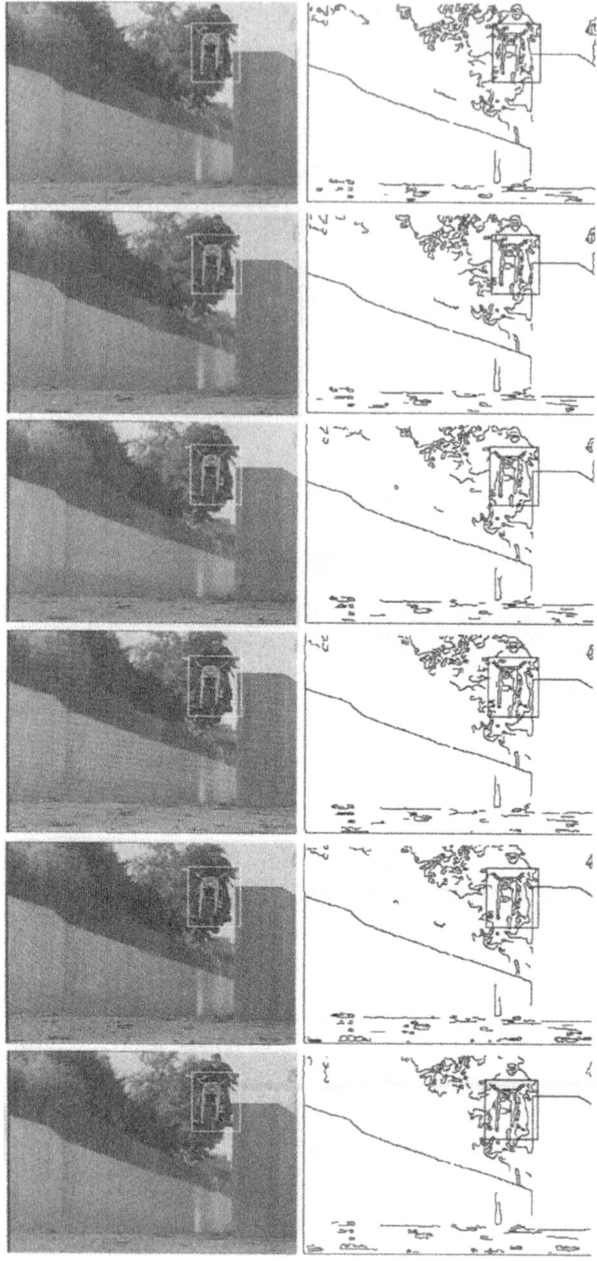

Fig. 2. Every 30th frame from a 150 frame motion sequence (see text).

time t is used to select a subset of the image to form the new model. This subset can either be selected by simply taking all the feature points that lie inside the bounding box of the transformed model from time t, or a "reverse match" can be done to select only those image feature points that are close to transformed model points. We have investigated both such methods, and currently simply use the transformed bounding box.

The tracking method is based entirely on comparing two-dimensional geometric structures, as represented by binary image models, between successive frames. There are no three-dimensional representations or model fitting (e.g., splines, etc.) and there is no computation of local motion such as the optical flow. Our experiments find that the method can successfully track non-rigid objects (such as people walking), even in relatively cluttered scenes, with other moving objects, poor lighting conditions, specular reflections, unknown camera motion, and large displacements in the image. We show some additional examples illustrating such results below.

The two major aspects of our approach are:

1. Decomposing the image of a moving 3D object into two parts: a 2D motion (translation and scale) and a 2D shape change. The shape change is assumed to be relatively small from one frame to the next, but the motion can be arbitrarily large (i.e., there is no 'local search window' in the image).
2. Capturing the 2D shape change between successive images with 2D geometric models that evolve across time. These models provide nonlocal spatial information which is important for accurately tracking an object.

One of the motivations of our work is investigating the extent to which techniques such as the one presented here, which are based solely on two-dimensional image matching, can be used for interacting with the three-dimensional world. We have developed a number of applications of the tracking technique, including interactive remote operation of a robot, mobile robot navigation, and segmenting video sequences based on when moving objects leave the field of view. This suggests that a large range of tasks are supported by two-dimensional image matching.

The remainder of this paper is organized as follows. First we review the generalized Hausdorff measure, which is used to find the transformation mapping the model at time t to the image at time $t+1$. Then we discuss the tracking method and its applications to robot navigation and image guided remote operation.

2 Comparing Images

The generalized Hausdorff measure is a means of determining the degree to which two point sets differ from one another. We use this measure to find the best transformation mapping the model at one time frame to the image at the next time frame, by searching over possible transformations for one which minimizes the measure. Given two point sets P and Q, with m and n points respectively, the generalized Hausdorff measure is defined in [5] as

$$h_k(P, Q) = \underset{p \in P}{k^{\text{th}}} \min_{q \in Q} ||p - q||, \tag{1}$$

where $k^{\text{th}}{}_{p \in P} f(p)$ denotes the k-th ranked value of $f(p)$ over the set P. That is, if we consider the points in P to be in sequence ordered by their values $f(p_1) \leq \ldots \leq f(p_n)$, the k-th element in this sequence is the k-th ranked value. For example, the n-th ranked value is the maximum (the largest element), and the $n/2$-th ranked value is the median. Equation (1) generalizes the classical Hausdorff distance, which maximizes over $p \in P$. In other words, the classical distance uses the maximum rank rather than some chosen rank.

We refer to $f = k/m$ as the *Hausdorff fraction*, because it is the fraction of the m points of P that are within distance d of points in Q. A Hausdorff fraction less than 1.0 allows for part of P to not correspond to anything in Q (as occurs, for example, when an object P is partly occluded from view in some image Q). For instance, if we compute $h_k(P, Q)$ with $f = .75$ ($k = \lfloor .75m \rfloor$), then up to 25% of the points of P need not be near any points of Q. That is, the distances of the 25% of points in P that are farthest from points in Q are not included in the computation of $h_k(P, Q)$ when $f = .75$.

When comparing two point sets P and Q using the generalized Hausdorff measure, it is possible to specify a fixed fraction f and then determine the smallest distance d such that $k = fm$ of the points of P are within d of points of Q. We call this finding the distance for a given fraction. Intuitively, this measures how well the best subset of size $k = fm$ of P matches Q. Alternatively, it is possible to specify a fixed distance d and then determine the resulting number (or fraction) of points that are within that distance. Intuitively, this measures what portion of P is near Q, for some fixed neighborhood size, d.

In using the generalized Hausdorff measure we generally specify both a fraction and distance, finding the largest fraction greater than the given one and the smallest distance less than the given one. That is, we require the distance to be less than some threshold δ (generally a pixel or two) and the fraction to be larger than some threshold τ. This puts constraints both on how much occlusion (the fraction) and how much perturbation (the distance) is allowed. Matches which do not meet both of these constraints are rejected.

The generalized Hausdorff measure is asymmetric (as is the classical distance). Given a model set M with m points and an image set I with n points, in general $h_{k_1}(M, I)$ and $h_{k_2}(I, M)$ will attain very different values. Intuitively, $h_{k_1}(M, I)$ measures how much of the model M is close to the image I, whereas $h_{k_2}(I, M)$ measures how much of the image I is close to the model M. When comparing a model to an image, we generally refer to $d_f = h_{k_1}(M, I)$ as the *forward* distance and $d_r = h_{k_2}(I, M)$ as the *reverse* distance. The corresponding fractions $f_f = k_1/m$ and $f_r = k_2/n$ are referred to as the forward and reverse fractions respectively (where M has m points and I has n points). The forward distance and fraction measure how much of the model is present in the image, whereas the reverse distance and fraction measure how much of the image is accounted for by the model.

In cluttered scenes, the reverse measure will generally have a large distance and/or low fraction because most of the image has nothing to do with a particular object model. Thus we use the forward distance and fraction to hypothesize possible instances of a (transformed) model in an image, and then use a *modified* version of the reverse distance and fraction to "verify" that transformation. The modification is to only compute the reverse measure for those image points that are contained in the bounding box of the (transformed) model. This excludes points in the image that are far from the object, but provides an added check on the match (e.g., it helps prevent models from matching to very dense image regions, where every transformed model point is near some image point but most of the image points are not near transformed model points).

Methods for computing the generalized Hausdorff measure efficiently for a number of different transformations (translation, translation and scale, affine) have been reported in [5, 6, 11, 12]. These methods are all based on rasterization of the space of possible transformations of the model. This transformation space is then searched using various techniques to prune the search. The most efficient search methods for transformations having more than two parameters are hierarchical, using branch and bound search techniques on a $2^d - tree$ (for a d-dimensional transformation space). These methods are described in [6, 11, 12].

For the tracker we use the branch and bound search of a 16-tree, as the transformation has four parameters (two translations and two scales). The search method makes it straightforward to search for matches with the largest fraction and smallest distance, because whenever a match is found the search can be modified to consider only those matches with an equal or larger fraction and equal or lesser distance. We generally use a "best fraction" mode, which returns those matches with the largest fraction and a distance less than or equal to some fixed amount (generally one pixel). That is, the search finds those transformations of the model that bring the largest fraction of model features close to (within one pixel of) image features, and vice versa.

3 Tracking an Object

The tracking system operates by finding matches of the model at one time frame, M_t, to the image at the next time frame, I_{t+1}. At each time frame, the system searches for transformations of the form $T = (p_x, p_y, s_x, s_y)$, where each point $u = (u_x, u_y)$ of M_t is mapped to $T(u) = (s_x u_x + p_x, s_y u_y + p_y)$. The quality of a given transformation is measured using both the forward generalized Hausdorff measure, $h_{k_1}(T(M_t), I_{t+1})$, and the modified reverse measure, $h_{k_2}(I'_{t+1}, T(M_t))$. Recall that the modified reverse measure considers just those points of the image that are contained in the bounding box of the transformed model (i.e., I'_{t+1} is the subregion of I_{t+1} contained in the bounding box of $T(M_t)$).

There are four parameters that specify the quality of a given transformation T of M_t: the forward fraction f_f and distance d_f, and the reverse fraction f_r and distance d_r. Recall that these parameters measure the fraction f_f of the transformed model points that lie within distance d_f of image points, and the fraction f_r of image points (within the bounding box) that lie within distance d_r of transformed model points. We search for the best transformations (largest fractions and smallest distances) that have forward and reverse distances of at most 1, thereby allowing feature points to be perturbed by one pixel. The search is also constrained to report only matches having forward and reverse fractions larger than certain thresholds. These thresholds on the fractions are determined from the previous match, so that the system dynamically adjusts its notion of a "good match" based on how good the prior match was.

The search is done over a local region of the transformation space, using the transformation values from the previous match. That is, we initially assume that the object translated and scaled by approximately the same amount as in the previous pair of frames. If there is no good match of the model to the image in this local range of transformations (no match with fractions larger than the thresholds determined from the previous frame) then the search region is enlarged and the fractions are lowered. The process of increasing the region and lowering the fractions continues until either a match is found, or the entire image has been searched with a scale range of 0.8 to 1.2 and with forward and reverse fractions of 0.5. If no match is found at this point then the tracker gives up. Note that such a failure means that there is no match which transforms at least half the model features from the previous frame to be within 1 pixel of image features (and vice versa) when translating the model to any location in the image and scaling it by up to $\pm 20\%$.

The Hausdorff matcher returns all the transformations that are found with forward and reverse distances less than or equal to the specified thresholds (1 pixel), and with forward and reverse fractions equal to the largest fractions that were found, greater than or equal to the thresholds. Thus there are often several matches returned, with similar translations and scales, as slightly different transformations of the model produced the same distances and fractions. We average these values together in determining the transformation for a given match. As the search region gets large compared to the size of the model, it is possible to get multiple distinct matches of the same quality. This can be checked for by simply looking at the range of parameters reported for the matches. When more than one distinct match is found, the one with parameters closer to the match at the previous time frame is preferred.

The overall search strategy is thus to find a transformation of the model that is as close as possible to the transformation found for the previous time frame. When no good match is found in a local region of transformation space, the search region is expanded. Thus if the model changes its motion, the system should not be confused unless there is an impostor object which is both near the location where the object was predicted to be and is a good match (of course in such cases it is not clear what would be the correct interpretation anyway). This search strategy provides efficiency when the object is changing location and scale slowly, or in a manner similar to the previous frame. It also makes the system more immune to impostor objects which might be located elsewhere in the field of view.

Our current implementation of this tracking method runs in about 2.5 seconds per frame on a Sparc-10 for 320×240 images. This is the time when the model is found in the initial search, which is true most of the time (generally at least 90%). If the full range of scales from 0.8 to 1.2 must be searched then the time is at least 10 seconds. Note that these are total times, including edge detection (a version of the Canny operator [2]) as well as the model matching using the Hausdorff method. There is still considerable work to be done in speeding up the processing. For example, this implementation does not exploit the local search region as much as possible, as the feature extraction and various computations for the Hausdorff matcher are performed on the entire image, rather than a local region of the image.

For the biker sequence shown in Figures 1 and 2, the worst match of a model at one time frame to the image at the next time frame had a Hausdorff fraction of 79% (i.e., that percentage of the transformed model feature points were within one pixel of image feature points at the next frame and vice versa). In general the matches were around 90% in both directions.

Figure 3 and Figure 4 show two additional examples of the tracking output. Figure 3 shows every sixth frame from a one second (30 frame) sequence of a football game. There are relatively large displacements and changes in the 2D shape between adjacent frames. The number "18" on the back of one of the players was selected for tracking. The view changes from being at an extreme angle to being more or less straight on in just a few frames. Note that nearly half of the "18" leaves the field of view a little after the frame that is shown in the fourth row of the figure. This means that subsequent frames are only able to track that portion of the object which remained visible, because the system only uses the model from the previous frame.

Figure 4 shows one frame every second from a five second sequence (every 30th frame of 150 frames) of a person walking in our laboratory. There is a significant change in viewpoint over this sequence, and the person being tracked turns around completely during the last half of the sequence. The tracker is able to follow the person despite the high degree of clutter and the large changes in the object being tracked (because the changes are gradual over about 75 frames).

4 Visually Guided Robot Motion

One application in which we have used the tracking method is a system for remote guidance of a robot using visual landmarks. This system allows a mobile robot to be operated remotely by indicating landmarks in an image. The system operates in two modes. The interactive mode enables a user to specify a landmark in the visual field (by marking a bounding box in the image), and to have the robot move towards that object. The offline mode uses a stored series of such visual landmarks, that were previously input in the interactive mode, in order to follow a route that the user has previously shown to the system.

In the interactive mode a user can specify an object in the visual field by marking it with a rectangular bounding box. The robot then begins moving

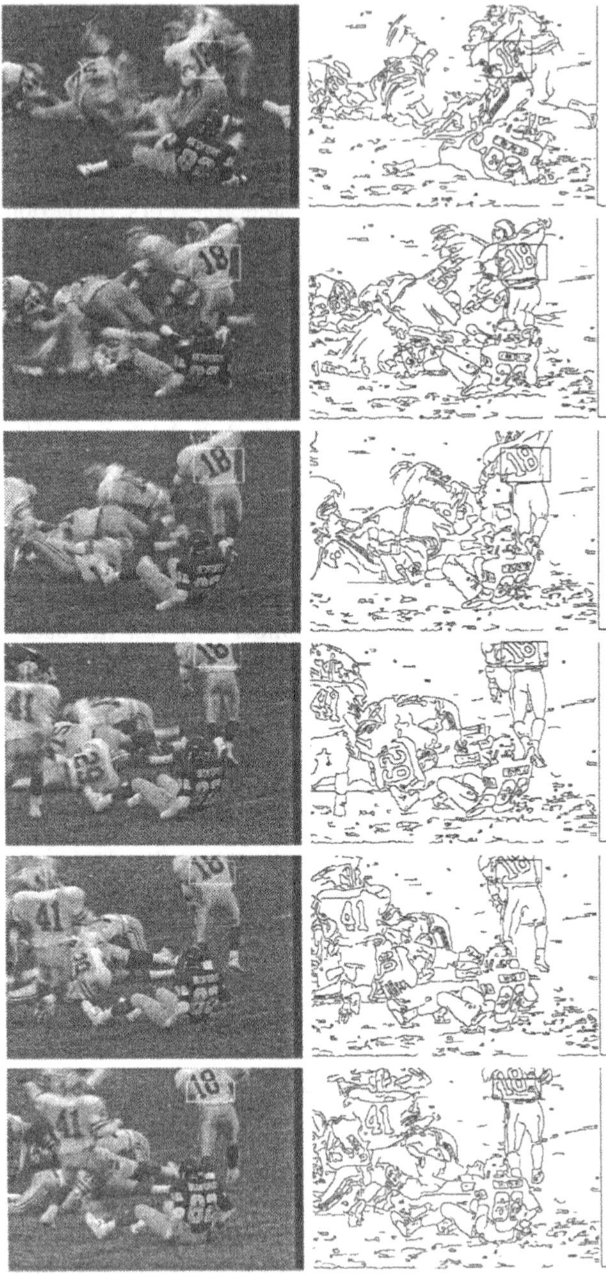

Fig. 3. Every 6th frame from a 30 frame motion sequence (see text).

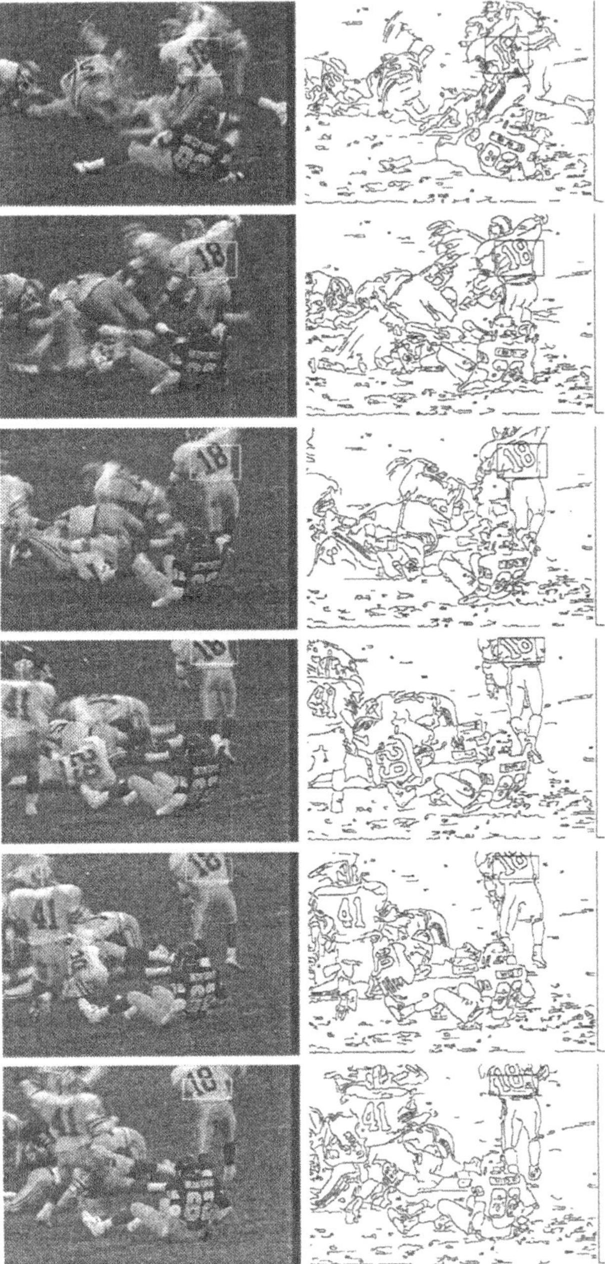

Fig. 4. Every 30th frame from a 150 frame motion sequence (see text).

forward and uses the tracker to keep the object in the field of view. That is, the user-specified landmark is located in the image at the next time frame and a new model is constructed. The robot is then commanded to rotate (if necessary) so as to center the landmark in the image. The robot continues moving and the tracking continues until the landmark reaches a particular size in the image (specified by the user, and changeable while the system is running), or until the user specifies that the robot should stop.

The sequence of landmarks input by the user can be stored for future reference. Thus it is possible to show the robot a path, which is specified as a series of landmarks to navigate to (stopping and locating the next landmark when each landmark reaches a certain size in the image). In addition to commanding the robot to navigate to a series of landmarks, the user can specify direct commands such as rotations or translations by a given amount. This sequence of tracking commands and other commands can be stored as a "route" which the robot can later be commanded to follow from a starting position in which the initial landmark is visible. The search for the initial landmark is often relatively slow, because it may be over a large range of possible locations and scales (as the robot will generally not start out in the same position as when the route was initially specified). After the first landmark is located, however, the tracking runs at about the same speed as normal (a couple of seconds per frame).

The tracking method is used to keep a landmark centered in the field of view as the robot moves with respect to the landmark (the landmark may also be moving). Centering the landmark in the image is straightforward given an estimate of the focal length of the camera. We assume that the camera is mounted looking forward so that the x-axis of the image is approximately parallel to the ground plane and the focal point of the camera is near the center of rotation of the robot. Small inaccuracies in these assumptions are not an issue (i.e., we do not do any careful alignment or calibration of the camera with respect to its mount on the robot).

With this camera configuration, a rotation of the robot primarily changes the x-coordinate of an object in the image. Thus if the landmark is centered at some image location $c = (c_x, c_y)$, a rotation by

$$\theta = \tan^{-1}\left(\frac{c_x}{f}\right) \tag{2}$$

will place the landmark at the image center ($x = 0$), where f is the focal length of the camera. We do not assume that f is known, but rather estimate it from the motion of the landmark in the image as a function of the rotation of the camera. The focal length, f, is estimated by rotating the camera a fixed, known, amount.

Figure 5 illustrates the manner in which we compute an approximation to f by rotating the camera some given amount about the focal point. Consider a (stationary) point X in the world that projects to a point x in the image. If the camera is rotated about the focal point by some given amount θ, then X will project to some new location x' in the resulting image. Let φ be the angle

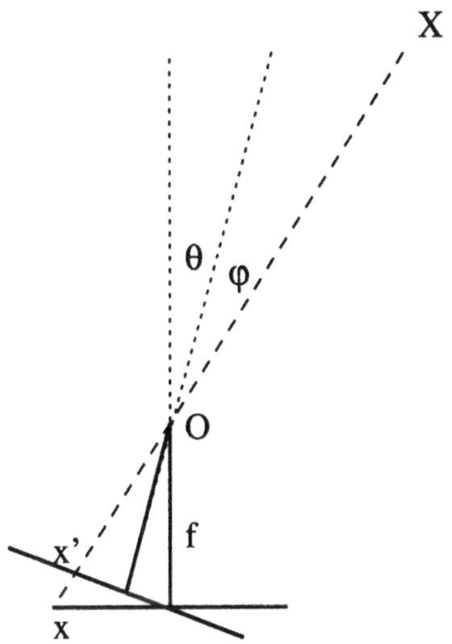

Fig. 5. Computing focal length.

between the optic axis of the rotated camera and the line $\overline{XO}$ (where O is the focal point of the camera). Let $\psi = \theta + \varphi$ be the angle between the optic axis of the original camera and the line $\overline{XO}$.

We know that $\tan(\psi) = x/f$ and $\tan(\varphi) = x'/f$, and hence

$$\tan(\psi) - \tan(\varphi) = \frac{x - x'}{f}.$$

This can be used to approximate f by noting that since $\theta = \psi - \varphi$, when $\tan(\psi) - \tan(\varphi) \approx \tan(\psi - \varphi)$ then

$$f \approx \frac{x - x'}{\tan(\theta)}. \tag{3}$$

This approximation is close when $\cos(\psi - \varphi) \gg \sin(\psi)\sin(\varphi)$ which clearly holds when ψ and φ are both small, but which also holds relatively well over the range of rotations that occur with our camera (which has a field of view of about $\pm 15°$). We choose the rotation of the camera to be $\theta = \pm 8°$ (where the sign of the rotation is in the direction that will move the object towards the image center). Thus if the initial angle is, for example, $\psi = 12°$, then $\varphi = 4°$ and $\cos(\psi - \varphi) \approx .99 \gg .01 \approx \sin(\psi)\sin(\varphi)$. In an image with a visual angle of $\pm 15°$ the initial angle between the optic axis and the image of the landmark is rarely more than $12°$.

Figure 6 shows every tenth frame from a 50 frame sequence of images taken by a mobile robot that was using the visual guidance mechanism just described in order to track a moving object (another mobile robot). The robot being tracked was moving down a hallway, zig-zagging across the hall from one wall to the other. The tracking robot was getting closer across time, because its path was

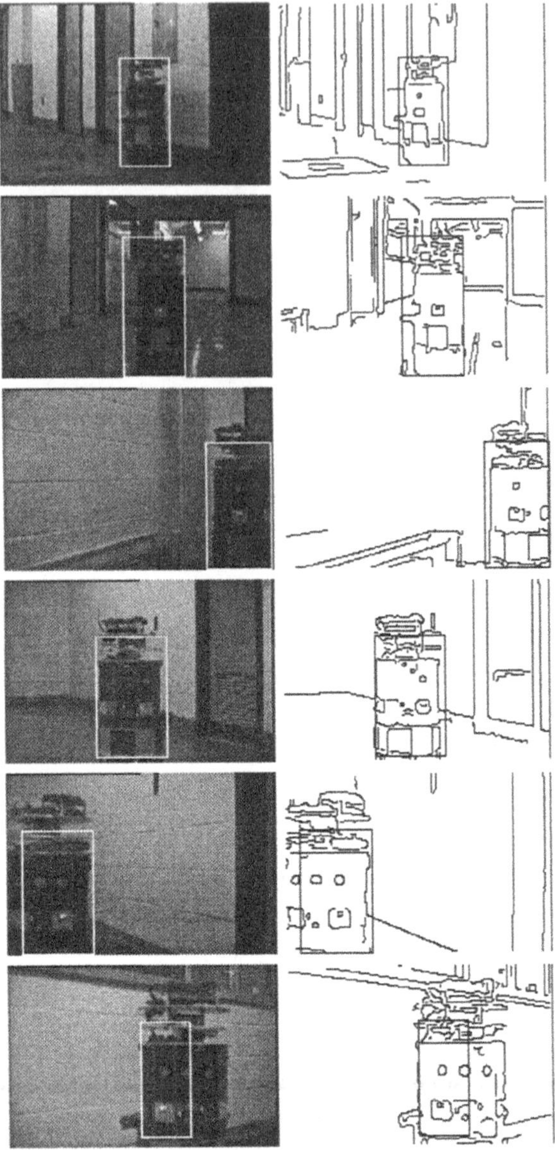

Fig. 6. Every 10th frame from a 50 frame image sequence (see text).

shorter. These images were only taken every few seconds, thus the 50 frame sequence corresponds to several minutes of motion by the robots.

5 Summary

This paper has presented a method for tracking a moving object in an image, when the camera motion is unknown and other moving objects may be visible in the image. The shape of the object in the image may change substantially across time, but the change between successive frames must be relatively small. One restriction of the current method is that it cannot track objects that become significantly occluded. We are investigating techniques for extending the method to allow for occlusion, both by employing models from previous timeframes, and by identifying points near the model that are moving rigidly with it.

The method is based on matching two-dimensional geometric structures between successive frames of an image sequence. A bitmap representing the object being tracked at one time frame is matched to features extracted from the image at the next time frame. The matching allows for translation and separate scale factors along the image axes. This enables the tracker to follow an object that moves anywhere in the image, and that changes size and aspect ratio (due to foreshortening). We have illustrated the method on a number of image sequences, and have used the same parameter settings for a wide range of different sequences. The parameters to the method are the feature extraction parameters (for the Canny edge operator), and the bounds on the maximum amount of shape change that will be tolerated between two successive images (we have used a translation anywhere in the image and a scale change of $\pm 20\%$ that brings at least half of the model edges to within one pixel of image edges and vice versa).

This work is part of our broader investigation of using techniques based solely on two-dimensional image representations for problems that involve interpreting events in the three-dimensional world. We have demonstrated a tracking method that uses just two-dimensional geometric information in order to track solid objects as they move in the three-dimensional world. This method has been used for visual guidance of mobile robots, in addition to tracking people in videos.

References

1. P. Bouthemy and E. Francois, Motion Segmentation and Qualitative Dynamic Scene Analysis from an Image Sequence, *Intl. Journal of Computer Vision*, 10(2), pp. 157–182, 1993.
2. J.F. Canny. A computational approach to edge detection. *IEEE Trans. Pat. Anal. and Mach. Intel.*, 8(6):34–43, 1986.
3. E. Dickmanns and V. Graefe. Dynamic monocular machine vision. *Machine Vision Applications*, 1:223–240, 1988.
4. D. Gennery. Tracking known three dimensional objects. In *Second National Conf. on Artificial Intelligence*, pages 13–17, 1982.

5. D.P. Huttenlocher, G.A. Klanderman and W.J. Rucklidge, Comparing Images Using the Hausdorff Distance, *IEEE Trans. Pat. Anal. and Mach. Intel.*, 15(9), pp. 850–863, 1993.
6. D.P. Huttenlocher and W.J. Rucklidge, A Multi-Resolution Technique for Comparing Images Using the Hausdorff Distance, *Proceedings of the IEEE Computer Vision and Pattern Recognition Conference*, pp. 705–706, 1993.
7. M. Kass, A. Witkin, and D. Terzopoulos. Snakes: Active contour models. *International Journal of Computer Vision*, 1(3):321–331, 1988.
8. D. Koller, K. Daniilidis, T. Thórhallsson, and H.-H. Nagel. Model-based object tracking in traffic scenes. In *Proc. 2nd European Conf. on Computer Vision*, pages 437–452, Santa Margherita Ligure, Italy, May 1992.
9. D.G. Lowe. Robust model-based motion tracking through the integration of search and estimation. *International Journal of Computer Vision*, 8(2):113–122, 1992.
10. N. Ueda and K. Mase. Tracking moving contours using energy-minimizing elastic contour models. In *Proc. 2nd European Conf. on Computer Vision*, pages 453–457, Santa Margherita Ligure, Italy, May 1992.
11. W.J. Rucklidge, *Efficient Computation of the Minimum Hausdorff Distance for Visual Recognition*, PhD thesis, Computer Science Dept., Cornell Univ., 1995.
12. W.J. Rucklidge, Locating Objects Using the Hausdorff Distance, Proc. Fifth Intnl. Conf. on Computer Vision, 1995.
13. G. Verghese, K. Gale, and C.R. Dyer. Real-time, parallel motion tracking of three-dimensional objects from spatiotemporal image sequences. In Kumar et. al., editor, *Parallel Algorithms for Machine Intelligence and Vision*, pages 340–359, New York, 1990. Springer-Verlag.
14. J. Woodfill and R.D. Zabih. An algorithm for real-time tracking of non-rigid objects. In *Proc. American Association for Artificial Intelligence Conference*, 1991.

Representations for Recognizing Complex Curved 3D Objects

David J. Kriegman[1] and Jean Ponce[2]

[1] Center for Systems Science, Yale University, New Haven, CT 06520-8267
[2] Beckman Institute, University of Illinois, Urbana, IL 61801

Abstract. *While there have been great strides in the development of systems to recognize 3D objects in images using viewpoint independent features, we have yet to develop algorithms for recognizing complex curved objects from large model databases; in part, the difficulty arises because image features are viewpoint dependent. Consequently, we must either develop viewer centered representations or have methods for directly relating viewpoint dependent features to 3D models. Aspect graphs, which enumerate all topologically distinct line drawings, may be too weak by themselves to support recognition. However, they can be used to control the search for image-model correspondences when coupled with the constraints afforded by viewpoint dependent features. When objects are represented by algebraic surfaces, these constraints can be expressed as systems of polynomial equations which can be solved using well established techniques. Alternatively, a new representation has been proposed called HOT Curves. Like representations based on geometric invariance, HOT Curves encode the relationship of image features for a particular 3D object. The representation can be constructed directly from a set of images, the features are viewpoint dependent, and indexing schemes are supported.*

1 Introduction

One of the primary goals of computer vision is object recognition; that is, given image data as input, determine the identity of the objects in a scene. In the purest form of this problem, three dimensional objects are observed from a single, arbitrary viewpoint; the recognition system has a database of object models and must determine which of these models is most compatible with some subset of the image. Ideally, object models are acquired directly from images rather than being manually encoded.

In classical approaches to pattern recognition, objects are represented as points in some finite dimensional model or feature space $\mathcal{M}$. Some separatrix distinguishes them, and it may be represented explicitly as a hypersurface in $\mathcal{M}$ or implicitly in terms of some discriminant function [8]. When measurement statistics are known, a Bayesian decision rule is an optimal discriminator. In computer vision, projection geometry has been emphasized, and simple minimum distance classifiers are typically used.

Unfortunately, when observing the projection of a 3D object into an image, measurements do not directly yield estimates of points in $\mathcal{M}$. Instead, images are also a function of the object's location, lighting, material characteristics and camera characteristics. To some extent, the process of forming an image of an object can be viewed as a mapping

$$F : \mathcal{M} \times \mathcal{T} \times \mathcal{P} \rightarrow \mathcal{I},$$

where $\mathcal{T}$ is the space of all possible object poses, $\mathcal{P}$ captures photometric effects such as lighting, material characteristics and camera characteristics, and $\mathcal{I}$ is the space of all possible images. (Note: this does not account for inter-object effects such as occlusion and inter-reflection.) Since this map is not invertible, image measurements do not directly yield points in $\mathcal{M}$; consequently, recognition approaches either explicitly account for the dependency of $\mathcal{I}$ on $\mathcal{T} \times \mathcal{P}$ as well as $\mathcal{M}$ or establish simplified relationship between models and images.

Appearance-based methods may try to represent all possible images of an object; to make this manageable, $\mathcal{I}$ is projected to a much lower dimensional space, often based on principal component analysis (eigenspace) [42]. For example, Hallinan explicitly accounted for different lighting conditions in face recognition [14]. Furthermore, Murase and Nayar allowed differences in viewpoint and lighting by representing the images of an object as a low dimensional manifold in the eigenspace [28]. Since these methods are based on appearance, the representation is learned by gathering images over the full range of parametric variation of $\mathcal{L} \times \mathcal{T}$. While this process is precise, it is cumbersome and may be impractical over the full range of variation.

Alternatively, the domain of the mapping F can be simplified by deriving features from the image measurements that are independent of lighting and material types so that F maps $\mathcal{M} \times \mathcal{T}$ onto $\mathcal{I}'$ where $\mathcal{I}'$ is the space of derived features. Eliminating the photometric effects $\mathcal{P}$ is usually accomplished by using image intensity discontinuities (edges) which arise from occlusion boundaries, surface normal discontinuities and albedo discontinuities. When the model space $\mathcal{M}$ is suitably restricted to $\mathcal{M}' \subset \mathcal{M}$, it may be possible to derive a set of features $\mathcal{I}''$ from $\mathcal{I}'$ which is invariant under all pose transformations $\mathcal{T}$. This yields a new map $\mathcal{M}' \to \mathcal{I}''$; the geometric invariance methods articulated in [27] rely on restricting the model space. Because of the direct relationship between object models and measured features, some of the methods from statistical pattern classification can be applied [8].

Unfortunately, since such a relationship does not exist in general [5, 17], the pose space and model space cannot be completely decoupled for all objects. This becomes rather clear for curved objects since image features such as the occluding contour are viewpoint dependent. Consequently, standard techniques from statistical pattern classification cannot be applied directly.

Instead, objects are generally modelled in some finite dimensional space (e.g., the number of parameters characterizing the model); a particular object is represented as a single point in $\mathcal{M}$ while an object class can be represented as a subset of $\mathcal{M}$. A correspondence between a measured feature and a model feature places a constraint on the possible models and poses. This constraint defines a surface in $\mathcal{M} \times \mathcal{T}$. Given enough measurements and hypothetical correspondences, these surfaces will intersect at a point in $\mathcal{M} \times \mathcal{T}$. Recognition approaches based on the Hough transform or pose clustering try to compute this intersection. Alternatively, alignment approaches exploit the limited number of models; for a hypothesized model $\mathbf{m} \in \mathcal{M}$ and hypothesized set of correspondences, the pose $\mathbf{T} \in \mathcal{T}$ is estimated; the quality of a hypothesis is determined by computing the image features $\mathbf{I}' \in \mathcal{I}'$ for model $\mathbf{m}$ observed in pose $\mathbf{T}$ and measuring the distance between $\mathbf{I}'$ and the measured features.

Due to measurement uncertainty (e.g. random noise, bias, specularities, etc.), the set of model points satisfying the constraints afforded by measurements will not necessarily correspond to a particular model $\mathbf{m} \in \mathcal{M}$. Also, because the modelling process is imperfect, the model is not a precise representation of the object; part of this approximation error is due to modelling an arbitrary surface (a member of an infinite dimensional space) with a finite number of parameters. While we can often find real objects that our finite dimensional models can approximate (sometimes this is challenging), perhaps we should consider infinite dimensional model spaces along with a suitable metric for determining matches.

Two difficulties in recognition systems are the density of models in $\mathcal{M}$ and unmodelled objects. When two models are nearby in $\mathcal{M}$, image data are more likely to be misclassified due to image noise, even when the noise statistics are known and optimal classifiers are used. Secondly, even when the number of models is limited, the constraint surfaces produced from unmodelled objects or random combinations of features may intersect at points in $\mathcal{M}$ that are close to some modelled object, and these features are falsely identified. Given the combinatorics of most matching algorithms, this can happen fairly often.

One successful area of recognition is CAD-based vision where the model domain is limited. Because of the generative nature of CAD systems, the model space can be easily articulated, objects are well approximated by the models, there is little variation amongst instances, and the relationship between image data and models can be ascertained. While objects in the database may be similar, images are typically free of clutter, and so few unmodelled objects will be encountered.

In the rest of this paper, we will consider tools for recognizing curved 3D objects from monocular image data. In particular, we consider two representations, algebraic surfaces and HOT curves. We also show how a view-based representation can be coupled with the constraints afforded by a relatively simple 3D object model. While algebraic surfaces can model a 3D object with high fidelity, a representation based on HOT curves can be directly constructed from images and used to model complex objects.

2 Algebraic Surfaces

Approaches to recognizing three dimensional objects differ in their object representation, image features, and search strategies. Broadly speaking, there are three classes of geometric representations: Explicit 3D descriptions such as collections of 3D points [17], polyhedra [15, 24], algebraic surfaces [21], and structural descriptions of volumetric primitives (e.g. generalized cylinders [29]). Secondly, objects are represented by their appearance either as collections of views of the 3D object (e.g. an aspect graph [3, 20, 35, 38] or as an appearance model capturing the components of the appearance which are independent of viewpoint [28]); each view may be represented by Fourier coefficients, moments, a line drawing, collection of features, topological description, etc. Finally, an object may be represented by a collection of geometric invariants, which can be computed directly from image features and are independent of viewpoint [27, 46]. Depending upon the representation, only certain image features (vertices, straight lines, silhouettes, etc.) can be used for recognition. Specific recognition algorithms are a consequence of these choices and may tradeoff computational time, memory, and robustness.

So, why use algebraic surfaces? The problem of choosing the right shape representation can be addressed from many points of view [2, 4, 25]; here, we consider the expressiveness of the representation and its adequacy for solving vision problems involving curved surfaces.

Most explicit 3D representation schemes proposed so far have relied on very simple surfaces, namely polyhedra and quadric surfaces [10] and, more recently, on superquadrics [1, 13, 30]. Alone, these surfaces are not expressive enough to represent complex curved objects: for example, a quadric surface only has three shape parameters, and a superquadric surface only has five shape parameters (eight if bending and tapering are allowed). By contrast, a quartic surface has twenty eight shape parameters! Each of these surfaces actually has six more coefficients that specify its position and orientation in space, but do not change the surface shape. Even when more complex objects are modelled by Boolean combinations of these primitives, they cannot be smoothly stitched together; higher degree surface patches are needed. Note that when collections of primitives are used, they can approximate a surface to arbitrary precision given enough primitives; however, while the surface is well approximated [11], the number of shape parameters and their values are unstable and are unlikely to be useful for recognition. Generalized cylinders [2] form a very general shape representation, but many of them, for example straight homogeneous generalized cylinders [36], can be modelled conveniently by algebraic surfaces [21]. Furthermore, several recent papers have shown how algebraic surfaces could be used for recognizing and locating object models in range and video images [21, 33, 39, 40, 41].

A parametric algebraic surface patch is defined by:

$$\mathbf{x}(u,v) = (x(u,v), y(u,v), z(u,v))^t, \quad (u,v) \in I \times J \subset \mathbb{R}^2$$

where x, y, z are (rational) polynomials in u, v. Examples include Bezier patches and non-uniform rational B-splines (NURBS).

An implicit algebraic surface of degree d is the set of points $\mathbf{x} = (x, y, z)^T$ where some trivariate polynomial

$$P(\mathbf{x}) = P(x,y,z) = \sum_{i+j+k \leq d} a_{ijk} x^i y^j z^k$$

of degree d vanishes. In other words, the surface is the zero set of the polynomial.

For most vision problems, the geometric constraints associated with algebraic surfaces are themselves algebraic, i.e., they can be expressed by polynomial equalities and inequalities. In turn, these constraints can be manipulated and solved using extremely powerful methods, such as elimination theory, curve tracing, cell decomposition and homotopy continuation.

Most importantly, these tools can be used to relate the 3D object model to the image observables. For algebraic surfaces and their occluding contour, it was shown in [21] that resultants could be used to construct an implicit polynomial equation

$$c(\mathbf{p}, \mathbf{T}) = 0 \tag{1}$$

relating image coordinates $\mathbf{p}$ of points on the occluding contour to the pose $\mathbf{T}$. That is, up to occlusion, (1) fully describes all possible image contours of the algebraic surface. The polynomial function c can be used in an alignment approach to recognition: for a given hypothesized model j and set of image measurements corresponding to the projection of the occluding contour, the pose can be estimated by minimizing with respect to $\mathbf{T}$

$$\mathcal{O}_j(\mathbf{T}) = \frac{1}{n}\sum_{i=1}^{n} d^2(\mathbf{p}_i, c_j(\mathbf{p}, \mathbf{T})) \tag{2}$$

where c_j is the contour polynomial corresponding to model j, and d is a (possibly Euclidean) distance measure. Note that $\mathcal{O}_j$ is a nonlinear objective function; local minimization techniques (e.g. gradient descent, Levenberg-Marquadt) require some initial estimate of the pose, whereas global techniques are computationally prohibitive. For the optimal pose, $\mathcal{O}_j(\mathbf{T})$ provides a measure of the error of fit. The recognized model is taken to be the one which minimizes $\mathcal{O}_j(\mathbf{T})$.

This method was proposed in [21], then extended in a number of ways in [33]: rotations were decomposed from translations leading to an order of magnitude speed improvement; the relationship between the model and other types of image measurements was established for range data, contour curvature, and image intensity. Note that different distance measures can be used in the minimization, for example the algebraic distance or Taubin's approximate distance function [40]. One way to obtain an initial estimate of $\mathbf{T}$ is to exploit the relationship between viewpoint dependent point features and certain surface curves [22, 34]. For example, contour inflections are the projection of points on parabolic curves [19], and bitangent lines are the projection of the limiting bitangent developable surface [18, 43]. The constraints imposed from a few correspondences can be expressed as a system of polynomial equations which can be solved using the global method of homotopy continuation [26].

3 Viewer Centered Representations and Aspect Graphs

Because of either self occlusion or inter-object occlusion, 3D recognition systems cannot utilize global shape properties; instead, they use collections of identifiable image features such as vertices, line segments, inflections, maxima of curvature, etc. This results in the combinatorial matching problem of identifying the correspondence between image features and object features. While simple graph matching techniques can be used, this approach is exponential in the number of features. Alternatively, it has been shown that object rigidity or the viewpoint consistency constraint [24] can be exploited to effectively reduce matching to polynomial complexity in practice [7].

For polyhedra, the matching combinatorics are easily enumerated since image junctions (except T-junctions) are the projection of object vertices, and line segments are the image of object edges. For the viewpoint dependent image features of curved objects, such point-to-point or curve-to-curve matches cannot be easily established. These types of correspondences are only possible for viewpoint independent features such as surface markings or singular points of the surface. In general, a point on the occluding contour may arise as the image of any surface point up to occlusion, and the differential geometry of contour points does not aid in establishing matches. However, for certain special types of feature points, a correspondence between a contour point and specific surface curves can be established. For example, it is well known that the projection of a point on a parabolic line is an image contour inflection, and that points on a limiting bitangent developable curve project to a contour bitangent [19] (Figure 1).

Whereas the number of possible image-to-model correspondences can be readily established for polyhedra, it is not even obvious *a priori* how many image features can arise from the projection of a single surface patch. For certain classes of surfaces, it is possible to derive an upper bound. For example, the Bezout number [45] can be used to determine an upper bound on the number of roots of a polynomial system characterizing features for algebraic surface. Unfortunately, such an upper bound may not be particularly tight since the Bezout number includes complex roots, multiple roots, and roots at infinity. Alternatively, the maximum number of occurrences of each feature type over all possible viewpoints can be determined from a viewer centered representation of a particular surface. Having a tight upper bound can greatly reduce the maximum number of hypothetical correspondences during recognition.

A natural next step is to note that certain sets of features are never mutually visible while other sets are visible together over a range of viewpoints; so, it is pointless to hypothesize certain sets of correspondences. This leads us to consider multiple view representations that explicitly characterize which groups of features are mutually visible; using such a representation, only correspondences involving these feature groups are hypothesized during recognition. Popular multiple-view representations such as aspect graphs [20] or characteristic views [6] explicitly enumerate all topologically distinct line drawings of an object, and algorithms have been implemented for constructing aspect graphs from 3D object models. (See [3] for a review.)

For an object modelled by an algebraic surface, an algorithm was presented in [32] for constructing its aspect graph. Using an aspect graph along with the viewpoint consistency constraints afforded by a 3D surface representation, only possibly valid correspondences will be hypothesized, and they can be validated using the algebraic constraints mentioned in section 2 and presented in [22, 34]. An interpretation tree can be used to organize the search for correspondences [12], and the likelihood of observing a particular grouping of features (perhaps derived by viewpoint volume of a particular aspect) can be used to organize this tree.

One oft repeated criticism of aspect graphs is that the representation of an object may contain an enormous number of views, many of which are irrelevant. One approach to reduce the size has been to note that many of the inconsequential views result because sensor resolution is never considered; using resolution models [9, 37] or better sensor models [16] during aspect graph construction, the size can be reduced. It should be noted that any recognition system that fully enumerates all possible feature correspondences would necessarily include all of the possible matches represented in an aspect graph.

The aspect graph representation can also be simplified to account for the matching process. An "aspect tree" can be constructed. Like an interpretation tree, the level of the tree indicates the number of features that are matched. In particular, rather than enumerating all mutually visible features, the representation would enumerate all n-tuples of features which are mutually visible at level n of the tree. The children of a

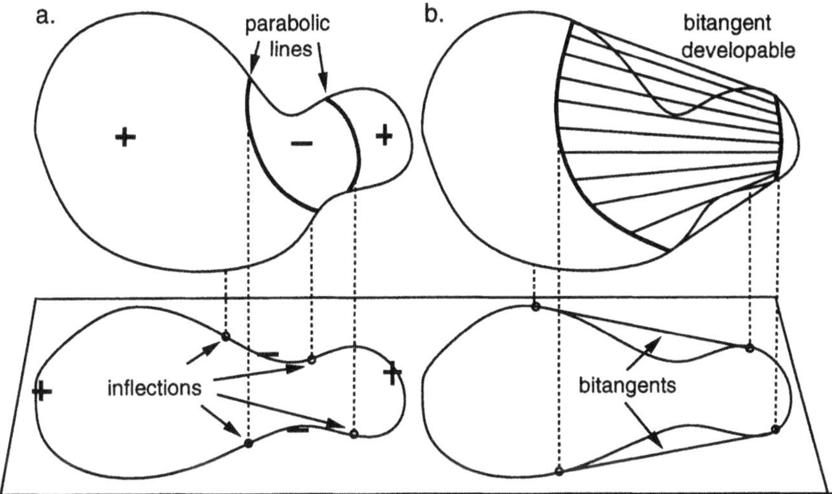

Fig. 1. Parabolic curves, limiting bitangent developables, and their projections.

node at level n would include all of the features at level n plus an additional feature which is mutually visible. For a node near the root of the tree, the set of features will be visible over a large range of viewpoints; however, as one goes deeper in the tree, the features will only be visible over a smaller and smaller set of viewpoints. Ignoring NIL matches, the tree would only be constructed to d levels where verifying d correspondences using the viewpoint consistency constraint adequately identifies the object. Such a representation would be much smaller than a full aspect graph, yet the system would only hypothesize valid correspondences. The same set of visual events used for constructing aspect graphs would be used, but the resulting representation would differ. To the best of our knowledge, algorithms for constructing such a representation and exploiting such a control strategy have not been implemented.

4 HOT Curves and Invariant Features in Images

We have implemented components of the recognition methods for curved objects outlined in the previous two sections. Objects are modelled either by hand crafting a CAD model or by fitting a single algebraic surface to range data or a set of images. Aspect graphs have been constructed from these models. Pose estimation using a small number of viewpoint dependent features as well as the entire occluding contour has been implemented, and recognition has been performed from a small model database. To handle more complex objects and a large model database, a number of enhancements are required: First, the object would have to be represented by multiple surface patches, and methods for automatically constructing multi-patch models must be further developed. While the aspect-tree and point feature matching approach outlined in the previous section can be easily extended, it is not entirely obvious how to extend the objective function of equation (2) to handle multiple patches. For a large database, indexing mechanisms are needed to access models in sub-linear time.

Building on the recognition constraints and geometric notions developed earlier, but not on the algebraic representation, it was noted that the image of a point lying on certain surface curves (termed HOT curves for high order tangency) can be easily identified in an image [18, 43]. In particular, points on parabolic curves generally project to inflections of the occluding contour, and points on the limiting bitangent developable curves project to contour bitangents (Figure 1). Thus, we have a means for establishing a correspondence between certain image features (inflections and bitangents) and the 3D surface curves (parabolic and limiting bitangent curves) that project to those features.

Now, consider parallel projection of a single limiting bitangent developable (parabolic points are handled similarly). Up to occlusion, this 3D bitangent developable will fall on the occluding contour for any viewing direction that lies in the tangent plane common to the two endpoints. Now consider the surface points whose tangent planes are parallel to that of the limiting bitangent. Since all of these planes are parallel, the tangents to the image contour at their projections are also parallel (Figure 2). We call these feature points *parallel tangents*; they are stable and easy to detect in images. Since all of these features have parallel tangent planes, up to occlusion they will all be visible from a great circle of viewing directions defined by rotating a vector about the common surface normal direction [43]. Kutulakos and Dyer also considered four parallel tangents and camera motion about a great circle to define a 3D affine basis [23]. Consider the distance d_i between parallel tangent i and the bitangent line; for a set of n parallel tangents, the projective coordinates of the distances $(d_1, \cdots, d_n)$ are invariant under affine image transformations and under rotation of the 3D surface about an axis parallel to the features' common surface normal [43]. If a bitangent and n parallel tangents can be detected in an image, $n-1$ invariants can be computed. Thus, a limiting bitangent and its n parallel tangents can be described by a point in an $n-1$ dimensional projective space, itself conveniently represented by an $(n-1)$ dimensional sphere embedded in an n dimensional Euclidean space.

Considering the entire limiting bitangent curve, the set of parallel tangent points defines a family of surface curves. When modeling an object, nearly any camera motion (except in the common tangent plane) will reveal a new pair of points on the limiting bitangent curve. Consequently, the entire family of curves can be observed from a trajectory (a curve) of viewpoints. From each viewpoint, an invariant can be computed. Taken over the entire trajectory, a curve in the $n-1$ dimensional invariant space is defined. Note that this curve is independent of the particular trajectory, as long as new points on the limiting bitangent curve are revealed at each step. Thus, an object can be modelled by a set of invariant curves, one for each limiting bitangent or parabolic curve. Note that these curves can be constructed from a sequence of images without knowledge of the camera motion – the 3D structure of the object is not needed, nor is reconstruction performed. Thus, model construction is greatly simplified. A set of models can then be represented by a set of curves.

To recognize an object in a single image, the bitangents, inflections, and corresponding parallel tangents are detected in the image; the projective coordinates of the distances are computed. For a particular bitangent (inflection) and measured invariants, the closest point on a model invariant curve is found. If the distance between the point and the curve is within some threshold determined from expected image noise, the

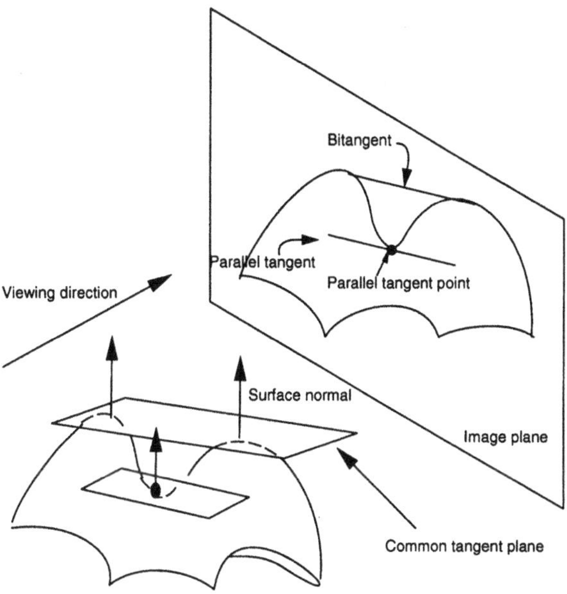

Fig. 2. Parallel tangents.

returned point provides evidence as to the object's identify. Since false matches may be returned for a single invariant, a voting procedure is used to finally identify the object. Note that the measured invariants can be used as an index similar to geometric hashing to retrieve a model in constant time. Furthermore, only one component of the image coordinates of each feature is used to construct the invariant. So, there is additional information available at each feature point. As shown in [44], the other component can be used to verify a match returned through invariant-based indexing.

We have implemented this approach and illustrate it with an example. Consider two modelled objects, a mallard decoy shown in Figure 3.a and a pintail duck in Figure 3.c; the detected image contour and a bitangent with its associated parallel tangents are shown in Figure 3.b while Figure 3.d shows all of the bitangents used to model the pintail duck. While n parallel tangents define a point in P^{n-1}, we cannot render curves in P^{n-1}, so we show all groupings of three parallel tangents which yield invariant curves in P^2. For the bitangents of both ducks, Figure 3.e shows these invariant curves drawn on the sphere; because there are more than three parallel tangents associated with some bitangents, there are more invariant curves than bitangents. Additionally, since the features can be ordered by increasing distance, it is unnecessary to hypothesize all match permutations for a given set of features. Applying this ordering constraint, the invariant curves only occupy a portion of the sphere. Figure 3.f shows the contour extracted from an image with both ducks as well as all of the bitangents extracted form the contour. The model base is indexed with the invariants derived from each bitangent and all combination of the associated parallel tangents. When more than three parallel tangents are measured during recognition, indexing can occur at a higher dimension than illustrated in Figure 3.e. Subsequent verification eliminates false matches [44]. For the image in Figure 3.f, four correct matches were returned, and one of these is illustrated

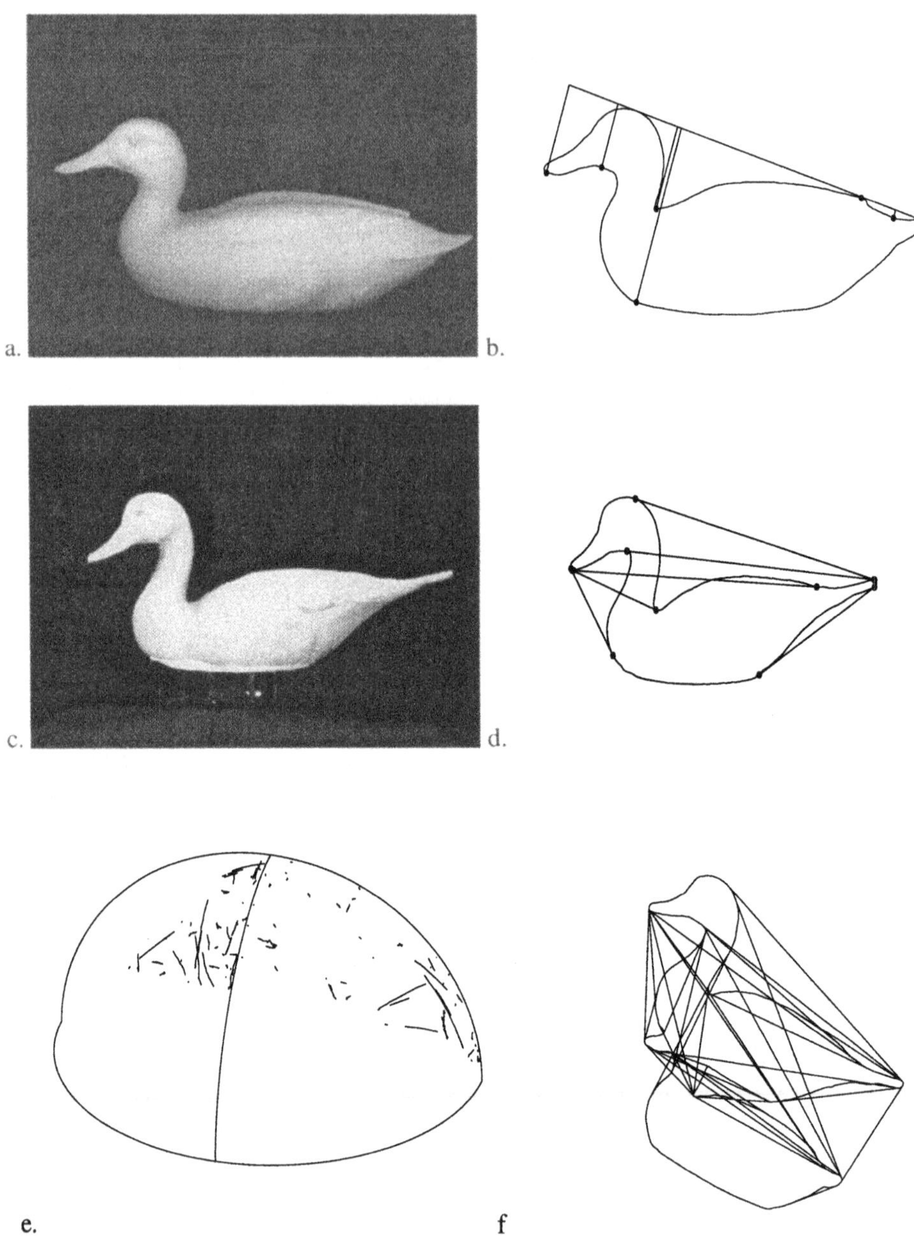

Fig. 3. Invariant-based recognition of curved objects using HOT curves: a,c. Images of a mallard and a pintail duck decoy; b. A bitangent and one set of parallel tangents; d. The bitangents of the pintail included in the model base; e. All of the invariant curves used for modelling drawn on a section of the sphere; f. The silhouette of an image with both ducks and all of the measured bitangents.

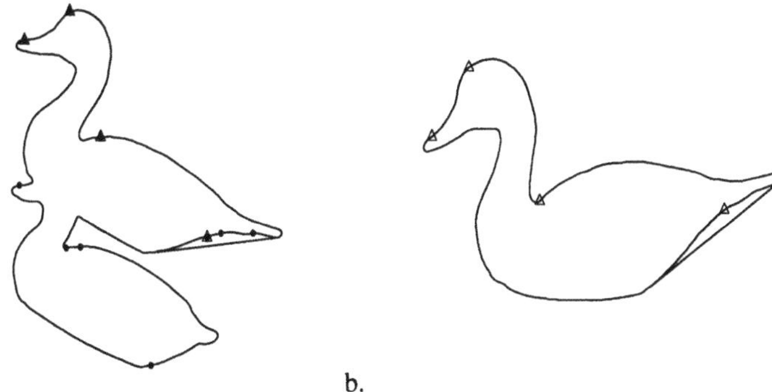

Fig. 4. The result of indexing and verification: Of the bitangents shown in Figure 3.f and associated parallel tangents, indexing and verification returned four correct matches. For one of these matches, (a) shows a bitangent and all of the parallel tangents for this bitangent; the four matching ones are drawn as triangles. The set of corresponding features returned from indexing and verification is shown in (b). The contour is from one of the images used during modelling.

in Figure 4. Figure 4.a shows a bitangent and associated parallel tangents. The ones that matched a database entry are indicated as triangles. Figure 4.b shows the matched features and image contour returned from the database after indexing.

5 Discussion

Beyond duck decoys, consider how birds are identified. While everyone can identify a bird as a bird, most of us can only identify a few species that are either particularly distinctive (penguins, flamingos, ostriches) or ubiquitous (pigeons and sea gulls). Others may simply be lumped into the category of birds. Now, with study or exposure, we can identify many other types of birds. Birding books, for example, provide a number of different recognition cues. First is geographic scope or range. We do not expect to find either a Rosegate Spoonbill or an Arctic Tern in Santa Monica, California. Second is habitat; if we are at the beach, it would be surprising to find quail. Thus, context is a method for reducing the set of models to be considered. In birding, the next step is usually to identify a rough category. Is it a plover, sandpiper, cormorant or pelican? Often gross morphology can be used. For example, size and overall color can be used to distinguish these four shore birds. Finally, when there are similar birds within the same family (e.g., Peterson's guide lists thirteen different plovers [31]), specific field marks (a black ring around the neck, a white belly, wing shape, etc.) are used to identify the particular species. General recognition, like bird identification, should exploit a number of different features including context, size, shape, color, and markings. Alone, none of these is strong enough for complete categorization, but together they are sufficient.

In this paper, we have focussed on the geometry of rigid objects and neglected these other features. Fortunately, many these features are either directly observable (e.g. color) or are external to the actual recognition system (e.g. context), and are relatively easy to incorporate in geometric recognition paradigms. Furthermore, the approaches described in this paper only apply directly to rigid objects though they can be used for the components of articulated objects. When an object is partly nonrigid such as a true duck (as opposed to a decoy), the viewpoint consistency constraint will not directly apply to some parts of the object. The recourse seems to be either to soften this constraint or to segment the object into multiple rigid parts.

In some circumstances, all of features mentioned above can be employed in recognition. However, this is not always true. Lighting conditions may not permit colors to be readily distinguished, objects may be too far away to discern fine morphological detail, or an object may be seen out of context. A general purpose recognition system, should utilize all available information; however, with less information, performance should degrade continuously, not catastrophically. As such, a recognition system might return a set of possible models instead of one. With a suitably organized database, it may still be feasible to recognize an object up to a particular class or category under degraded conditions.

Acknowledgments: This work was supported in part by the National Science Foundation under Grants IRI-9224815 and IRI-9015749. D. Kriegman was supported in part by a National Science Foundation NYI Grant IRI-9257990. J. Ponce was partially supported by NASA Contract NAG 1 613 and by the Campus Research Board of the University of Illinois at Urbana-Champaign. Thanks to A. Hoogs, S. Petitjean, L. Sandford, and S. Sullivan for their work on algebraic surfaces, to B. Vijayakumar and T. Joshi for their work on HOT Curves and T. Gallagher for her comments on birding.

References

1. R. Bajcsy and F. Solina. Three-dimensional object representation revisited. In *Int. Conf. on Computer Vision*, pages 231–240, London, U.K., June 1987.
2. T. O. Binford. Visual perception by computer. In *IEEE Conf. on Systems and Control*, 1971.
3. K. Bowyer and C. R. Dyer. Aspect graphs: An introduction and survey of recent results. *Int. J. of Imaging Systems and Technology*, 2:315–328, 1991.
4. M. Brady. Criteria for representations of shape. In J. Beck, B. Hope, and A. Rosenfeld, editors, *Human and Machine Vision*, pages 39–84. Academic Press, 1983.
5. B. Burns, R. Weiss, and E. Riseman. The non-existence of general-case view-invariants. In *Geometric Invariance in Computer Vision*, pages 120–131. MIT Press, 1992.
6. I. Chakravarty. The use of characteristic views as a basis for recognition of three-dimensional objects. Image Processing Laboratory IPL-TR-034, Rensselaer Polytechnic Institute, October 1982.
7. C. Chen and A. Kak. A robot vision system for recognizing 3-d objects in low-order polynomial time. *IEEE Systems, Man, and Cybernetics*, 19:1535–1563, 1989.
8. R. Duda and P. Hart. *Pattern Classification and Scene Analysis*. Wiley, New York, 1973.
9. D. Eggert, K. Bowyer, C. Dyer, H. Christensen, and D. Goldgof. The scale space aspect graph. *IEEE Trans. Pattern Anal. Mach. Intelligence*, 15(11):1114–1131, 1993.
10. O. Faugeras and M. Hebert. The representation, recognition, and locating of 3-D objects. *Int. J. Robot. Res.*, 5(3):27–52, Fall 1986.
11. F. Ferrie, J. Lagarde, and P. Whaite. Darboux frames, snakes, and super-quadrics: Geometry from the bottom-up. In *Workshop on Interpretation of 3D Scenes*, pages 170–176, 1989.

12. W. E. L. Grimson. *Object Recognition by Computer: The Role of Geometric Constraints.* MIT Press, 1990.
13. A. Gross and T. Boult. Error of fit measures for recovering parametric solids. In *Int. Conf. on Computer Vision*, pages 690–694, Tampa, Fl., December 1988.
14. P. Hallinan. A low-dimensional representation of human faces for arbitrary lighting conditions. In *Proc. IEEE Conf. on Comp. Vision and Patt. Recog.*, pages 995–999, 1994.
15. D. Huttenlocher and S. Ullman. Object recognition using alignment. In *Int. Conf. on Computer Vision*, pages 102–111, London, U.K., June 1987.
16. K. Ikeuchi and T. Kanade. Automatic generation of object recognition programs. *Proceedings of the IEEE*, 76(8):1016–35, August 1988.
17. D. W. Jacobs. Space efficient 3D model indexing. In *Proc. IEEE Conf. on Comp. Vision and Patt. Recog.*, pages 439–444, 1992.
18. T. Joshi, J. Ponce, B. Vijayakumar, and D. Kriegman. HOT curves for modelling and recognition of smooth curved 3D shapes. In *Proc. IEEE Conf. on Comp. Vision and Patt. Recog.*, June 1994. In Press.
19. J. J. Koenderink. *Solid Shape.* MIT Press, Cambridge, MA, 1990.
20. J. J. Koenderink and A. J. Van Doorn. The internal representation of solid shape with respect to vision. *Biological Cybernetics*, 32:211–216, 1979.
21. D. Kriegman and J. Ponce. On recognizing and positioning curved 3D objects from image contours. *IEEE Trans. Pattern Anal. Mach. Intelligence*, 12(12):1127–1137, 1990.
22. D. Kriegman, B. Vijayakumar, and J. Ponce. Constraints for recognizing and locating curved 3D objects from monocular image features. In *European Conf. on Computer Vision*, pages 829–833, 1992.
23. K. Kutulakos and C. Dyer. Occluding contour detection using affine invariants and purposive viewpoint adjustment. In *Proc. IEEE Conf. on Comp. Vision and Patt. Recog.*, pages 323–329, 1994.
24. D. G. Lowe. The viewpoint consistency constraint. *Int. J. Computer Vision*, 1(1):57–72, 1987.
25. D. Marr and K. Nishihara. Representation and recognition of the spatial organization of three-dimensional shapes. *Proc. Royal Society, London*, B-200:269–294, 1978.
26. A. Morgan. *Solving Polynomial Systems using Continuation for Engineering and Scientific Problems.* Prentice Hall, Englewood Cliffs, 1987.
27. J. Mundy and A. Zisserman. *Geometric Invariance in Computer Vision.* MIT Press, Cambridge, Mass., 1992.
28. H. Murase and S. Nayar. Illumination planning for object recognition in structured environments. In *IEEE Conference on Computer Vision and Pattern Recognition*, pages 31–38, Seattle, WA, 1994.
29. R. Nevatia and T. Binford. Description and recognition of complex curved objects. *Artificial Intelligence*, 8:77–98, 1977.
30. A. Pentland. Perceptual organization and the representation of natural form. *Artificial Intelligence*, 28:293–331, 1986.
31. R. Peterson. *A Field Guide to Western Birds.* Houghton-Mifflin, Boston, 1990.
32. S. Petitjean, J. Ponce, and D. Kriegman. Computing exact aspect graphs of curved objects: Algebraic surfaces. *Int. J. Computer Vision*, 9(3):231–255, 1992.
33. J. Ponce, A. Hoogs, and D. Kriegman. On using CAD models to compute the pose of curved 3D objects. *Computer Vision, Graphics, and Image Processing: Image Understanding*, 55(2):184–197, Mar 1992.
34. J. Ponce, D. Kriegman, S. Petitjean, S. Sullivan, G. Taubin, and B. Vijayakumar. Representations and algorithms for 3D curved object recognition. In *3D Object Recognition Systems*, pages 327–352. Elsevier Press, 1992.

35. M. Seibert and A. Waxman. Adaptive 3-D object recognitin from multiple views. *IEEE Trans. Pattern Anal. Mach. Intelligence*, 14(2):107–124, Feb. 1992.
36. S. Shafer. *Shadows and Silhouettes in Computer Vision*. Kluwer Academic Publishers, 1985.
37. I. Shimshoni and J. Ponce. Finite resolution aspect graphs of polyhedral objects. In *IEEE Workshop on Qualitative Vision*, June 1993.
38. L. Stark, D. Eggert, and K. Bowyer. Aspect graphs and non-linear optimization. In *Int. Conf. on Computer Vision*, pages 501–507, 1988.
39. J. Subrahmonia, D. Cooper, and D. Keren. Reliable object recognition using high-dimensional implicit polynomials for 2D curves and 3D surfaces. Technical Report LEMS-94, Brown University, 1991.
40. G. Taubin. Estimation of planar curves, surfaces and nonplanar space curves defined by implicit equations, with applications to edge and range image segmentation. *IEEE Trans. Pattern Anal. Mach. Intelligence*, 13(11):1115–1138, 1991.
41. G. Taubin, F. Cukierman, S. Sullivan, J. Ponce, and D. Kriegman. Parameterized families of polynomials for bounded algebraic curve and surface fitting. *IEEE Trans. Pattern Anal. Mach. Intelligence*, 16(3):287–303, 1994.
42. M. Turk and A. Pentland. Face recognition using eigenfaces. In *Proc. IEEE Conf. on Comp. Vision and Patt. Recog.*, pages 586–591, 1991.
43. B. Vijayakumar, D. Kriegman, T. Joshi, and J. Ponce. Towards invariant-based recognition of complex 3D curved objects in images. In *Yale Workshop on Adaptive and Learning*, June 1994.
44. B. Vijayakumar, D. Kriegman, and J. Ponce. Invariant-based recognition of complex 3D curved objects. In *Int. Conf. on Computer Vision*, 1995. Submitted.
45. R. Walker. *Algebraic Curves*. Princeton University Press, 1950.
46. A. Zisserman, D. Forsyth, J. Mundy, and C. Rothwell. Recognizing general curved objects efficiently. In Mundy and Zisserman, editors, *Geometric Invariance in Computer Vision*, pages 228–251. MIT Press, 1992.

On Representation and Invariant Recognition of Complex Objects Based on Patches and Parts

David B. Cooper and Zhibin Lei

Laboratory for Engineering Man/Machine Systems
Division of Engineering, Brown University, Providence, RI 02912

Abstract. This paper presents an outline of our view of $3D$ object modeling and recognition. The problem of interest is the recognition of which *free-form* object from a large possible set of articulated deformable objects in arbitrary positions is present in sensed data when clutter is also present and the objects are partially occluded. The modeling approach is to represent complex objects by patches or parts that individually capture significant local information and to geometrically relate them in order to provide the complete structure of an object. A representation that we use generally consists of implicit polynomial curves and surfaces for $2D$ data and $3D$ data, respectively. Recognition is based on geometric invariants—functions of patches or parts that capture the shape but are invariant to the geometric transformations. Since an object to be recognized may be a member of a class, or since the invariants for an object may take values in a class, recognition of class membership is necessary, and for this we use Bayesian recognizers. The technology of self and mutual invariants and Bayesian recognizers for implicit polynomial patches or parts is touched on in the paper. Another topic briefly discussed is representation by generalized cylinders within the framework of algebraic curves and invariants.

1 Introduction

1.1 The Simplest Problem: Rigid Objects

An object, which can be any of L rigid objects stored in a database, is sensed in arbitrary position. A system is to make a minimum probability of error recognition as to which one of the L objects stored in the database is sensed in the data. Our approach is to partition the complex object being sensed into patches or parts, fit an implicit polynomial of modest degree greater than two to each of the patches or parts, compute a vector of Euclidean or affine invariants, where each invariant is a function of the coefficients of a polynomial fit to one patch or part or a function of the polynomials fit to a pair of patches or parts, and use a Bayesian recognizer based on comparing the vectors of invariants for the patches or parts in the sensed data with those stored in the database. The goal is a recognizer that functions with close to the minimum probability of error and incurs modest computational cost. The approach is especially effective for recognizing partially occluded articulated objects, since recognition can then be based on self and joint recognition of one or more patches or parts, each of which is unoccluded or is minimally occluded. The approach is also useful for distinguishing whether or not an individual patch or part belongs to an object in the database.

Data can be $3D$ x, y, z points on the surface of a $3D$ object, or data can be $2D$ points which are measurements of contours in a $2D$ image. A contour can be an occluding contour of a $3D$ object or it can be a contour of markings on a $3D$ surface, etc. In $3D$, since an object being sensed will be in arbitrary position, the surface being sensed will be a Euclidean transformation of the surface stored in the database. In $2D$, we assume that the curve being sensed is a projection of a *roughly* planar curve in $3D$ in arbitrary position with respect to the viewing camera. Hence, the curve being sensed will be *roughly* an affine transformation of the curve stored in the database.

1.2 A More Complicated Problem: Deformable Objects and Classes of Objects

Recognition of *classes* arises for three reasons, the first two because objects may belong to classes. Object classes occur for two reasons. First, we may wish to recognize that an object of interest is a pear, and there are a variety of shapes for a pear. Second, we may wish to recognize a particular person who may be laughing or crying or angry at the time so that the face of the individual takes various forms. Or we may wish to base recognition of the individual partially on the body, and the upper torso, e.g., can be in a variety of positions with respect to bending at the waist and may undergo a variety of deformations associated with these positions, e.g., sagging. This latter class is due to deformations of a particular object. Both of these *object class problems* result in the necessity for recognizing class membership of the data in measurement or representation space. Since our representation is usually implicit polynomial curves or surfaces or algebraic invariants, recognition of classes is in the space of polynomial coefficients or in the space of algebraic invariants of these polynomials. The third reason for the necessity of class membership recognition in data or representation space is because of *noise* or because certain assumptions about the data formation are approximations: a measured *invariant* for an object may differ somewhat from one data set to another and must therefore be treated as a member of a class.

The challenge in Bayesian recognition of classes in data or representation space is how to put a priori probability measures on these spaces for these classes. We touch on this in Section 5.

2 The Approach

2.1 An Approach That Has Been Used Previously for Simple Shapes

When an object shape is simple, each object can be represented by a single implicit polynomial of modest degree — typically of degree three or four. Then each of the L objects to be recognized can be stored in the database as the coefficient vector for a single implicit polynomial. A single 2nd degree implicit polynomial can represent any of the commonly used conics — a right circular cone, an ellipsoid, a cylinder of circular cross section, a sphere, etc. Fourth degree implicit polynomials can represent much more interesting, useful complicated shapes. For example, each of the lower case hand printed letters can usually be represented by a 2nd, 3rd, or 4th degree polynomial in x, y, and shapes such as a pear, an eggplant, computer keyboard mice can each be represented by a single 4th degree implicit polynomial in x, y, z [4]. Then a number of invariants are computed for these polynomials. Typically, three to seven arranged as a vector have been used [4]. One such vector is stored for each object in the database, and recognition is then done by fitting an implicit polynomial to the data to be recognized, and comparing a vector of invariants with each stored vector of invariants. Bayesian comparisons are then made in order to realize minimum probability of error recognition. The advantages of this approach are the following: (a) Data are processed only once — to fit a single polynomial to the data to be recognized; (b) Recognition is based on invariants — hence, is object position and camera position invariant; and (c) Recognition is Bayesian — hence, robust to missing data due, for example, to occlusion, and also robust to other variations from the stored model. For example, if the $2D$ curve being recognized is the projection of a curve in $3D$ that is not planar, then the *affine invariants* used for recognition are not really constant under change in camera viewing position. Hence, such an *invariant* is really only locally invariant and can take any value in a set of values and must therefore be treated as a class of invariants. Then recognition is recognition of the class association of a measured vector of invariants.

2.2 Complicated Shapes: The Problem of Interest in This Paper

Here, an object of interest is too complicated to be represented by a polynomial of modest degree, e.g., 4th degree. The approach, then, is to decompose a complex object into patches or parts such that each patch or part can be represented by a polynomial of 4th or lower degree. An object is now represented by a set of implicit polynomials — one polynomial for each patch or part in the object. We would now like to implement object recognition as Bayesian recognition based on invariants of the polynomials representing an object. There are two formidable obstacles to this approach: (a) How can one capture the joint geometry of an object represented by many patches or parts, e.g., ten, each being a 4th degree polynomial? (b) It is necessary to determine patches or parts in the data that correspond *roughly* to patches or parts in the objects stored in the database. If there is data missing, due, e.g., to partial occlusion, then the number of patches or parts of an object in the data will likely be significantly smaller than the number in the database.

The preceding two issues are the major challenges, but others exist as well. An important observation at this point concerns the benefit of using high degree polynomials, e.g., 4th degree per patch rather than using smaller patches requiring only 2nd or 3rd degree polynomials. Since the data around an object being sensed may be over only part of the object boundary, due, e.g., to partial self occlusion or to partial occlusion by some other object, the portion of the object boundary being sensed will often be less than that which is represented in the database. Furthermore, what may then be unclear is which group of patches in the sensed data correspond to a group of patches for a single object in the database. Using patches that are fairly complex helps to solve this problem because checking to see which object patches stored in a database match a patch in the data can help to quickly eliminate many objects in the database as being candidate matches for the object being sensed in a particular patch. A 4th degree polynomial has much more discriminatory power for this purpose than does a 2nd degree polynomial.

3 Bayesian Recognition in Terms of Patches and Parts

3.1 Parts

What are parts and patches, and why are they useful? There is not general agreement on the concept of *parts*, but there are a number of different approaches to the problem, and we touch on a few that we find appealing. Our inclination at present is to decompose data into parts when parts are easily detectable invariantly, and otherwise decompose data into patches. Our view of parts is that they are simply chunks of $2D$ curves or $3D$ surfaces that can be found reliably in any data in which they are visible and irrespective of sensor viewing direction. Among the bases for doing this are: (a) general shape features; (b) geometric primitives; (c) features pertinent to the restricted set of objects under consideration in a particular application; (d) area or volume scale; (e) complexity.

1). Limb and neck based parts [11]. They feel that there is a general multiscale criterion for decomposing silhouettes into parts motivated both by computational requirements for recognition as well as by psychophysical constraints. Essentially, objects are to be partitioned at limbs, or pairs of negative curvature minima whose boundaries show evidence for good continuation, and necks, or locally thinnest regions. Note that this decomposition is invariant to Euclidean and affine transformations of the data. (See Figure 5 for their multiscale decomposition of an airplane into parts). They handle occlusion by exploiting two features. First, the intersections of the boundaries of an occluding silhouette and an object are high curvature points, and second, if a long smooth boundary curve of an object is partially occluded, the visible portion of the curve to either side of the occluded portion can often be recognized as having roughly the same direction and therefore being part of the same continuous curve. This is illustrated in the rockets that are partially occluded by the wing, but the fore and aft portions of a rocket are inferred to be portions of the same object. Multiscale is handled by approximating the shape in scale (e.g., the reaction-diffusion space) to remove small scale structure and performing the decomposition hierarchically, first on large scale structure, then on smaller scale structure, etc. Scale here is distance across the silhouette in at least one direction.

2). Many manufactured parts are composed of 3d surface patches that are planar, spherical, cylindrical, conical, or hyperellipsoidal. It is easy find the portions of $3D$ space occupied by these primitive surfaces and to fit quadric models to the sensed $3D$ data. (Examples of approaches that could be further refined for this purpose are [7, 14, 13]). A desirable property of primitives is that it is not necessary that there be tangent discontinuities where they meet in order to segment them. Two cylindrical patches having parallel axes can intersect in a line such that the tangent is continuous in the neighborhood of the line of intersection, yet the cylindrical patches can be segmented and the parameters estimated in an unsupervised mode.

3). High curvature points are often used as markers at which to decompose a curve into segments. These would serve, e.g., to delimit the ends of the curve segment that represent the wing of an airplane or perhaps intersection of the occluding contour of the spout of a pitcher with the body of the pitcher. In $3D$ data, surface curvature would be the basis for segmenting a pitcher spout from the pitcher body, though other size and orientation properties can also be used for this purpose.

4). One concept of scale is illustrated by the $2D$ boundary or $3D$ surface of a tiger's tail. The tail is long but its width in $3D$ or $2D$ is very small, in which case it would be considered to be small scale and can be detected as such using morphology or scale-spaced filtering [10].

5). The complexity of a curve segment or of a surface patch is a good criterion for decomposition into parts. This measure is especially useful if it is in terms of the representation being used for objects. One measure of complexity is based on the combination of region size and the minimum degree of the implicit polynomial needed to represent the data in that region. For example, in Figure 9, 4th degree implicit polynomial surfaces are needed to represent the lips. chin, and nose in the regions shown, but degree 2 is probably all that is needed to represent the cheeks or most of the head over regions of the sizes used for the lips, nose and mouth. Hence, the lips, nose and chin are relatively complex distinctive regions. How can these regions, e.g., the lips region, be automatically segmented out? We suggest segmentation based on this complexity measure, or on curvature, or using other methods [6, 12, 13, 14].

3.2 Patches.

Patches are chunks of data which are chosen in very simple ways incurring little computational cost and such that the extent of the region for each patch is somewhat but not heavily dependent on the shape of the data. We partition an object boundary into patches of size L. For $2D$ curves that are subject to Euclidean transformations, this means that a curve is partitioned into curve segments having arc length L. For $2D$ curves that are subject to affine transformations, if we are dealing with a closed curve we rescale the curve to fit into a box of standard size and partition the curve into curve segments having arc length L. This rescaling is ok because it is an affine transformation, and affine invariants are used in the recognition. For $3D$ surfaces, we intersect the surfaces with spheres of diameter L, and the data patches are the data subsets that lie within the intersecting spheres. The spheres will need to overlap somewhat, so that the patches contain most of the data that is given. Since recognition involves comparing patches found in the sensed data with those stored in the database, it is necessary that for a patch found for an object in the data there is a patch in the database for that object that covers roughly the same region on the object. These chunks of data must be found *invariantly*. The general technique to break data into patches and compare them with those stored in the database is sketched in [16]. The basic idea is that for curves, overlapping patches of length $\frac{3}{2}L$(or less) are stored for an object in the database. Then a patch of length L in the data will always lie within a stored patch of length $\frac{3}{2}L$ (see Figure 2). The same ideas apply to $3D$. Multiscale can be a significant issue here. For example, suppose there were more detail in the panther's face in Figure 11, or in the engine exhaust area in Figure 10. Then either these regions would have to be represented by fourth degree polynomials over intervals of shorter arc lengths [16], or these regions could be smoothed and, the smoothed data could then be well represented by curve segments of length L as is used everywhere else. These types of small scale structure can be smoothed well by Gaussian smoothing, or, more simply, by letting the polynomial fit to an arc of length L do the smoothing.

3.3 Joint Geometry of Pairs of Patches or Parts and Their Mutual Invariants

It is shown in [16] that if the relative geometry is known for the two patches or parts in each pair of patches or parts for a complex object, then the geometry of the entire object is determined. Hence, recognition of complex objects based on recognition of Euclidean *mutual invariants* that capture the joint geometry in each pair of patches or parts should be essentially as reliable as recognition based on Euclidean *self invariants* of a single polynomial that represents the entire object. So the recognition of a complex object can be reduced to first breaking the data into simpler patches or parts and then comparing the mutual invariants of pairs of patches or parts with those of stored models. Two approaches to finding mutual algebraic invariants are those in [7, 8] and [16], the latter based on symbolic computation to find relative invariants and then linear algebra to find absolute weight and rank invariants. A simple example illustrating concepts of self and mutual algebraic Euclidean invariants is for the pair of elliptic arcs in Figure 3. Self invariants for each elliptic arc are the lengths of the major($R1$ and $R2$) and minor axes, which are polynomial functions of the coefficients of the second degree terms in the implicit polynomial representation for the ellipse. Mutual geometric invariants are the distance, d, between ellipse centers and the angle, Θ, between their major axes. Here d is determined by the coefficients of the linear and constant terms of each ellipse and Θ is determined by the coefficients of the second degree terms for each ellipse. An example of a mutual relative algebraic Euclidean invariant for a pair of 4th degree implicit polynomials is

$$\begin{array}{c} -9*a_{31}^2*b_{04}+24*a_{22}*a_{40}*b_{04}+3*a_{22}*a_{31}*b_{13}-18*a_{13}*a_{40}*b_{13} \\ -2*a_{22}^2*b_{22}+3*a_{13}*a_{31}*b_{22}+24*a_{04}*a_{40}*b_{22}+3*a_{13}*a_{22}*b_{31}- \\ 18*a_{04}*a_{31}*b_{31}-9*a_{13}^2*b_{40}+24*a_{04}*a_{22}*b_{40} \end{array} \quad (1)$$

where a_{ij} is the coefficient for the monomial x^iy^j in one polynomial and b_{ij} is the coefficient in the other polynomial.

3.4 Bayesian Object Recognition

Consider the simplest recognition scenario where there is a set of L objects, labeled by $l = 1, 2, \ldots, L$, each modeled by a single polynomial of the same degree n in x, y and z. Let α_l, $l = 1, 2, \ldots, L$, denote the parameter vector for object l. Then, given a sequence of data points $Z_1, Z_2, \ldots, Z_N$ from a part of one of the objects, the minimum probability of error recognition of object type is: "choose l for which (2) is maximum".

$$p(\mathbf{Z} \mid \alpha_l) \quad (2)$$

This, however, requires considerable computation because the raw data $\mathbf{Z} = \{Z_1, Z_2, \ldots, Z_N\}$ is processed a total of L times in order to compute (2) for each l. Using an asymptotic approximation [4] gives a computationally attractive recognition rule: "choose l for which (3) is maximum".

$$p(\mathbf{Z} \mid \boldsymbol{\alpha}_l) \approx p(\mathbf{Z} \mid \hat{\boldsymbol{\alpha}}_N) \exp\left\{ -\frac{1}{2}(\boldsymbol{\alpha}_l - \hat{\boldsymbol{\alpha}}_N)^t \Psi_N (\boldsymbol{\alpha}_l - \hat{\boldsymbol{\alpha}}_N) \right\} \tag{3}$$

Here, $\hat{\boldsymbol{\alpha}}_N$ is the maximum likelihood estimate of $\boldsymbol{\alpha}$, and is the set of parameter values of the unconstrained polynomial of degree n fit to the data. Ψ_N is an information matrix that depends on the specific data set $\mathbf{Z}$ being recognized. Since $p(\mathbf{Z} \mid \hat{\boldsymbol{\alpha}}_N)$ is the same for all l, maximizing the function in (3) is equivalent to minimizing the quadratic form $(\boldsymbol{\alpha}_l - \hat{\boldsymbol{\alpha}}_N)^t \Psi_N (\boldsymbol{\alpha}_l - \hat{\boldsymbol{\alpha}}_N)$. Thus, the recognition rule becomes: "choose l for which (4) is minimum".

$$(\boldsymbol{\alpha}_l - \hat{\boldsymbol{\alpha}}_N)^t \Psi_N (\boldsymbol{\alpha}_l - \hat{\boldsymbol{\alpha}}_N) \tag{4}$$

Equation (4) is a low computational cost recognition rule because the data set $\mathbf{Z}$ is involved just once to compute $\hat{\boldsymbol{\alpha}}_N$ and the information matrix Ψ_N. This minimum Mahalanobis distance recognition rule gives approximately the same result as does (2) when the number of data points N is at least a few times larger than the number of polynomial coefficients.

For object l, let G_l be the vector of invariants stored in the database, and let $\hat{G}$ be the vector of invariants for the best fitting polynomial to the data. Note that $\hat{G}$ is a function of $\hat{\boldsymbol{\alpha}}_N$. Then (3) extends to recognition based on invariants: "choose l for which (5) is maximum [4]".

$$p(\mathbf{Z} \mid G_l) \approx p(\mathbf{Z} \mid \hat{G}) \exp\left\{ -\frac{1}{2}(G_l - \hat{G})^t \Psi^G (G_l - \hat{G}) \right\} \tag{5}$$

where Ψ^G is computed from Ψ_N and the function $\hat{G}(\hat{\boldsymbol{\alpha}}_N)$; or, equivalently: "choose l for which $(G_l - \hat{G})^t \Psi^G (G_l - \hat{G})$ is minimum".

Finally, to extend this to patches or parts, let lm specify a subset indexed by m of a partition of object l into patches or parts, and let $\boldsymbol{\alpha}_{lm}$ specify the vector of polynomial coefficients for lm of object l, and G_{lm} specify the set of mutual invariants for pairs of patches or parts in lm. Then (5) expends to: "choose object l and subset lm of a partition of it for which (6) is minimum".

$$(\hat{G}_{m^*} - G_{lm})^t \Psi^{G_{m^*}} (\hat{G}_{m^*} - G_{lm}) \tag{6}$$

where $\hat{G}_{m^*}$ is the set of mutual invariants for the object to be recognized in partition m^* of the data [16].

4 Generalized Cylinders: A New Representation for Them

Though implicit polynomial surfaces of degree 4 can be quite complicated shapes, e.g., the lips in Figure 9, other shapes such as the spout of a tea kettle or a rotated forearm may require a higher degree polynomial or patches of two or more polynomials of degree 4. For the spout of a tea kettle, if the axis has an S shape, then even if the cross section were approximately ellipsoidal of constant shape, the required 3d implicit polynomial representation would be of degree 5. The forearm is interesting, because if the hand is rotated around the axis of the forearm bone, the roughly elliptic cross section of the surface of the arm twists around the forearm bone, thus requiring a description of the rotation of the elliptic cross section around the axis. Again, all of this leads to the need for patches of polynomials. An alternative is to use generalized cylinders for elongated objects. Perhaps we can capture the required recognition and position information with fewer parameters. But we do want to preserve the benefits of implicit polynomials, specifically, the use of invariants, computational ease of estimating the parameters for the models, and the easy use of Bayesian recognizers in order to have robustness. We can have it all by using implicit polynomial curves to represent generalized cylinders in two and three dimensions in new ways. Among possible approaches to segmenting generalized cylinders from a data set [15] would be adaptation of the maximum aposteriori estimation approach taken for roads in aerial images (Figure 4) where a road is modeled by its geometry, i.e., axis and cross section, and by its data formation [3].

4.1 3D Generalized Cylinders Represented by Algebraic Curves and Invariants

A generalized cylinder can be specified by its axis and its cross section in a plane perpendicular to the axis. For elliptic cross sections, our representation for the generalized cylinder is a set of 5 or fewer one dimensional curves which are invariant to Euclidean transformations of the data. Furthermore, each of these curves can be represented as an implicit polynomial, so that algebraic translation invariants of these curve polynomials are then invariants for the generalized cylinder. The details are as follows.

First, the axis of the generalized cylinder can be specified by its curvature and torsion as functions of arc length. These are two one dimensional curves that are invariant to Euclidean transformations of the $3D$ data space. Each of these curves can be represented as an implicit polynomial curve. For example, in the case of the spout of the pitcher, the axis is planar, so the torsion curve is 0 and the curvature curve probably has an S like shape and is therefore well represented by a 3rd degree implicit polynomial curve. Thus, the coefficients of this implicit polynomial curve are Euclidean invariants for the $3D$ generalized cylinder axis. Unfortunately, part of the generalized cylinder may be occluded, in which case the starting point for our measured axis may not be the same as the starting point for the axis stored in the database. We can get around this by using algebraic *translation* invariants of our implicit polynomial curve representation. These algebraic invariants are then Euclidean invariants of the generalized cylinder axis. If the generalized cylinder axis has both curvature and torsion, then we find algebraic invariants for both curves. However, using translation invariants for each curve separately does not capture the fact that both curves translate in the same way. Hence, our ultimate recognizer is the use of *mutual algebraic translation* invariants for the pair of curves.

Second, suppose the generalized cylinder cross section is a polynomial curve, and, in particular, an ellipse. The ellipse is partially specified by the lengths of its major and minor axes. Consider the curve of lengths of the major axes of the ellipses as a function of arc length along the generalized cylinder axis, and similarly for the sequence of values of the minor axes of the ellipses (Figure 1). These two *eigen-curves* are Euclidean invariant features of the generalized cylinder, and, as in the preceding paragraph, we can represent each as an implicit polynomial curve, and algebraic mutual *translation* invariants of these implicit curves will be Euclidean invariants of the original generalized cylinder. Now suppose the generalized cylinder axis lies in a plane, i.e., the torsion is 0. Then a complete description of the sequence of elliptic cross sections can be had by also including the angle between the major axis of the elliptic cross section with the plane of the generalized cylinder axis. The one dimensional curve of this sequence of angles is another Euclidean invariant of the generalized cylinder, so that we now have a total of four Euclidean invariant curves that describe a generalized cylinder with arbitrary planar axis and twisting-varying-in-size elliptic cross section. When this is extended to a generalized cylinder axis that twists, i.e., has nonzero torsion, we have 5 invariant curves that completely specify the generalized cylinder, and these curves in turn can be represented by mutual algebraic invariants which will be Euclidean invariants even if the generalized cylinder is partially occluded.

The final generalization is to more complicated cross section curves, for example, 4th degree implicit polynomial cross section curves. Then the eigen-curves of the 4th degree terms of the cross section curves are a generalization of the eigen-curves for the elliptic cross section curves in the preceding paragraph. And, as with the ellipses, we can talk about the twisting of the cross section curves about the generalized cylinder axis. For this purpose, we would use the concept of an intrinsic coordinate system for an implicit polynomial curve that was developed in [7, 8].

4.2 2D Representation of Generalized Cylinders

By a $2D$ axis for a generalized cylinder we will mean the curve midway between a pair of curves which are the occluding contours in an image of an elongated $3D$ surface. If the axis is the projection of a planar generalized cylinder axis in $3D$, then the $2D$ axes in two views, each view by a camera in different position, will be related by an approximately affine transformation. Hence, the axis can be represented by an implicit algebraic curve and it can be recognized based on algebraic affine invariants.

More generally, assume there is at least one view of the generalized cylinder $3D$ cross section in the image, and assume that the $3D$ axis is planar and the cross section curve keeps its shape but can vary in size as a function of location along the axis. There are then interesting invariants for the sequence of projections into the image of the $3D$ cross section curves, but space precludes discussing them here.

5 Recognition of Classes of Objects and Deformable Objects

If instead of recognizing individual objects, there are classes of objects where each class consists of two or more objects, then in order to distinguish to which of L classes an object present in the data belongs, the Bayes solution for recognition based on polynomial curves or surfaces is: "choose the class j for which (7) is maximum".

$$\int_{whole\ \alpha_j\ space} p(\mathbf{Z} \mid \alpha_j, class\, j) p(\alpha_j \mid class\, j) P(class\, j) d\alpha_j \tag{7}$$

where α_j is the parameter vector for class j. It is a random vector here because there is more than one element in the pattern class, e.g., a pear in the class of pears. Practical use of this recognition function depends on having a simple closed-form expression for the integral. Because of the simple form that $p(\mathbf{Z} \mid \alpha_j, class\, j)$ takes, (7) can be evaluated if $p(\alpha_j \mid class\, j)$ is simple. We suggest a mixture of Gaussians for the purpose. The recognizer will then be a mixture of Gaussians. If recognition is to be done in terms of invariants, the mechanism is similar, where asymptotic approximations are used and the expressions are similar to (7) but are in terms of invariants rather than polynomial coefficients. See [4, 5] for details.

For deformable objects, in preliminary experiments it appears that an appropriate prior locally is often the convolution of a multivariate Gaussian probability density function (pdf) with a smooth manifold. In particular, in Figure 7 each surface is the plot of one Euclidean algebraic invariant of the polynomial for a tank silhouette as a function of elevation and azimuth angle in 42 views of a model tank by a moving camera. A few views within the range of views used, and associated occluding contours, each of which is then represented by a 4th degree polynomial, are shown in Figure 6 [5]. The Gaussian that would be used in the convolution is due to measurement noise and other phenomena, and is the Gaussian that arises in the asymptotic distributions. Note, if deformations are over a limited range of angles, then recognition based on polynomial coefficients may be appropriate because a smooth manifold may be an appropriate mean value function for the range of coefficients values for the deformations of an object. Otherwise, Euclidean or affine algebraic invariants, which are shape descriptors, appear to be the recognition features for which the deformation manifolds should be studied and used.

References

1. D. Keren. Some New Invariants in Computer Vision. *IEEE Transactions on Pattern Analysis and Machine Intelligence,* November 1994.
2. M. Barzohar, D. Keren and D.B. Cooper. Recognizing Groups of Curves Based on New Affine Mutual Geometric Invariants, with Applications to Recognizing Intersecting Roads in Aerial Images. *Proceedings, 12th International Conference on Pattern Recognition,* Jerusalem, Israel, October 1994.
3. M. Barzohar and D.B. Cooper. Automatic Finding of Main Roads in Aerial Images by Using Geometric Stochastic Models and Estimation. *Proceedings, IEEE Conference on Computer Vision and Pattern Recognition,* N.Y.C., June 1993.
4. J. Subrahmonia, D. Keren, and D. B. Cooper. An Integrated Object Recognition System Based on High Degree Implicit Polynomials, Algebraic Invariants, and Bayesian Methods. *Proceedings, Image Understanding Workshop,* Washington DC, April 1993.
5. J. Subrahmonia. Practical Reliable Bayesian Recognition of 2D and 3D Objects Using Implicit Polynomials and Algebraic Invariants. *PhD thesis,* Brown University, May 1993.
6. G. Taubin. Estimation of Planar Curves, Surfaces and Nonplanar Space Curves Defined by Implicit Equations, with Applications to Edge and Range Image Segmentation. *IEEE Transactions on Pattern Analysis and Machine Intelligence,* November 1991.
7. G. Taubin. Recognition and Positioning of Rigid Object Using Algebraic and Moment Invariants. *PhD thesis,* Brown University, May 1991.

8. G. Taubin and D.B. Cooper. 2D and 3D Object Recognition and Positioning with Algebraic Invariants and Covariants. *Symbolic and Numerical Computation for Artificial Intelligence,* B.R. Donald, D. Kapur and J.L. Mundy editors, Academic Press, N.Y.C., 1992.
9. F. Mokhtarian and A. Mackworth. Scale-based Description and Recognition of Planar Curves and Two-dimensional Shapes. *IEEE Transactions on Pattern Analysis and Machine Intelligence,* January 1986.
10. B.B. Kimia, A.R. Tannenbaum and S.W. Zucker Entropy Scale-Space. *Visual Form: Analysis and Recognition,* C. Arcelli, ed. Plenum Press, New York, 1991.
11. K. Siddiqi and B.B. Kimia. Parts of Visual Form: Computational Aspects. *IEEE Transactions on Pattern Analysis and Machine Intelligence,* March 1995.
12. O.D. Faugeras, M. Hebert and E. Pauchon. Segmentation of Range Data into Planar and Quadratic Patches. *Proceedings, IEEE Conference on Computer Vision and Pattern Recognition,* Washington D.C., June 1983.
13. P.J. Besl and R.C. Jain. Segmentation Through Variable-order Surface Fitting. *IEEE Transactions on Pattern Analysis and Machine Intelligence,* March 1988.
14. J.F. Silverman and D.B. Cooper. Bayesian Clustering for Unsupervised Estimation of Surface and Texture Models. *IEEE Transactions on Pattern Analysis and Machine Intelligence,* July 1988.
15. M. Zerroug and G. Medioni. The Chanllenge of Generic Objects Recognition. *NSF/ARPA Workshop on 3D Object Representation for Computer Vision,* N.Y.C., December 1994.
16. Z. Lei, D. Keren and D.B. Cooper. Recognition of Complex Free-Form Objects Based on Mutual Algebraic Invariants for Pairs of Patches of Data. Lems report 140, Division of Engineering, Brown University, January 1995.

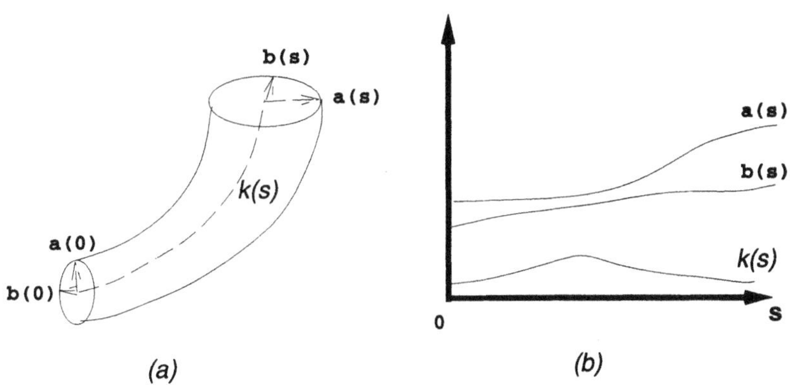

Fig. 1. The generalized cylinder and its central axe and cross-section representation

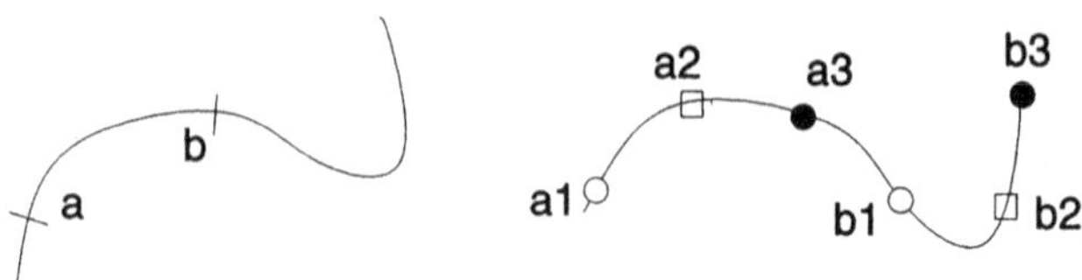

Fig. 2. A patch ab of length L in data and three overlapping patches of length $\frac{3}{2}L$ in the database

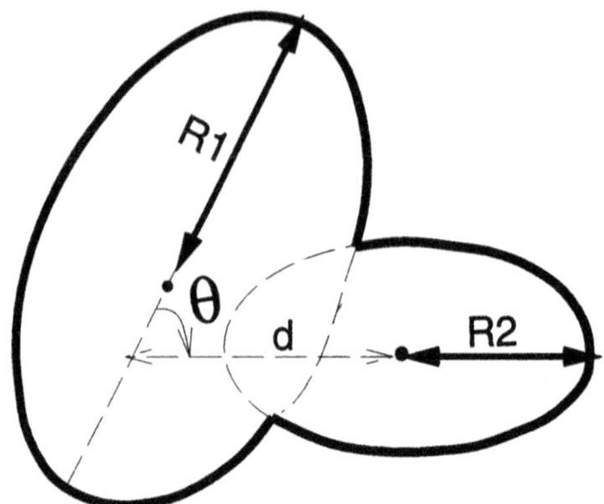

Fig. 3. Geometry illustrating the self and mutual algebraic invariants

Fig. 4. The original and the estimated roads

Fig. 5. The part decomposition of an original and a partially occluded plane

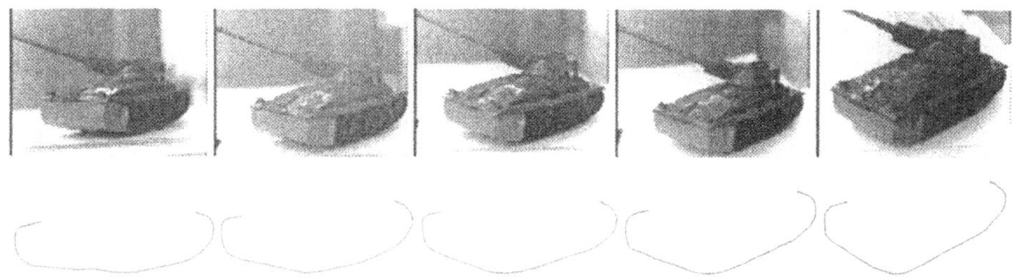

Fig. 6. Silhouettes extracted from the images of a tank viewed from different directions

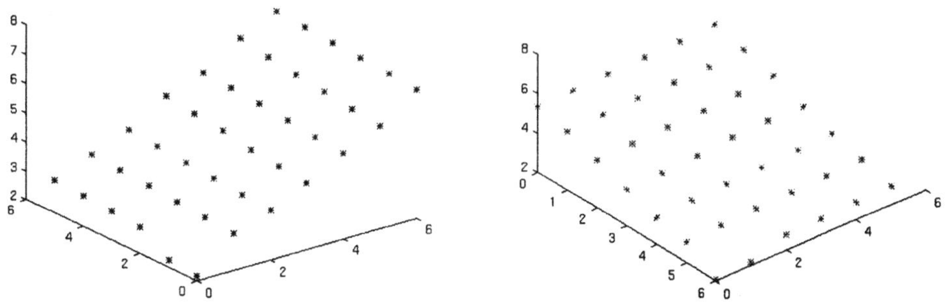

Fig. 7. Plots of each of two invariants, for the polynomial fits to the silhouettes of the tank, as functions of camera azimuth and elevation angle.

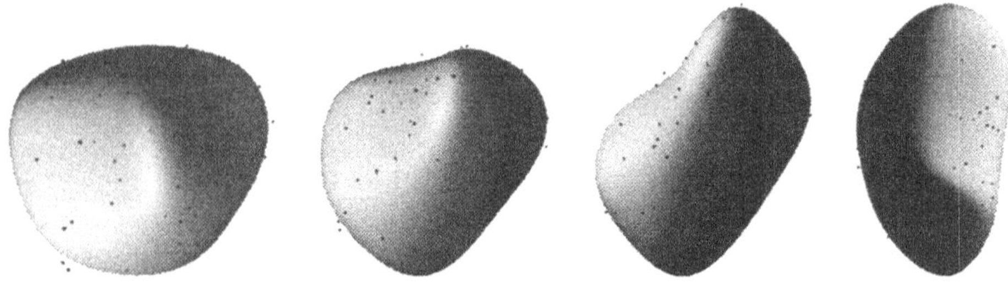

Fig. 8. Four views of one 4th degree implicit polynomial surface fit to dog's heart

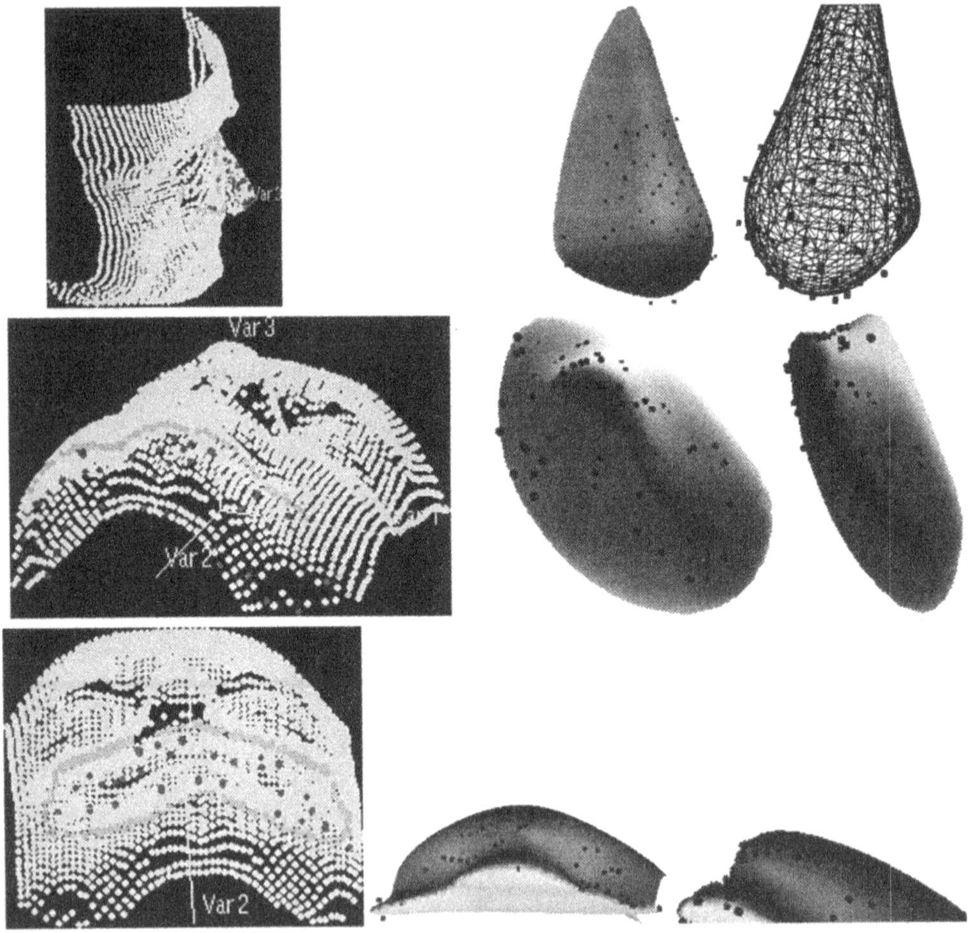

Fig. 9. Two views of each single 4th degree implicit polynomial surface fit to a nose, a chin and lips

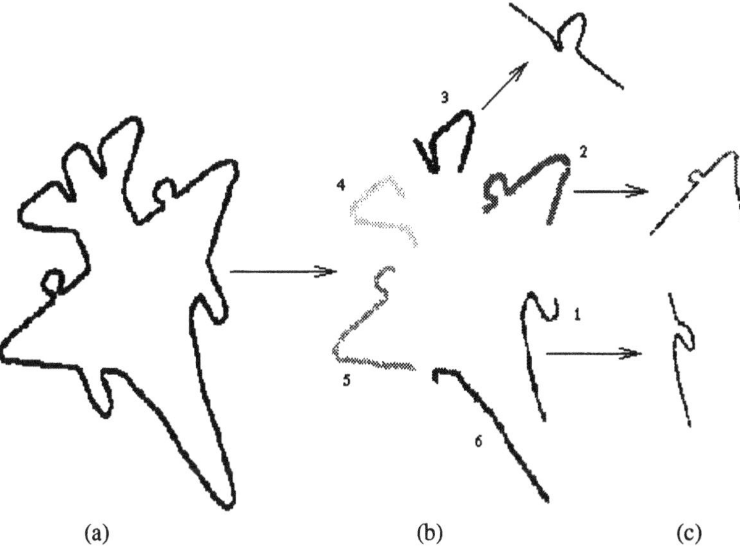

Fig. 10. (a) Sensed data of an affine transformation of a plane in the database; (b) its decomposition into patches; (c) 4th degree polynomial fits to several patches.

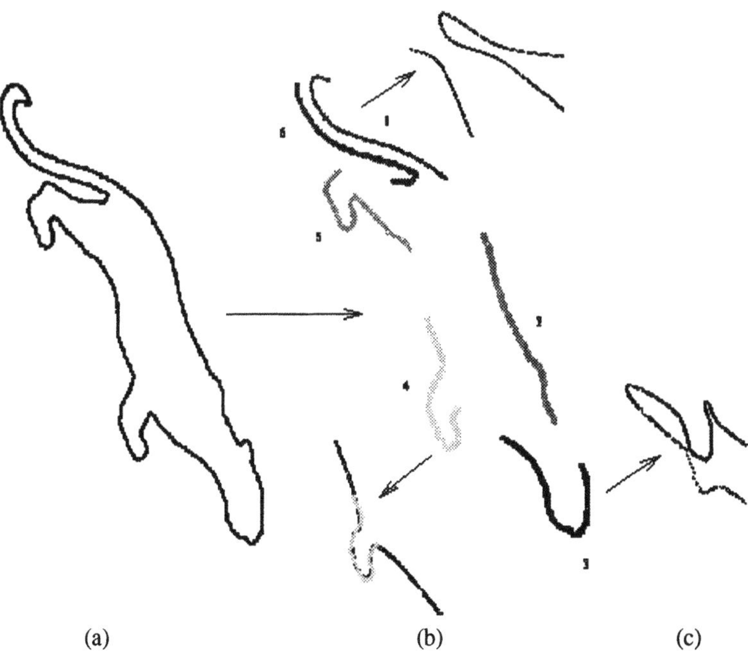

Fig. 11. (a) Sensed data of an affine transformation for an animal in the database; (b) its decomposition into patches; (c) 4th degree polynomial fits to several patches.

Algebraic Geometry and Object Representation in Computer Vision

Sylvain Petitjean

CRIN / INRIA Lorraine
B timent LORIA
Campus Scientifique
BP 239
54506 Vandœuvre-les-Nancy cedex, France
e-mail: petitjea@loria.fr

Abstract. The goal of algebraic geometry is to gain an understanding of the behaviour of functions related by polynomial relationships. Algebraic curves and surfaces having considerable advantages as objects of study in computer vision, the relevance of algebraic geometry when dealing with the representation of polynomial objects seems only natural. However, due to its high level of abstraction, this mathematical field has seen only few applications in computer vision in comparison to its huge achievements, by contrast with differential geometry for instance. In this paper, we examine several methods that we think could see some developments in object representation. An example of their use is presented for the construction of aspect graphs of smooth surfaces and for the understanding of the complexity of this representation.

1 Introduction

Up to now, most 3D recognition systems have used polyhedra as object models, which have the advantage for instance that the structure of their observed line-drawing is easily understood and that most of its features are viewpoint-independent. Unfortunately, the world is full of curved objects, smooth and piecewise-smooth, with viewpoint-dependent object features. Recent years have seen a broadening of the scope of recognition systems to encompass more expressive shape representations. Algebraic surfaces are proving increasingly useful because of their flexibility and because the geometric constraints associated with them are themselves algebraic.

Algebraic geometry is the part of mathematics which specifically deals with algebraic varieties and the solutions of polynomial systems, which makes it a good candidate for possible interactions with vision. As a summer school entitled "Algebraic Geometry and Computer Vision" (Nice, June 1993) made clear [1], the classical literature of algebraic geometry is still very useful to people in vision, and putting these old ideas in today's framework could benefit to many.

In this paper, we examine two directions in which algebraic geometry has developed that should find increasing applications in object representation. In fact, we illustrate two different aspects of an algebraic representation: its construction and its complexity. Real algebraic geometry is concerned with symbolic methods to characterise the solutions of systems of polynomial equalities and inequalities. Enumerative geometry shows how to "count" the number of solutions of polynomial systems, something that is usually achieved using intersection theory, a generalisation of the well-known Bezout's theorem to higher-dimensional varieties.

The rest of this presentation is organised as follows. After a general introduction of algebraic geometry, Section 2 presents real algebraic geometry, with a sample of methods to compute solutions of polynomial systems exactly or to count the number of real solutions of such systems, and enumerative geometry (along with intersection theory). Section 3 applies those ideas to the aspect graph representation. In particular, it is shown that since all entities entering in the construction of aspect graphs are algebraic or semi-algebraic, aspect graphs may be computed entirely using algorithms from real algebraic geometry. An indication of how the complexity of aspect graphs may be approached via enumerative geometry is given, in connection with the problem of lines and planes having specified contacts with a projective surface, before concluding.

2 Algebraic Geometry and Object Representation

Algebraic geometry is aimed at providing an algebraic characterisation of the geometry of solutions of systems of polynomial equations. Perhaps the first issue to address is the relevance of algebraic geometry in vision. First, it is not necessary to elaborate much on the fact that geometry is the basis for computer vision. Projective geometry and differential geometry have seen numerous applications in the past years. The main reason why methods developed in the context of algebraic geometry have not yet, for the most part, penetrated vision is because of their high level of abstraction, though one may note that projective geometry has mostly been relevant in machine vision in it algebraic form (introduce coordinates right at the beginning and proceed by defining geometric entities as equations in the coordinates). Second, algebraic entities have significant advantages as objects of study. Line systems have been investigated in connection with the problem of reconstruction from line correspondences [2]. Algebraic curves are of interest in the following respects: as curves on surfaces, such as the shadow boundary or the intersection between two interpenetrating surfaces, or as descriptions of certain physical phenomena which are used to interact with shapes, like light rays for instance. Recently, attention has focused on algebraic surfaces [3, 4], and it is known that many man-made surfaces are well approximated by patches of algebraic surfaces.

It is thus clear that tools from algebraic geometry can definitely help in vision. We now turn our attention to the examination of two domains that we think should prove increasingly useful in the field of object representation. We start with real algebraic geometry, a recent theory whose goal is to design efficient algorithms for solving problems related to systems of polynomial equalities and inequalities. We then move on to enumerative geometry and its modern sibling intersection theory, a powerful tool to assess the complexity (one aspect of which is the degree) of an algebraic representation. Readers interested in other interactions between algebraic geometry and computer vision should refer to [5].

[6, 7, 8] are general sources covering the whole spectrum of algebraic geometry, while [9, 10] cover invariant theory from an algebraic standpoint and [11] deals with projective techniques.

2.1 Real and Computational Algebraic Geometry

In recent years, the search for efficient algorithms for handling problems related to systems of polynomial equations has received renewed attention because of their demonstrated importance to a variety of problems of both practical and theoretical interest. The need to count the number of real solutions or to find exact or approximate solutions to such systems has arisen in a wide range of practical efforts, including robotics, solid modelling, and of course computer vision. *Real algebraic geometry* [12] is concerned with *semi-algebraic sets* (subsets of some real Euclidean space that are defined by polynomial equalities and inequalities) and algorithms on these sets. *Computational algebraic geometry* [13] deals with computer implementations of these algorithms.

In this section, we briefly go through the different methods that have been developed in this framework, and we shall see later (Section 3.1) how these methods have been used in connection with aspect graphs.

Solving Systems of Polynomial Equations: Exact Methods. Many different methods have been developed for computing solutions of polynomial systems. In this section, we turn our attention to "exact" or "symbolic" methods.

Resultant-based methods are useful in algorithms which solve problems by recursion on dimension. Briefly, the *resultant* (see [14] for instance) is an algebraic criterion for determining when two polynomials have common roots. One very important application of resultants is in the construction of a description of the projection of an algebraic set. For example, to solve a geometric problem in $n + 1$ dimensions, in which a geometric object is defined by polynomial equalities, we may first construct the projection of the given set onto n-dimensional space, then recursively solve the problem there, and "lift" this solution back to the higher dimensional space.

Collins' *cylindrical algebraic decomposition* [15] is a well-known algorithm which also makes use of this general scheme, though in an indirect manner. The idea here is to obtain the partition of a n-dimensional real space into subsets of a specified character, so that the given polynomials have the same sign at every point of each such subset.

In the limited case of homogeneous polynomials, [16] gives an algorithm using *multivariate resultants*, a multivariate analogue of the classical resultants which can be employed to eliminate a set of variables simultaneously. Note also the methods based on *generalised characteristic polynomials*, introduced in [17], which have the advantage of working for systems of inhomogeneous equations but require to have exactly n polynomials in n variables.

Related to the problem of finding solutions to systems of equations is the problem of deciding whether any such solution exists. One way of tackling this issue is by using *Gröbner bases* [18], which have been heavily used in a wide variety of algebraic and geometric algorithms.

Finally, let us mention that there is a strong connection between the problem of finding the solutions to a system and finding the factors of a multivariate polynomial, in the sense that both essentially amount to finding the irreducible components of an algebraic set. The latter problem of identifying and describing each of these components has been studied in [19].

Counting Real Solutions. In 1835, Sturm gave an algorithm to compute the number of real solutions of a single polynomial equation, which was later extended to the case of a system formed by an equation and an inequation by Sylvester. *Generalised Sturm sequences* [20] give the means to count the exact number of real roots of univariate polynomials satisfying a finite number of polynomial inequalities. *Hermite's method* [21] generalise these results to multivariate polynomials. These techniques, coupled with new algorithms computing Gröbner bases, give the means to characterise the semi-algebraic subsets of the space of parameters on which the system has a fixed number of real solutions. Then, a coding based on *Thom's lemma* [22] allows to characterise the solutions that have been counted this way by examining the signs of the derivatives of the defining polynomials at the solutions.

Finally, without going into a detailed description, note that some criteria have been given to characterise the magnitude of roots and the distance between roots of multivariate polynomial equations. [23] contains a nice introduction to these notions.

Computational Aspects. The last few years have seen numerous computer implementations of the above ideas. *Macaulay* [24] is a computer algebra system whose main task is the efficient calculation of Gröbner bases. It has solved many problems and provided many examples to the algebraic community. The *IF* Maple package [25] is aimed at handling systems of polynomial inequations, and methods like generalised Sturm sequences have been implemented.

As a final comment, the interested reader should consult [26] for a survey of what computers can currently help compute in algebraic geometry and of the different complexity bounds involved in those computations.

2.2 Intersection Theory and Enumerative Geometry

Let k be a field and P, Q two homogeneous polynomials in $k[x_0, x_1, x_2]$ of degrees p and q respectively. Then it is well-known by Bezout's theorem that if the zero-set of P and Q is 0-dimensional, then it contains exactly pq points counted with proper multiplicity. *Intersection theory* [27] generalises this relation between degree and multiplicity to obtain higher-dimensional analogues of Bezout's theorem. *Enumerative geometry* [28, 29] is concerned with enumerating the solutions of certain intersection problems over the field of complex numbers.

Cycles and Intersection Rings. To be able to generalise the notion of intersection to higher dimensions, one has to introduce the language of *cycles.* A *n-cycle* is a formal element of the free Abelian group generated by the irreducible subvarieties of dimension n. A cycle is an element of the direct sum of the groups of n-cycles.

Now, to give the intersection theory on a particular variety is to be able to predict, given two polynomial subvarieties P and Q or equivalently their associated cycles $[P]$ and $[Q]$, the components of the intersection of P and Q, or $[P]$ and $[Q]$, with their proper multiplicities. If the zero-set of P and Q is r-codimensional, then this intersection, denoted $[P].[Q]$ or simply $[P][Q]$, is a r-cycle. If $r = 0$, then $[P].[Q]$ is a 0-cycle, i.e. a finite set of points, and taking the *degree* of this 0-cycle, denoted $\int$, gives the number of these points.

Enumerative Geometry. Taking the degree of a 0-cycle is the same as counting the number of solutions of a systems of equations whose zero-set if 0-dimensional. Enumerative geometry and intersection theory are thus siblings. On a more general level, enumerative geometry is a fascinating subject the goal of which, according to Schubert, is to answer all questions of the following kind: how many (algebro-) geometric figures of a fixed type satisfy certain given conditions. The finesse of enumerative geometry comes in finding the number of figures without finding the figures themselves. Solving enumerative problems has led to a deeper understanding of basic geometry, and to the development of new tools, like the theory of multiple points of maps that we will develop later.

Parameter Spaces. *Parameter spaces* may be thought of as generalisations of projective spaces, i.e. spaces parameterising algebraic entities other than points. Examples of these are the *Grassmannians*, fundamental compact complex manifolds parameterising linear spaces, and the *flag manifolds*, which parameterise for instance points in lines or points in planes (so-called *flags*). They are very useful in connection with the study of the geometry of line and plane systems.

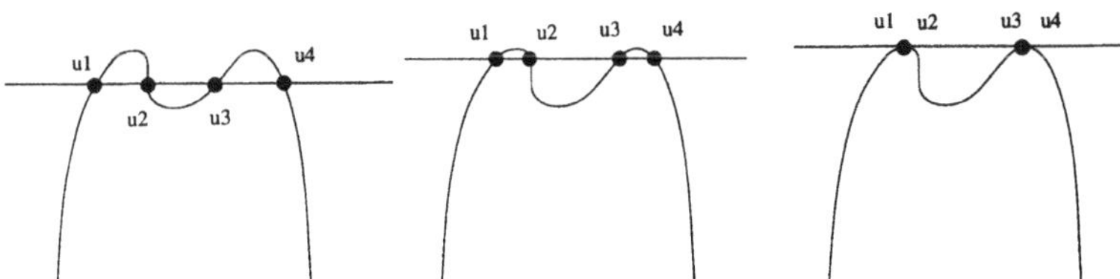

Fig. 1. Deformation of an ordinary 4-fold point into a (2,2)-point.

Mappings and Multiple-Point Theory. One of the ways in which enumerative geometry has developed is through the use of multiple-point theory. Given a map $f : U \to V$, an *r-fold point* of f is a point u_1 of U such that there exist points $u_2, \cdots, u_r$ of U all having the same image under f. If $u_1, \cdots, u_r$ are all distinct points of U, then we shall say that the point is *ordinary*. On the contrary, if some of the u_i lie *infinitely near* each other, then it is called *stationary*. We can rephrase this a little bit. Let $\mathbf{a}$ be a partition of r, i.e. $\mathbf{a} = (a_1, \cdots, a_k)$ is a k-tuple of positive integers such that $\sum_{g=1}^{k} a_g = r$. We consider the multiple point locus $S_f(\mathbf{a})$ described as follows. $S_f(\mathbf{a})$ consists of points u_1 of U such that there are $r-1$ other points $u_2, \cdots, u_r$ of U all having the same image under f and also such that the first a_1 points lie infinitely near each other, the next a_2 points lie infinitely near each other, and so on.

The precise definition of "infinitely near" need not worry us here and this notion is best described by an example. Let C be a curve of the projective plane $\mathcal{P}^2$. Let U be

$$U = \{(x, l) \in C \times \check{\mathcal{P}}^2 \;/\; x \in l\},$$

where $\check{\mathcal{P}}^2$ is the dual projective plane, i.e. the space parameterising lines in $\mathcal{P}^2$. Both U and $\mathcal{P}^2$ are 2-dimensional spaces. Now, we consider multiple points of the map $f : U \rightarrow \check{\mathcal{P}}^2$, induced by projection onto the second factor. Suppose for instance that we look at the multiple-point locus corresponding to the partition $\mathbf{a} = (1, 1, 1, 1)$. Pictorially this locus is as indicated in the first drawing of Figure 1. Now, if we allow some points to coalesce (i.e. to lie infinitely near each other), we may get a different picture. Figure 1 illustrates the case where the first two points coalesce and the last two points coalesce (it is the case $\mathbf{a}$=(2,2)). This example shows that if we have a satisfactory theory to "count" multiple points, then we may be able in this case to tell how many bitangents the curve C has. So we see that in this example, the notion of stationarity is very close to that of order of tangency.

In [30], a procedure for computing a cycle $l(\mathbf{a})$ whose geometric support is the multiple-point locus $S_f(\mathbf{a})$ is defined. Thus, modern enumerative geometry, from the standpoint of multiple-point theory, consists in the following steps: identify parameter spaces U and V suitable for the problem under consideration, find the intersection theories on U and V, and apply multiple-point theory to the map $f : U \rightarrow V$. An example of this will be given in Section 3.2, in connection with the understanding of lines having specified contacts with a projective surface.

Computational Aspects. The *Schubert* Maple package [31] deals with all sorts of enumerative problems, from a modern point of view. Another program called *Singular* is especially suited for computing information about singularities. And in [32], we have described a first attempt at automating the construction of stationary multiple-point classes.

3 Aspect Graphs and Algebraic Geometry

The *aspect graph* is a viewer-centered representation that enumerates all the topologically distinct appearances of an object. The range of possible viewpoints can be partitioned into maximal connected sets that yield identical aspects. The change at the boundary between regions is called a *visual event*.

It is a well-known fact [33] that these boundary points (*accidental* viewpoints) belong to at least one of five visual event surfaces. There are three local events (see Figure 2): lip and beak-to-beak transitions occur for viewpoints belonging to the parabolic surface (made of asymptotic lines - lines having contact of order 3 - at parabolic points), and swallowtail transitions occur when looking from a point of the flecnodal surface (made of asymptotic tangents at flecnodal points - points having a tangent of order 4). In addition, there are three multilocal events, as shown in Figure 3. The tangent crossing surface is made of lines supporting the points of contact of bitangent planes. The cusp crossing surface is made of lines having contact of order 2 at one place and order 3 at another place. The triple point surface is made of lines triply tangent.

In the next two sections, we shall see two ways in which algebraic geometry has recently helped researchers working on aspect graphs: Section 3.1 shows how tools from real algebraic geometry have been used for computing "exact" aspect graphs and

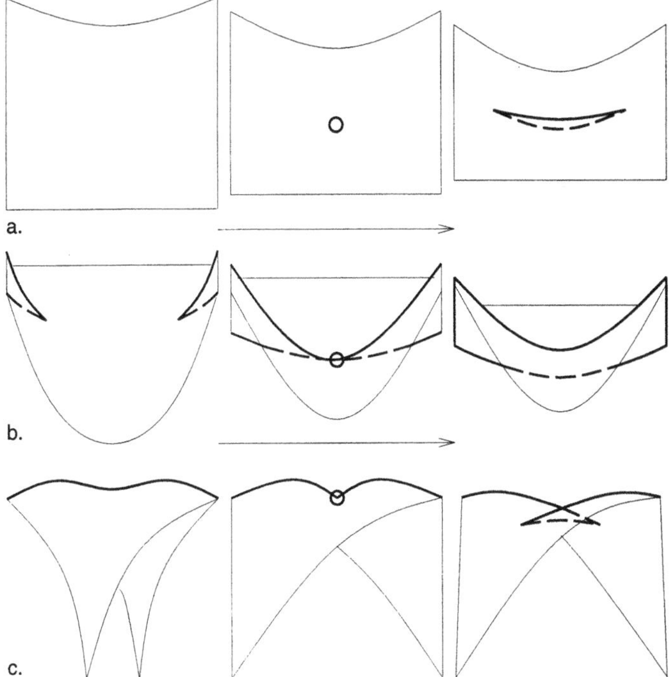

Fig. 2. Local events. a. Lip. b. Beak-to-beak. c. Swallowtail. The transitions take place from left to right, the singular points being indicated by small circles. The contours actually seen by the observer are shown as thick lines, and their hidden parts are dashed.

Section 3.2 shows how enumerative geometry can be used to understand the complexity of this representation.

3.1 Aspect Graphs and Real Algebraic Geometry

When the surface boundaries are semi-algebraic (as is the case for CAD objects), then the occluding contours, the image contours, the visual event curves and the visual event surfaces are all semi-algebraic sets. In turn, this means that aspect graphs may be constructed entirely using algorithms from real algebraic geometry (see [34] for bodies of revolution and [35] for algebraic surfaces). Making exact computations, using symbolic methods, is of key importance to avoid the numerical instabilities inherent to polynomial computations, but at the cost of memory usage.

Assume $\mathcal{V}$ is the viewspace considered. We let $\mathcal{B}_i \subset \mathcal{V}$ be the visual event surfaces and $\tilde{\mathcal{B}}_i \subset \mathcal{V} \times \mathcal{R}^\alpha$ be the recognition equations, $i = 1, \cdots, 5$. For instance, if ω is a viewpoint, p a point of the surface and $l(t) = p + \omega t$ the viewline, then the set $\tilde{\mathcal{B}}_2$ corresponding to the flecnodal event surface is:

$$\tilde{\mathcal{B}}_2 = \{(\omega, p) \in \mathcal{V} \times \mathcal{R}^3 \; / \\ l(0) = 0, l'(0) = 0, l''(0) = 0, l'''(0) = 0\}.$$

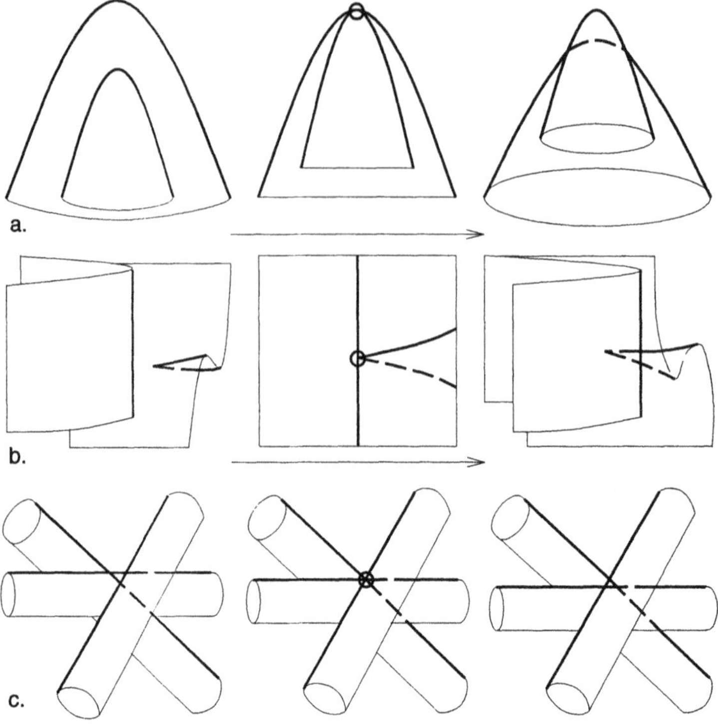

Fig. 3. Multilocal events. a. Tangent crossing. b. Cusp crossing. c. Triple point.

The construction of the aspect graph then goes as follows:

- Eliminate the surface parameters (i.e. the components of p) between the defining equations of the $\tilde{\mathcal{B}}_i$, using resultants or Gröbner bases. The resulting equations are singular curves $\breve{\mathcal{B}}_i \subset \mathcal{V}$.
- Compute the connected components of $\mathcal{V} - \cup_{i=1}^{5} \breve{\mathcal{B}}_i$ using Collins' cylindrical algebraic decomposition.
- Obtain the components $\mathcal{B}_i$ of the bifurcation set by deleting certain branches of the $\breve{\mathcal{B}}_i$ which are not projections of real points of $\tilde{\mathcal{B}}_i$ (the projections of the $\tilde{\mathcal{B}}_i$ onto the first factor are the $\mathcal{B}_i$) and by merging the components separated by such branches.
- Compute sample viewpoints for the remaining connected components.

3.2 Aspect Graphs and Intersection Theory

In [36], we have described how the degrees of the different visual event surfaces could be obtained in an exact manner, and we have deduced a bound on the complexity of the aspect graph, i.e. on the number of views of the object. This is what we intuitively describe in this section.

We have seen above that the visual event surfaces determining the partitioning of viewpoint space in the construction of aspect graphs of smooth surfaces correspond to lines and planes having specified contacts with the surface. For instance, the triple point visual event surface is made of lines triply tangent to the surface. We illustrate now how contacts of lines may be understood. Very similarly to what we have seen for plane curves, we let X be a projective surface of degree d and we consider the map $q : F_X \to G$, where

$$F_X = \{(x, l) \in X \times G \;/\; x \in l\},$$

and where G is the 4-dimensional Grassmannian of lines in $\mathcal{P}^3$. Suppose we look at the multiple-point locus $S_q(2,2,2)$ of q. This locus is the set of pairs (x_1, l_1) such that there exist pairs $(x_2, l_2), \cdots, (x_6, l_6)$, in such a way that all those pairs have the same projection under q and also such that $(x_1, l_1), (x_2, l_2)$ are infinitely near each other, $(x_3, l_3), (x_4, l_4)$ are infinitely near other and $(x_5, l_5), (x_6, l_6)$ are infinitely near each other, i.e. it is the locus of pairs (x, l) such that l is tritangent to X and one of its points of contact is x.

Now, projecting this multiple-point locus onto G, one obtains the set of lines tritangent to X. This set is a one-dimensional line system S and thus represents a ruled surface R. The degree of a surface in $\mathcal{P}^3$ being the number of its intersections with a line in general position, the degree of R is the number of lines of S meeting a line in general position. This shows that if we are able to give the intersection theory on G (or on F_X actually), then we are able to construct the cycle $l(2,2,2)$ whose geometric support is $S_q(2,2,2)$. Projecting $l(2,2,2)$ onto G, we obtain the degree of the triple point visual event surface by intersecting this cycle with the cycle of lines meeting a given line.

The above setup allows to compute almost mechanically any formula concerning lines having specified contacts with a surface. When dealing with planes (as is the case for the tangent crossing surface), a slightly different framework allows to draw conclusions as well. This way, we are able to obtain the degrees of the 5 visual event surfaces. The largest of these degrees is the degree of the triple point surface, which is bounded by $O(d^6)$. In turn, this means that the global bifurcation set may be included in a variety of degree $O(d^6)$. We may conclude on the number of views of this object with Thom-Milnor's bound [37], which says that the number of connected components of a semi-algebraic set of $\mathcal{R}^m$ that is defined by a polynomial equation of degree p is upper bounded by $O(p^m)$. Thus X has at most $O(d^{12})$ aspects in the orthographic case and $O(d^{18})$ aspects in the perspective case.

As a final comment, note that singularities of the outline may also be studied in this framework. Singularities of a stable outline occur when viewing the surface along either an asymptotic ray or a bitangent ray. These two systems of rays are both doubly infinite, and are therefore *line congruences* [38].

4 Conclusion

In this paper, we were concerned with the possible use in object representation of tools developed in algebraic geometry. In particular, we have introduced real algebraic geometry, with methods to characterise the solutions of systems of polynomial equalities and inequalities, and enumerative geometry, to determine the complexity of algebraic representations. We have illustrated some of these points in the context of a multiview representation, the aspect graph.

To enlarge a little bit on the possible future interactions between algebraic geometry and computer vision, let us mention some promising areas of research. Algebraic invariant theory is currently receiving renewed interest and questions like how many linearly independent invariants of a given configuration can one find or are there algebraic relations (*syzygies*) between basic invariants are beginning to receive satisfactory answers. The study of linear systems of curves may also prove useful. In particular, the linear system of contour generators and its global properties are as of now not fully understood. Its relative simplicity in comparison to the complexity of the linear system of outlines is something that should be investigated.

References

1. D. Laksov, "Report on "Algebraic Geometry and Vision School"," *Europroj News*, vol. 10, July 1993.
2. S. Maybank, "The Critical Line Congruence for Reconstruction from Three Images," 1993. Lecture Notes from "Algebraic Geometry and Vision" Summer School, June 1-4, 1993, Nice, France.
3. D. Kriegman and J. Ponce, "On Recognizing and Positioning Curved 3D objects from Image Contours," *IEEE Transactions on Pattern Analysis and Machine Intelligence*, vol. 12, pp. 1127–1137, December 1990.
4. G. Taubin, "Estimation of Planar Curves, Surfaces, and Nonplanar Space Curves Defined by Implicit Equations with Applications to Edge and Range Image Segmentation," *IEEE Transactions on Pattern Analysis and Machine Intelligence*, vol. 13, pp. 1115–1138, November 1991.
5. O. Faugeras and S. Petitjean, "Algebraic Geometry and Theoretical Computer Vision: Past Interactions and Future Trends," 1995. In preparation.
6. J. Semple and L. Roth, *Introduction to Algebraic Geometry*. Clarendon Press, 1949.
7. R. Hartshorne, *Algebraic Geometry*. Springer-Verlag, 1977.
8. J. Harris, *Algebraic Geometry: A First Course*. Graduate Texts in Mathematics, Springer-Verlag, 1992.
9. J. Grace and A. Young, *The Algebra of Invariants*. Cambridge University Press, 1903.
10. B. Sturmfels, *Algorithms in Invariant Theory*. Texts and Monographs in Symbolic Computation, Springer-Verlag, 1993.
11. J. Semple and G. Kneebone, *Algebraic Projective Geometry*. Oxford University Press, 1952.
12. J. Bochnak, M. Coste, and M.-F. Roy, *Géométrie algébrique réelle*, vol. 12 of *Ergebnisse der Mathematik*. Springer-Verlag, 1987.
13. D. Arnon, *Computational Methods in Real Algebraic Geometry*. Academic Press, 1989.
14. B. van der Waerden, *Modern Algebra*. Frederick Ungar Publishing Co., 1970. Fifth Edition.
15. G. Collins, "Quantifier Elimination for Real Closed Fields by Cylindrical Algebraic Decomposition," in *Lecture Notes in Computer Science*, vol. 33, pp. 134–183, Springer-Verlag, 1975. Proceedings of the 2nd GI Conference on Automata Theory and Formal Languages.
16. D. Lazard, "Résolution des systèmes d'équations algébriques," *Theoretical Computer Science*, vol. 15, pp. 77–110, 1981.
17. J. Canny, "Generalized Characteristic Polynomials," *Journal of Symbolic Computation*, vol. 9, pp. 241–250, 1990.
18. B. Buchberger, "Gröbner Bases: An Algorithmic Method in Polynomial Ideal Theory," in *Multidimensional Systems Theory* (N. Bose, ed.), pp. 184–232, Dordrecht-Boston-Lancaster: Reidel, 1985.

19. D. Grigor'ev and A. Chistov, "Fast Decomposition of Polynomials into Irreducible Ones and the Solution of Systems of Algebraic Equations," *Soviet Math. Dokl.*, vol. 29, no. 2, 1984.
20. L. Gonzalez, H. Lombardi, T. Recio, and M.-F. Roy, "Sturm-Habicht Sequences," in *Proceedings of ISSAC'89 (International Symposium on Symbolic and Algebraic Computation)*, pp. 136–146, 1989.
21. P. Pedersen, M.-F. Roy, and A. Szpirglas, "Counting Real Zeros in the Multivariate Case," in *Proceedings of MEGA'92 (International Symposium on Effective Methods in Algebraic Geometry)* (F. Eyssette and A. Galligo, eds.), (Nice, France), pp. 203–224, Progress in Mathematics, Birkhäuser, 1992.
22. M. Coste and M.-F. Roy, "Thom's Lemma, the Coding of Real Algebraic Numbers and the Computation of the Topology of Semi-Algebraic Sets," *Journal of Symbolic Computation*, vol. 5, pp. 121–129, 1988.
23. M. Mignotte, "Some Useful Bounds," *Computer Algebra: Symbolic and Algebraic Computation*, pp. 259–263, 1982.
24. D. Bayer and M. Stillman, "Macaulay: A System for Computation in Algebraic Geometry and Commutative Algebra," 1992. Computer software available via anonymous ftp from zariski.harvard.edu.
25. F. Cucker and M.-F. Roy, "IF, A Package of Maple Programs for Computing with Real Algebraic Numbers and Working with Real Solutions of Equations and Inequalities," 1989. Version 0.1.
26. D. Bayer and D. Mumford, "What Can Be Computed in Algebraic Geometry?," May 1992. Preprint.
27. W. Fulton, *Intersection Theory*. Ergebnisse der Mathematik und ihrer Grenzgebiete, Springer-Verlag, 1984.
28. H. Schubert, *Kalkül der abzählenden Geometrie*. Leipzig: Teubner, 1879.
29. G. Salmon, *A Treatise on the Analytic Geometry of Three Dimensions*, vol. II. Dublin, 5th ed., 1915.
30. S. Colley, "Enumerating Stationary Multiple-Points," *Advances in Mathematics*, vol. 66, no. 2, pp. 149–170, 1987.
31. S. Katz and S. Stromme, "Schubert: A Maple Package for Intersection Theory," 1992. Available by anonymous ftp from ftp.math.okstate.edu in the directory pub/schubert.
32. S. Petitjean, "Automating the Construction of Stationary Multiple-Point Classes," in *Proceedings of ISSAC'94 (International Symposium on Symbolic and Algebraic Computation)*, (Oxford, United Kingdom), pp. 9–14, ACM Press, July 1994.
33. V. Arnold, "Singularities of Smooth Mappings," *Russian Mathematical Surveys*, vol. 38, no. 2, pp. 87–176, 1983.
34. M.-F. Roy and T. van Effelterre, "Aspect Graphs of Bodies of Revolution with Algorithms of Real Algebraic Geometry," in *Proceedings of AAGR'92*, (Linz, Austria), August 17-19 1992.
35. J. Rieger, "Computing View Graphs of Algebraic Surfaces," *Journal of Symbolic Computation*, vol. 11, pp. 1–14, 1993.
36. S. Petitjean, "The Complexity and Enumerative Geometry of Aspect Graphs of Smooth Surfaces," in *Proceedings of MEGA'94 (International Symposium on Effective Methods in Algebraic Geometry)*, (Santander, Spain), Progress in Mathematics, Birkhäuser, April 1994. To appear.
37. J. Milnor, "On the Betti Numbers of Real Varieties," in *Proceedings of the American Mathematical Society*, vol. 15, pp. 275–280, 1964.
38. J. Koenderink, "An Internal Representation for Solid Shape Based on the Topological Properties of the Apparent Contour," in *Image Understanding: 1985-86* (W. Richards and S. Ullman, eds.), ch. 9, pp. 257–285, Norwood, NJ: Ablex Publishing Corporation, 1986.

Discrete Surface Signal Processing: The Polygon as the Surface Element

Gabriel Taubin

IBM T.J.Watson Research Center
P.O.Box 704, Yorktown Heights, NY 10598

Abstract. Smooth surfaces are approximated by polyhedral surfaces both for display, and for other computational purposes. This is probably the most common surface representation in computer graphics, because polygons are what the current generation of graphics hardware can display efficiently. In computer vision this representation has not been very popular for a long time, probably because it was not seen as very appropriate for object recognition and positioning applications. But it is now being used in more applications, particularly in the medical domain, where polyhedral surface approximations of unknown surfaces are routinely generated with iso-surface construction algorithms. We look at polyhedral surfaces as a generalization of digital images, where the polygon takes the place of the pixel. The polygon is the fundamental surface element. Digital images are functions defined at the nodes of a regular rectangular grid. *Discrete surface signals* are functions defined at the vertices of a polyhedral surface of arbitrary topology. Signal processing operations, and Fourier analysis in particular, are the fundamental tools of low level computer vision. The lack of regularity and self-similarity of arbitrary polyhedral surface meshes, as opposed to the regular rectangular grids of digital images, complicates the analysis and processing of digital surface signals. Nevertheless, in this paper we generalize Fourier analysis to discrete surface signals. As a first application of this theory we consider the problem of surface smoothing, which corresponds to low-pass filtering within this framework. As in the classical cases of one-dimensional signals, and of digital images, the analysis of discrete surface signals is reduced to matrix analysis, and matrix computation techniques are used to achieve fast discrete surface signal processing operations. We intend to continue along this line of research in the near future, extending higher level computer vision operations to arbitrary polyhedral surfaces.

1 Introduction

The approach described in this paper was motivated by the problem of smoothing a polyhedral surface of arbitrary topology [11], such as those produced by iso-surface construction algorithms, so commonly used in three-dimensional medical imaging [8, 1, 5, 6]. Since these polyhedral surfaces usually have a very large number of vertices and faces – a million vertices is not unusual –, we decided very early on in our research that only algorithms with linear time and space complexity would be acceptable.

Many methods for smoothing closed curves, both in two and three dimensions, can be found in the computer vision literature. Since closed curves can be described parametrically by functions defined on the unit circle, i.e., by *one-dimensional periodic signals*, Fourier analysis was seen as a natural tool to solve the problem of smoothing. The space of one-dimensional periodic signals is decomposed into orthogonal subspaces associated with different frequencies, with the low frequency content of a signal regarded as subjacent data, and the high frequency content as noise. To smooth a closed curve it is sufficient to remove the noise from the coordinate signals, i.e., to project the coordinate signals onto the subspace of "low frequencies". This is what the method of *Fourier descriptors*, which dates back to the early 60's, does [13], To smooth a closed curve with the method of Fourier descriptors, the Fourier series associated with the coordinate signals are truncated. In the discrete domain this continuous process is replaced by computing the Discrete Fourier Transform of each sequence of vertex coordinates, setting the coefficients of the transform associated with high frequencies to zero, and back transforming the resulting sequence, obtaining a new sequence of vertex coordinates.

The problem with the method of Fourier descriptors for polygonal curves is that it is computationally expensive. Even using the Fast Fourier Transform algorithm, the computational complexity is in the order of $n \log(n)$ operations, where n is the number of vertices. An alternative is to do the projection onto the space of low frequencies only approximately. This is what a linear low-pass filter does. In a linear low-pass filter the filtered function is the convolution of the original function with a low-pass filter kernel. In the discrete case, a function defined on a closed polygon is given by a sequence of values, one for each vertex. The value of the filtered function associated with a vertex is computed as a weighted average of some neighbors. Perhaps the most popular convolution-based smoothing method for parameterized curves is the so-called *Gaussian filtering* method, associated with scale-space theory [12, 7]. It is well known though, that Gaussian filtering produces shrinkage, and this is so because the Gaussian kernel is not a low-pass filter kernel [9].

All of these methods can be generalized to functions defined on simple surfaces, such as the plane, the sphere, the cylinder, and the torus; where classical Fourier analysis can be applied. In all these cases, as in the case of closed curves, the decomposition of the space of functions defined on one of these surfaces into orthogonal subspaces of low and high frequencies is defined in advance, and in analytic form. The decomposition depends only on the type of domain.

The main reason why it is difficult to extend Fourier analysis from the simple cases described above to polyhedral surfaces of arbitrary topology is that there is no notion of shift-invariant functions defined on a general surface, and so, no notion of convolution. To define a filter, the convolution kernel must be replaced by a set of vertex-dependent smoothing kernels. Also, based on the usual representation of a polyhedral surface as a list of vertices and a list of faces, it is very difficult to access vertices that are far away from a given vertex without building special purpose data structures. Furthermore, for a general surface the neighborhood structure changes from vertex to vertex, consuming a significant amount of storage. All of this is very impractical, both in terms of the number of arithmetic operations and in terms of the amount of storage required to encode all this information.

The main idea behind the new theory described below, is the observation that the classical Fourier transform of a signal can be seen as the decomposition of the signal into a linear combination of the eigenvectors of the Laplacian operator. To extend Fourier analysis to surfaces of arbitrary topology we only have to define a new operator that takes the place of the Laplacian. In this paper we show how to extend the analysis to polyhedral surfaces of arbitrary topology, which can be done with elementary tools from matrix analysis. The mathematics of the continuum case will be treated elsewhere.

2 Going From Curves To Surfaces

As we mentioned above, this work was motivated by the problem of surface smoothing. Let us describe Fourier analysis of a discrete time n-periodic signal, i.e., a function defined on a regular polygon of n vertices, in relation to Gaussian smoothing. Let us denote $x_1, \ldots, x_n$ the values of the function at the vertices of the polygon, and let us group them together as a column vector $x = (x_1, \ldots, x_n)^t$. The discrete Laplacian of an n-periodic signal is the second order difference operator defined by the following formula

$$\Delta x_i = \frac{1}{2}(x_{i-1} - x_i) + \frac{1}{2}(x_{i+1} - x_i) \ , \tag{1}$$

where the indices are incremented and decremented modulo n. In the simplest possible implementation of Gaussian smoothing, a signal x is smoothed by adding its Laplacian multiplied by a scale factor $0 < \lambda < 1/2$ to it

$$x'_i = x_i + \lambda \, \Delta x_i \ . \tag{2}$$

This can be written in matrix form as follows

$$x' = (I - \lambda K) x \ , \tag{3}$$

where

$$K = \begin{pmatrix} 1.0 & -0.5 & & & -0.5 \\ -0.5 & 1.0 & -0.5 & & \\ & \ddots & \ddots & \ddots & \\ & & -0.5 & 1.0 & -0.5 \\ -0.5 & & & -0.5 & 1.0 \end{pmatrix} .$$

This is a *circulant matrix*, whose eigenvectors are the columns of the Fourier matrix [2]. Both eigenvectors and eigenvalues have well known analytic expressions, which are the basis of the Fast Fourier Transform algorithm. The elements of the Fourier matrix are complex numbers, but since the matrix K is symmetric, it also has real eigenvalues and eigenvectors, which we will order according to increasing value of the corresponding eigenvalue $k_1 \leq \cdots \leq k_n$. By looking at the analytic expressions of the eigenvalues, it is not difficult to check that in this particular case we have $0 \leq k_1 \leq \cdots \leq k_n \leq 2$, but a more general argument to justify this statement will be given below. To decompose the signal x as a linear combination

$$x = \sum_{i=1}^{n} \xi_i \, u_i \ , \tag{4}$$

of the real eigenvectors $u_1, \ldots, u_n$ of K is equivalent to computing the Discrete Fourier Transform of x. To smooth the signal x with the method of Fourier descriptors, this is exactly what we have to do. But if a linear convolution-based smoothing method is used, it is not necessary to decompose the signal.

This process can be generalized to polyhedral surfaces by just carefully redefining the discrete Laplacian as follows. First of all, a polyhedral surface is usually represented as a pair of lists $S = \{V, F\}$, a list of n vertices V, and a list of faces F, with each face being a sequence of non-repeated indices of vertices, and representing itself a closed three dimensional polygon, not necessarily flat. *Triangulated surfaces* are the most common, where all the faces are triangles. There are other popular representations for piece-wise linear shapes [3], but this one is the most appropriate for our purposes.

Both for curves and for surfaces, a *neighborhood* of a vertex v_i is a set $i^\star$ of indices of vertices. If the index j belongs to the neighborhood $i^\star$, we say that v_j is a *neighbor* of v_i. The *neighborhood structure* of a polygonal curve or polyhedral surface is the family of all its neighborhoods $\{i^\star : i = 1, 2, \ldots, n\}$. A neighborhood structure is *symmetric* if every time that a vertex v_j is a neighbor of vertex v_i, also v_i is a neighbor of v_j.

A particularly important neighborhood structure is the *first order* neighborhood structure, where for each pair of vertices v_i and v_j that share a face (edge for a curve), we make v_j a neighbor of v_i, and v_i a neighbor of v_j. For example, for a polygonal curve represented just by a list of consecutive vertices, the first order neighborhood of a vertex v_i is $i^\star = \{i-1, i+1\}$. The first order neighborhood structure is symmetric. and since it is implicitly represented in the list of faces of the surface, no extra storage is required to represent it.

A *discrete surface signal* is a function $x = (x_1, \ldots, x_n)^t$ defined on the vertices of a polyhedral surface. Gaussian smoothing can still be defined for a discrete surface signal by the smoothing step of equation (2), but with the discrete Laplacian defined by the following formula instead of (1)

$$\Delta x_i = \sum_{j \in i^\star} w_{ij}\,(x_j - x_i)\ , \tag{5}$$

where the weights w_{ij} are positive numbers that add up to one for each i. The weights can be chosen in many different ways taking into consideration the neighborhood structures. One particularly simple choice that produces good results is to set w_{ij} equal to the inverse of the number of neighbors $1/|i^\star|$ of vertex v_i, for each element j of $i^\star$.

If $W = (w_{ij})$ is the matrix of weights, with $w_{ij} = 0$ when j is not a neighbor of i, the matrix K can now be defined as $K = I - W$. With this definition the matrix formulation of Gaussian smoothing of equation (3) remains valid, ant the case of a curve becomes a particular case.

Although the method applies to general neighborhood structures, we will restrict our analysis here to those cases where the matrix W can be factorized as a product of a symmetric matrix E times a diagonal matrix D. Such is the case for the first order neighborhood of a shape with equal weights $w_{ij} = 1/|i^\star|$ in each neighborhood $i^\star$. In this case E is the matrix whose ij-th. element is equal to 1 if vertices v_i and v_j are neighbors, and 0 otherwise, and D is the diagonal matrix whose i-th. diagonal element is $1/|i^\star|$. Since in this case W is a *normal matrix* [4], because $D^{1/2}WD^{-1/2} = D^{1/2}ED^{1/2}$

is symmetric, W has all real eigenvalues, and sets of n left and right eigenvectors that form respective bases of n-dimensional space. Furthermore, by construction, W is also a *stochastic matrix*, a matrix with nonnegative elements and rows that add up to one [10]. The eigenvalues of a stochastic matrix are bounded above in magnitude by 1, which is the largest magnitude eigenvalue. It follows that the eigenvalues of the matrix K are real, bounded below by 0, and above by 2. Let $0 \le k_1 \le k_2 \le \cdots \le k_n \le 2$ be the eigenvalues of the matrix K, and let $u_1, u_2, \ldots, u_n$ a set of linearly independent unit length right eigenvectors associated with them. Seen as discrete surface signals, the eigenvectors of the matrix K can be considered as the *natural vibration modes*, and the corresponding eigenvalues as the associated *natural frequencies*. Note that the constant signal $u_1 = (1, \ldots, 1)/\sqrt(n)$ is an eigenvector of K associated with the frequency $k_1 = 0$, and the values of a natural vibration mode u_i associated with a low natural frequency k_i varies slowly when we move from a vertex to a neighbor vertex.

As opposed to the simple case of curves described above, where the eigenvectors and eigenvalues of K have analytic expressions, in the general case they do not, and if needed they have to be computed using numerical matrix computations at a computational cost even higher than $n \log(n)$. However, for filtering operations it is not necessary to compute the eigenvalues and eigenvectors explicitly. Besides, for the typical large surfaces that we are interested in processing, it is just practically impossible from a numerical point of view to compute all the eigenvalues and eigenvectors of K.

The decomposition of equation (4), of a discrete surface signal x as a linear combination of eigenvectors of K, remains valid with these definitions. Using this decomposition, the Gaussian smoothing step of equation (3) can be described as

$$x' = \sum_{i=1}^{n} (1 - \lambda k_i)\, \xi_i\, u_i \ ,$$

because $Ku_i = k_i u_i$. And after N iterations we have

$$x^N = \sum_{i=1}^{n} (1 - \lambda k_i)^N \xi_i\, u_i \ .$$

This is in fact a filter with transfer function $g(k)^N = (1 - \lambda k)^N$. The component of x of frequency k_i is attenuated or enhanced by the factor $g(k_i)$. In the case of Gaussian smoothing, all the frequency components are attenuated, except for frequency $k = 0$. This is why a closed polygon converges to its centroid after an infinite number of Gaussian smoothing steps

$$\lim_{N \to \infty} g(k)^N = \begin{cases} 1 & \text{for } k = 0 \\ 0 & \text{for } 0 < k \le 2 \end{cases} ,$$

i.e., why Gaussian smoothing produces shrinkage.

3 Surface Signal Smoothing

In [11] we solve the problem of shrinkage of Gaussian smoothing by replacing the Gaussian smoothing step of equation (3) by the following step

$$x' = (I - \mu K)(I - \lambda K)\, x \ , \tag{6}$$

where μ is a new parameter such that $0 < \lambda < -\mu$. That is, after we perform the Gaussian smoothing step of equation (2) for all the vertices, we then perform another similar step

$$x'_i = x_i + \mu \, \Delta x_i \tag{7}$$

for all the vertices, but with negative scale factor μ instead of λ. Figures 1, 2, 3 show some examples of smoothed surfaces produced by this algorithm

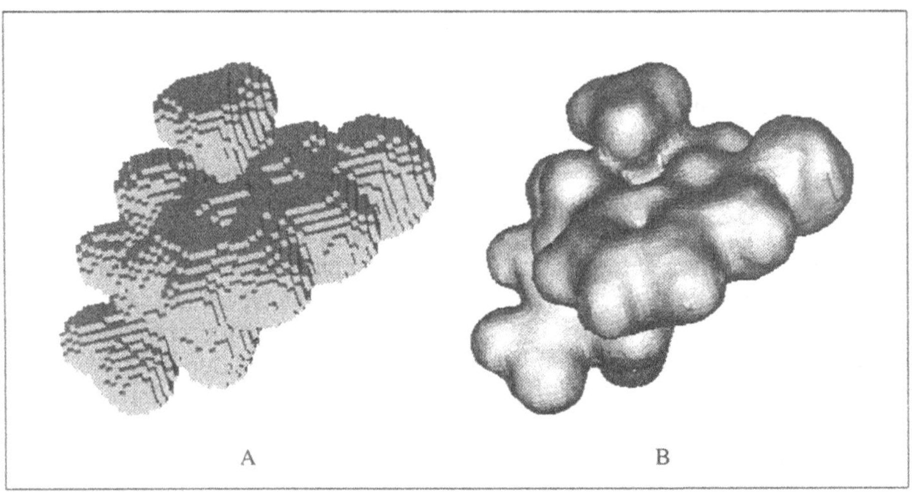

Fig. 1. (A) A very irregular polyhedral surface computed as the boundary of a set of volume elements. (B) The result of applying 90 steps of the smoothing method of [11] to A with parameters $\lambda = 0.33$ and $\mu = -0.34$.

The transfer function of the algorithm defined by equation (6) is $f(k) = (1 - \mu k)(1 - \lambda k)$ illustrated in figure 4. Figure 5 shows the resulting transfer function after N iterations of the algorithm, the graph of the function $f(k)^N$. The graph of the polynomial $f(k)$ is an inverted parabola with roots at $k = 1/\lambda > 0$ and $k = 1/\mu < 0$. The value of the polynomial $f(k)$ is positive for $1/\mu < k < 1/\lambda$, and negative for $k < 1/\mu$ and $k > 1/\lambda$. Furthermore, since $f(0) = 1$ and $\mu + \lambda < 0$, there is a positive value of k, let us denote it k_{PB} (the *pass-band frequency*), such that $f(k_{PB}) = 1$. The value of k_{PB} is

$$k_{PB} = \frac{1}{\lambda} + \frac{1}{\mu} > 0 \, .$$

The graph of the transfer function $f(k)^N$ displays a typical *low-pass filter* shape in the region of interest, i.e., from $k = 0$ to $k = 2$. The *pass-band region* extends from $k = 0$ to $k = k_{PB}$. For values of k in the pass-band region, $f(k)^N \approx 1$. The *transition region* extends from $k = k_{PB}$ to $k = k_{SB}$ (the *stop-band frequency*), where k_{SB} is a parameter defined by the user, that will be discussed in detail in the next section. The *stop-band region* extends from $k = k_{SB}$ to $k = 1/\lambda$. For values of k in the stop-band region, $f(k)^N \approx 0$ for large N.

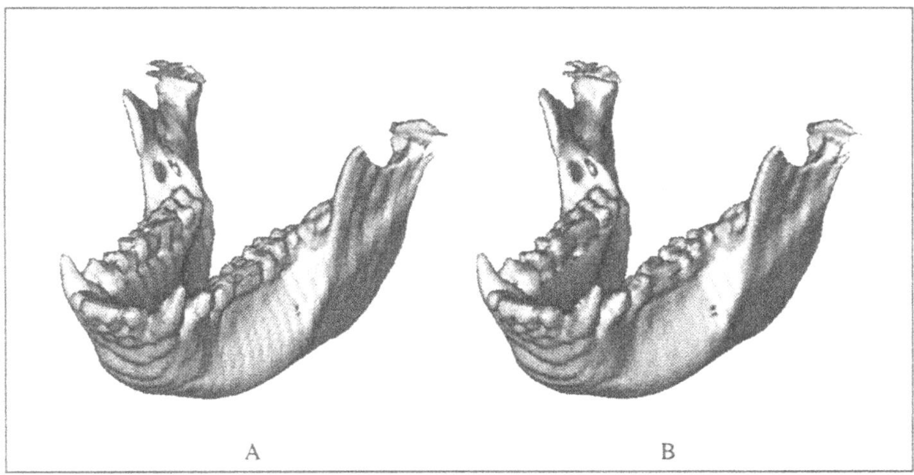

Fig. 2. Smoothing piece-wise linear surfaces created by a good iso-surface construction algorithm. (A) Iso-surface constructed with Kalvin's Alligator algorithm [6] from CT data (74,760 vertices, 149,776 faces). (B) The result of applying 100 steps of the smoothing method of [11] to A with parameters $\lambda = 0.33$ and $\mu = -0.34$. Note the faceting in A near the teeth produced by the greater separation between slices, has disappeared in B.

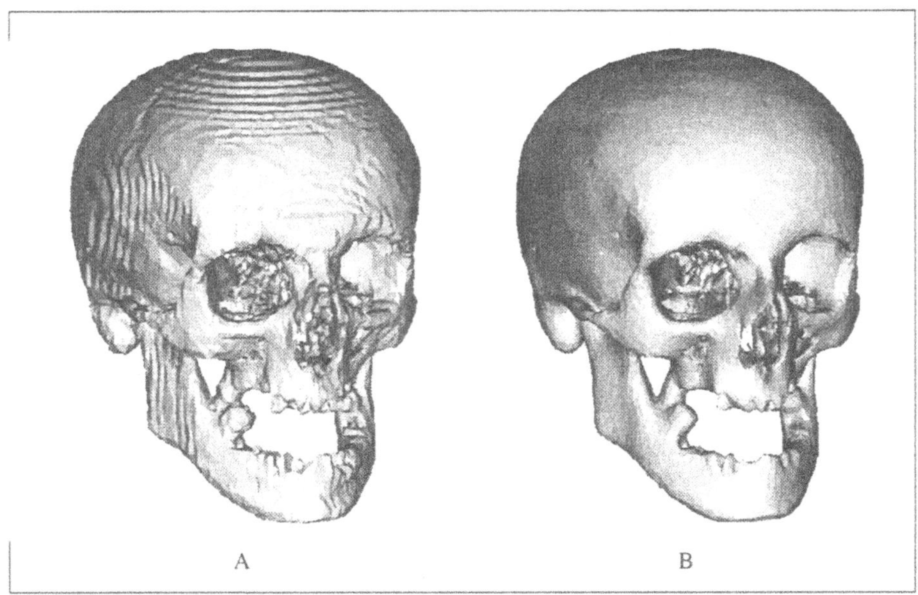

Fig. 3. Smoothing simplified piece-wise linear surfaces. (A) Simplified iso-surface constructed with Guéziec's Wrapper algorithm [5] from CT data (27,367 vertices, 55,310 faces). (B) The result of applying 60 steps of the smoothing method of [11] to A with parameters $\lambda = 0.330$ and $\mu = -0.331$.

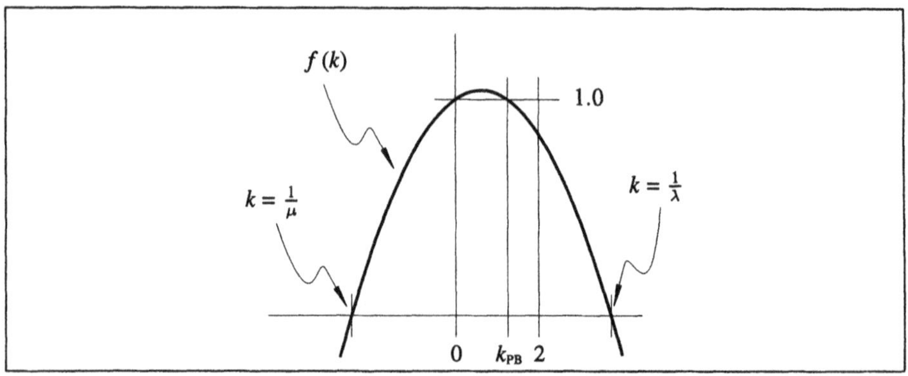

Fig. 4. Graph of transfer function of non-shrinking smoothing algorithm.

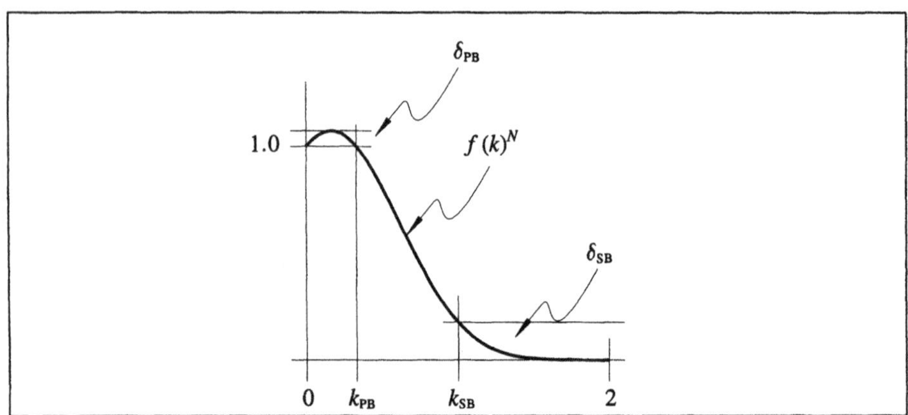

Fig. 5. Graph of the transfer function $f(k)^N$.

This algorithm is fast (linear both in time and space), extremely simple to implement, and produces smoothing without shrinkage. However, even faster algorithms can be achieved by choosing other polynomial transfer functions. Care should be taken to minimize space requirements, though. Smoothing is an approximate projection of the surface signal x onto the subspace of low frequencies of the matrix K, which in our typical case is large and sparse. There are well established matrix computation methods for doing so, which we are currently exploring. Note that because of the linear complexity constraint, only polynomial transfer functions are allowed. With rational transfer functions much better approximations of the ideal low-pass filter could be achieved, but a rational transfer function involves solving the sparse linear system $Ky = z$ a number of times, which is not a linear complexity operation.

4 Conclusions

Motivated by the problem of surface smoothing, we have shown how to generalize the classical Fourier analysis of one-dimensional discrete time signals to discrete surface signals, i.e., signals defined on polyhedral surfaces of arbitrary topology. The new theory is based on the observation that the classical Fourier transform of a signal can be seen as the decomposition of the signal into a linear combination of the eigenvectors of the Laplacian operator. Fourier analysis is extended to surfaces by defining an appropriate operator that takes the place of the Laplacian. In this paper we only treated the discrete case, because only elementary tools from matrix analysis are needed. We will describe the continuum case in more detail in a forthcoming article. As the first application of this theory we have described a very simple algorithm that produces surface smoothing without shrinkage. This algorithm is in fact a linear low-pass filter as a function of the natural frequencies of the surface. Other surface operations can also be seen as filtering with the methods introduced in this paper. For example, ridge curve detection can be looked upon as high pass filtering. The ideas introduced in this paper also open the door to non-linear filtering operations on surfaces, an area which we are currently exploring.

References

1. H. Baker. Building surfaces of evolution: The weaving wall. *International Journal of Computer Vision*, 3:51–71, 1989.
2. P.J. Davis. *Circulant Matrices*. John Wiley & Sons, 1975.
3. J.D. Foley, A. van Dam, S.K. Feiner, and J.F. Hughes. *Computer Graphics, Principles and Practice*. Addison-Wesley, Reading, MA, second edition, 1992.
4. G. Golub and C.F. Van Loan. *Matrix Computations*. John Hopkins University Press, 1983.
5. A. Guéziec and R. Hummel. The wrapper algorithm: Surface extraction and simplification. In *IEEE Workshop on Biomedical Image Analysis*, pages 204–213, Seattle, WA, June 24–25 1994.
6. A.D. Kalvin. *Segmentation and Surface-Based Modeling of Objects in Three-Dimensional Biomedical Images*. PhD thesis, New York University, New York, March 1991.
7. T. Lindeberg. Scale-space for discrete signals. *IEEE Transactions on Pattern Analysis and Machine Intelligence*, 12(3):234–254, March 1990.
8. W. Lorenson and H. Cline. Marching cubes: A high resolution 3d surface construction algorithm. *Computer Graphics*, pages 163–169, July 1987. (Proceedings SIGGRAPH).
9. J. Oliensis. Local reproducible smoothing without shrinkage. *IEEE Transactions on Pattern Analysis and Machine Intelligence*, 15(3):307–312, March 1993.
10. E. Seneta. *Non-Negative Matrices, An Introduction to Theory and Applications*. John Wiley & Sons, New York, 1973.
11. G. Taubin. Curve and surface smoothing without shrinkage. Technical Report RC-19536, IBM Research, April 1994.
12. A.P. Witkin. Scale-space filtering. In *Proceedings, 8th. International Joint Conference on Artificial Intelligence (IJCAI)*, pages 1019–1022, Karlsruhe, Germany, August 1983.
13. C.T. Zahn and R.Z. Roskies. Fourier descriptors for plane closed curves. *IEEE Transactions on Computers*, 21(3):269–281, March 1972.

Spline Representations in 3-D Vision

Visa Koivunen and Ruzena Bajcsy

GRASP Lab, University of Pennsylvania
301C 3401 Walnut St., Philadelphia, PA 19104, USA

Abstract. Spline representations are widely used in CAGD as well as in computer graphics. Splines are also useful in many computer vision tasks. In particular, in the manufacturing domain it is easier to share data and provide machine perception based services because similar representations are commonly used in both shape design and engineering tools.
In this paper, we study different spline representations and their applications in computer vision. The advantages and disadvantages of splines in various vision tasks are assessed. Different design options are described in detail and some guidelines for making appropriate choices are given. Examples on spline techniques using data from a 3-D imaging sensor are shown.

1 Introduction

Three common 3-D shape representations used in computer graphics, Computer Aided Geometric Design (CAGD), and computer vision are surface, volume, and sweep representations. Currently, there is no single representation that would be appropriate for all classes of shapes. Consequently, *multiple representations* have to be used in order to describe different shapes efficiently. Moreover, the representation should not be selected simply based on the types of shapes we are dealing with but also according to the way the shape description is subsequently to be used. The requirements for representing the shape of the same object may be different, for example, in obstacle avoidance, grasping, object recognition, or geometric model construction tasks. For further information about various 3-D shape representations, the reader is referred to [21, 2, 18, 8, 4, 31].

Spline representations are widely used in CAGD as well as in graphics. In this paper, we investigate applications of splines in computer vision. Splines are useful, in particular, if the focus is on describing the geometry of manufactured parts or free-form shapes. They are also compatible with commonly used representations in manufacturing automation. This facilitates the data sharing and makes it easier to provide machine vision based services to manufacturing industry. The paper is organized as follows. In section 2, different criteria for shape representation are addressed. In section 3, some important spline representations and their properties are described. Splines provide several options for a developer. The options in different vision tasks are described and some guidelines for making appropriate choices are given. In section 4, the pros and cons of splines are assessed in various applications including surface reconstruction and object recognition. Finally, examples on spline applications using range data are given in section 5.

2 Issues on shape representation

2.1 Criteria for shape representation

Marr [19] evaluated shape representations using the following criteria: *scope, accessibility, conciseness, uniqueness, stability* and *sensitivity*. Scope specifies what kinds of shapes the representation can describe. Accessibility defines how inexpensively the description can be obtained from the image. Conciseness describes the expressive power of the representation, i.e., the ability to describe shapes efficiently. Stability and sensitivity define how well the representation reflects similarity between different objects and differences between similar objects.

Nalwa [23] proposed that a good representation should exhibit the following characteristics: it should be *sufficient for the task, unique, unambiguous, stable, local* and *convenient*. Sufficiency corresponds to scope and accessibility from Marr's criteria, while convenience includes stability and sensitivity. A representation is stable if small perturbations in a described entity do not introduce large changes in the representation. A representation is unique if every distinct object has a single distinct representation. Unambiguity requires that, although an object may have multiple representations, no two distinct objects have a common representation. A representation is local if nonintersecting components have independent representations.

Clearly, it is difficult to satisfy fully all the above criteria simultaneously because there are trade-offs among them. The scope and the accessibility needed depend on the task domain. Conciseness is useful in general if a large family of shapes can be described using only few parameters. Unambiguity is required in most geometric modeling and computer vision tasks. Uniqueness is necessary to distinguish between objects easily, and to establish correspondences between the model and the data description. It is important in recognition tasks but not necessary in CAGD.

2.2 Surface representation forms

The mathematical representation of surfaces occurs in either parametric or nonparametric form. Nonparametric equations are further divided into explicit and implicit forms. The explicit graph form $z = f(x, y)$ is widely used in computer vision and image processing applications but it can't represent closed or multiple-value surfaces because only one z-value corresponds to each (x, y) pair. Implicit and parametric forms are not limited to single-valued surfaces.

The parametric form of a surface is defined as follows:

$$S(u, v) = (x = g(u, v), y = h(u, v), z = f(u, v))$$

where u and v are parameters on the surface and g, h, and f are smooth functions. The the parametric domain is often either rectangular or spherical which limits one's ability to represent surfaces of arbitrary topology. A surface expressed in implicit form $f(x, y, z) = 0$ is defined as a set of zeros of the smooth function f. The implicit form is very general because every rational parametric surface has a corresponding implicit form. The implicit form allows the determination of whether a point (x, y, z) lies on, above, or under a surface. Such a distinction is difficult when using the parametric form. Depending on the form of the surface representation, arrangement of input data and the definition of the error norm, different numerical techniques are used for fitting surfaces [31, 26, 28, 18].

3 Spline basics

3.1 General

In computer graphics and CAGD, surfaces are typically defined in a parametric form. Piecewise polynomials called splines [6, 8] are commonly used for designing sculptured free-form surfaces. Splines of degree p have a continuity of derivatives of order $(p - 1)$ with respect to their parameterization at the joints. As an example, cubic splines have C^2 continuity.

Various spline representations such as B-splines, cubic splines and splines in tension, have been employed, e.g., in CAGD, computer graphics, Finite Element applications, robot path planning, machine vision and data smoothing. In this paper, we address triangular Bezier-splines and tensor product Non-Uniform Rational B-Splines (NURBS) in more detail. Various design options are described as well. NURBS include Bezier and B-splines as special cases. They are widely used because they possess local control as well as good continuity properties. Furthermore, the coefficients have a direct geometric interpretation which is not the case for most other spline representations. NURBS are included in several CAD systems and the IGES [12] product data exchange standard. In industrial vision applications, it is easier to provide services and communicate with other manufacturing automation processes using standard representations. A thorough presentation of splines can be found in several computer graphics and geometric design related books and papers, for example, in [6, 8, 25].

3.2 Tensor product B-splines

It is often desirable to represent shapes using smooth surfaces. Commonly such shapes are defined using tensor product surfaces such as B-splines. B-spline surfaces have several desirable properties for geometric modeling and high quality surface approximation. In addition to the continuity property, they possess the local control property, so that the influence of a local change is limited to a small region. B-splines are invariant under affine or projective (NURBS) transformations. The surface lies within the convex hull formed from the control point mesh. B-spline surfaces have not been proved to have the same variation diminishing property as do B-spline curves.

NURBS are a rational generalization of B-splines. A NURBS surface is defined as a bivariate polynomial function of parameters u and v as follows:

$$S(u,v) = \sum_{i=1}^{n}\sum_{j=1}^{m} B_{i,j}^{h} N_{i,k}(u) M_{j,l}(v), \qquad (1)$$

where $N_{i,k}$ and $M_{j,l}$ are the basis functions of order k and l, and the $B_{i,j}^{h}$'s are the homogeneous coordinates $(x_{i,j}, y_{i,j}, z_{i,j}, h_{i,j})$ of the control points. A NURBS surface and its defining control point mesh using an open end condition are illustrated in Figure 1. The basis functions are defined recursively (see [25]) and

Fig. 1. A NURBS surface and its defining control point mesh. The surface and the control points coincide at the boundaries because of the open end condition.

the spacing of knots can be non-uniform. In order to introduce discontinuities, multiple knots have to be introduced in the same location.

It is important to have sufficient number of control points in surface fitting to be able to describe all the degrees of freedom of the underlying surface and to ensure convergence. On the other hand, if there are too many control points, unnecessary undulations may occur on the surface and the data compression rate is lower. The fewest number of control points tends to yield the fairest surfaces [28].

The mapping from physical coordinates to the parametric domain can be either uniform or reflect the spatial distribution of the data. Uniform parameterization generally produces unsatisfactory results because it does not adapt to the geometry of the data points [8].

3.3 Triangular spline patches

The rectangular arrangement of the parametric domain of tensor product surfaces and the strict C continuity definition do not allow for representing all classes of shapes. Triangular splines provide a convenient representation for surfaces of arbitrary topology and provide a tool for fitting surfaces if the input data are scattered. Furthermore, the differentiability property at the joints can be relaxed to geometric (G) continuity.

A smooth surface representation can be constructed on top of an optimal 2/3 triangulation [9, 29] or a constrained Delaunay triangulation such as [7]. The shape is represented by a collection of Bezier triangles where the vertices of the triangulation are the domain vertices for the Bezier patches. A Bezier triangle [8] of degree n is defined using barycentric coordinates (u,v,w) as follows:

$$\sum_{|\mathbf{i}|=n} b_{\mathbf{i}} B^n_{\mathbf{i}}(u,v,w) \tag{2}$$

where $|\mathbf{i}| = i + j + k$ so that $i + j + k = n$ and $i, j, k \geq 0$. $B^n_{\mathbf{i}}(u,v,w)$ are the Bernstein polynomials

$$B^n_{\mathbf{i}}(u,v,w) = \frac{n!}{i!\,j!\,k!} u^i v^j w^k,$$

where $u, v, w \geq 0$ and $u+v+w = 1$. The control net formed using coefficients $b_{\mathbf{i}}$ in (2) consists of $\frac{1}{2}(n+1)(n+2)$ control points forming a triangular structure. The control net in the case of cubic patch is depicted in Figure 2.

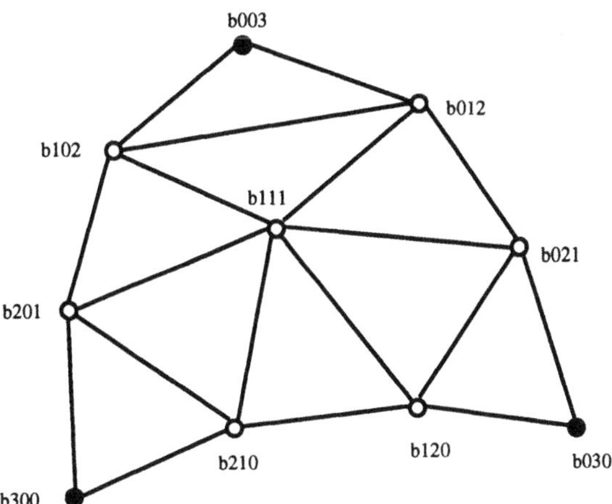

Fig. 2. The arrangement of control points for a cubic Bezier triangle, where b_{030}, b_{300} and b_{003} are the domain vertices from 2/3 triangulation.

For each patch, the coefficients $b_{\mathbf{i}}$ in (2) have to be estimated. This is accomplished by determining the 3 boundary curves of the triangle first. The intermediate coefficients along each boundary are computed through interpolation instead of approximation using the domain vertices and the tangent plane at each vertex. Therefore, the resulting surface passes through the domain vertices. The remaining coefficient can be set for example to $b_{111} = \frac{1}{3}(b_{300} + b_{030} + b_{003})$.

The continuity at the patch boundaries can be adjusted either locally or globally. As an example of a local technique, each patch is subdivided (through modified Clough-Tocher split) into 3 minitriangles at the centroid ($u = \frac{1}{3}, v = \frac{1}{3}, w = \frac{1}{3}$). Then, in each minitriangle, the coefficients on the next row parallel to the boundary are adjusted to meet the continuity requirement across the boundary. Subdivision allows for treating each boundary independently at the cost of increasing the number of coefficients. Convex combination schemes construct only one patch for each triangle. For a survey on various techniques, see [18].

3.4 Design options

In any application where splines are used, the developer has several design options. These include: type of spline; patch shape (triangular, rectangular or n-sided patches); parameterization; nonrational or rational basis functions; end condition; refinement techniques and handles; order of the spline; and the type of continuity. Various computational techniques have been developed for approximating and interpolating splines. As an example, different techniques are used for regularly spaced and scattered data.

In the case of tensor product spline patches, the rectangular parametric domain and strict differentiability requirement do not allow for representing surfaces of arbitrary topology. However, some engineering tricks such as trimmed surfaces have been developed in order to describe complex shapes with rectangular patches. Alternatively, triangular or n-sided patches can be used (see [8]). A trimmed NURBS surface depicted in Figure 3 is essentially a regular NURBS surface where certain parts of the surface are marked "invalid" [8]. A surface-

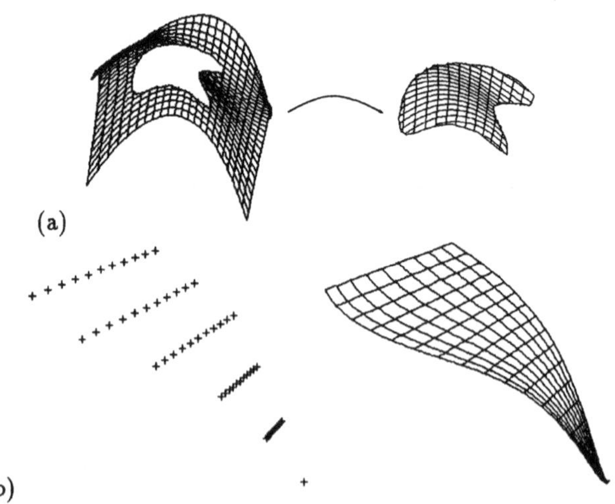

Fig. 3. (a) A trimmed NURBS surface and (b) a B-spline surface with degenerate control point mesh.

to-surface intersection algorithm is typically required for defining the trimming curve. Also, degeneracies can be introduced into the control point mesh by reducing the control points forming an edge of the mesh into a single point. The degeneracies, however, cause problems in the parameterization of the surface.

Tensor product surfaces can be used directly for data with rectangular arrangement. In the case of scattered data, however, such an arrangement has to be constructed. A more general technique is based on triangular spline patches. Such a representation is typically build on top of a triangulation such as Delaunay triangulation [29, 9, 7]. The coefficients of the splines are then estimated using vertices in the triangulation as corners of the spline patches.

Polynomial splines are a special case of rational splines. In the case of rational splines, homogeneous (4-D) coordinates are used and a weight is associated with each control point. They can describe a larger class of shapes and allow for defining conic sections and quadric surfaces exactly which is important because manufactured objects are typically bounded by such surfaces. Similar techniques as in the case of nonrational splines can be used for interpolation and approximation just by using 4-D coordinates.

The use of weights provides an additional handle for altering the shape of the spline. The curve or surface moves closer to the control point by increasing the weight. Other ways to alter the shape include nonuniform spacing of the knots, changing the order of the spline, moving the control points, changing the parameterization or adding or removing knots and, consequently, control points [5, 28, 8, 24].

Certain shapes cannot be defined to have C continuity at the joints. In such cases smooth surfaces can be constructed relaxing the differentiability condition to visual or *geometric continuity* (G). As an example, tangent planes are adjusted to be continuous at the joints (G^1) instead of being once differentiable with respect to the parameterization (C^1) [10]. Geometric continuity allows a larger class shapes to be defined by smooth surfaces. However, higher order patches are typically required. The concept of approximately G^1 continuous surface (ε-G^1) allows for using lower order patches [18].

4 Vision applications

In this section we investigate the use of spline representations in different vision tasks. In particular, the pros and the cons of splines in surface reconstruction, object recognition and CAD-based vision applications are assessed.

4.1 Surface reconstruction

In computer vision, splines have been used for surface reconstruction. Spline representations are convenient for reconstructing complete 3-D shapes, in particular, free-form shapes in medical imaging and CAGD. Tensor product splines are commonly used in CAD systems. Especially shapes in ship hulls and car bodies are defined using rectangular patches. Triangular splines provide a more general purpose representation for reconstructing surfaces of arbitrary topology. Vision systems can benefit from using similar shape representations with other manufacturing subsystems especially in *inspection* tasks and in rapid *geometric modeling* for free-form fabrication (FFF) purposes. A recent review on surface reconstruction techniques appears in [4].

Tensor product splines A rectangular parametric grid creating a mapping between physical measurements and parameterization have to be constructed for tensor product splines. Different techniques are required for reconstructing tensor product surfaces from regularly spaced or scattered data. In the case of scattered data, a rectangular arrangement has to be constructed, e.g., by interpolating the data between the observations and then extracting appropriate number of isoparametric lines in two directions and map them on a parametric grid. In addition, direction of the parameterization has to be chosen which is not necessarily a straight forward task. The main axes of inertia can be used as default directions. The rectangular arrangement of the parametric domain in tensor product splines precludes the representation of surfaces of arbitrary topology, e.g., surfaces with holes and branches.

The input data for the spline approximation can be given, for example, as a collection of sequences of 3-D points with an equal number of points in each sequence. Each sequence generates a 3-D space curve which corresponds to a row or a column in the parametric grid. The sequences are not allowed to cross and the network of curves in different parametric directions are then blended into a surface. Some 3-D imaging sensors, e.g. scanning range sensors with cylindrical or rectangular scan, often provide data that can be directly used for tensor product spline approximation. Such data can easily be projected on parameter plane or cylinder.

B-splines are employed in [22, 30] for reconstructing single valued surfaces ($z = f(x, y)$). In [30], outliers are first removed and missing data are interpolated through a local polynomial fit using the Least Median of Squares (LMS) error criterion. In the presence of non-impulsive noise, the LMS criterion performs poorly as demonstrated in [11]. Very dense knot vector and uniform parameterization are used which may cause unwanted oscillations.

B-spline approximation have been employed is ship and car design [28]. NURBS surfaces are used for 3-D shape estimation for constructing free-form surface models for CAGD [14]. Adaptive parameterization is used. An estimate for appropriate control point mesh size is computed automatically from the data using differential geometric properties of the surface. A parametric grid is constructed through a resampling procedure. The positions of the control points are estimated using the least squares error criterion. The obtained surface approximation is refined to meet a user-given tolerance value. Surface discontinuities are detected separately and dealt with using subdivision. Surfaces with holes and branches are approximated using trimmed surfaces. Surface boundaries are used to compute trimming curves that divide the surface into background and object components [14].

Triangular splines Triangular splines are not limited by the rectangular or spherical arrangement of the parametric space, and there is no requirement for choosing principal parameterization directions. These techniques use a surface triangulation as a starting point. The triangulation can be, for example, an optimal triangulation such as 2/3 Delaunay triangulation or a suboptimal constrained triangulation. The constrained triangulation provides a straight forward way to deal with discontinuities. The vertices of the triangles are used as domain vertices for triangular spline patches and the positions of the other control points are typically interpolated. Local techniques are preferred since they allow for reconstructing the surface incrementally and are not sensitive to occlusion or missing data. The error criteria used in the fitting procedure should include a term penalizing roughness because some degrees of freedom remain unset after adjusting the joints [15]. The domain vertices may unnecessarily appear as artifacts in the resulting interpolant because they lie on the obtained surface.

The key computational problem is to estimate the tangent plane accurately from noisy data at each vertex [15, 16]. Some techniques use second fundamental form coefficients for establishing the tangent plane continuity [18]. The coefficients are, however, difficult to estimate reliably from sparse data and are found to be sensitive to noise [1]. The estimation of geometric surface properties from noisy observations is typically neglected in graphics and CAGD literature and they are assumed to be known. These properties are, however, critical for obtaining a high-quality interpolant.

4.2 Object recognition

Spline representations are studied here using Nalwa's criteria. Splines are concise for describing free-form shapes if the order of the spline is appropriate. However, in the case of standard geometric primitives such as spheres, cylinders or faceted surfaces the number of parameters required is high. As an example, a sphere and the required control point mesh are shown in Figure 4. Rational splines have to be used to represent quadric surfaces exactly. Splines are very stable because the influence of a change is limited to a small region.

The knot vector of a tensor product spline surface in each parameter direction is basically the union of the individual curve knot vectors in that direction. As a consequence, there may be unnecessary knots and consequently control points on slowly varying areas of the surface. If there are several discontinuities and a high order spline is used, the compactness of the representation decreases because multiple knots are required at the discontinuity.

Unlike superellipsoids, for example, spline surfaces do not capture a large family of shapes with few parameters or convey much structural information about the object. Spline representations are unambiguous but not necessarily unique. These factors make splines less appealing as a representation for object recognition purposes.

Splines do, however, provide an intermediate representation that is useful in object recognition. A more compact, structured and unique description can easily be computed reliably from a spline surface. As an example, the second

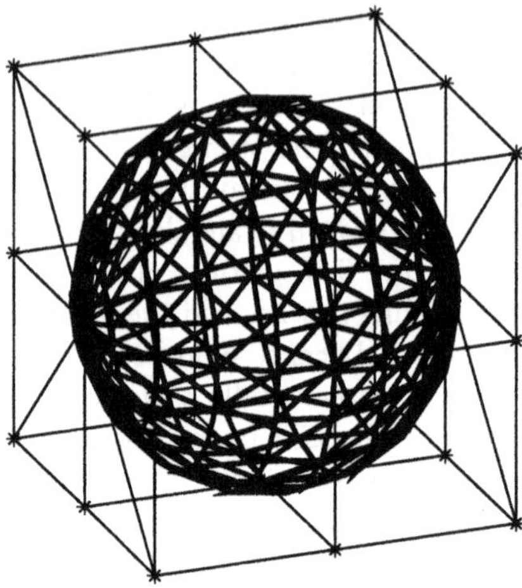

Fig. 4. A sphere defined by NURBS

fundamental form, mean and Gaussian curvatures [17] and Darboux frames are widely used in recognition applications. The estimation of these features from range data directly is very sensitive to noise [1]. Spline approximation or interpolation can be an intermediate step in the recognition process. The computation of derivatives, second fundamental form coefficients as well as mean and Gaussian curvatures and their sign maps is convenient from the smooth spline surface. In the case of CAD-based vision, a similar process can be done for the model shape which may be defined using splines as well. Moreover, spline representations have desirable invariance properties for object recognition from arbitrary viewpoint.

4.3 Other applications

In general, splines provide a flexible data smoothing tool and allow for regression with low-dimensional parametric models. In addition, spline representations are typically invariant under affine or projective (rational splines) transformations.

Shape registration is a very important step in many applications, in particular in medical imaging, where a complete data set have to be put together from several partial data sets. The transformation among multiple data sets from different viewpoints is estimated. Spline curves and surfaces provide a useful representation for free-form matching because of their continuity, local control and invariance properties. As an example, splines can be used as the underlying shape representation in the widely used Iterative Closest Point (ICP) technique [3].

An interactive sculpting technique for NURBS surfaces is developed for design and graphics purposes in [32] using the physics based framework introduced in [20]. NURBS control vertices and the associated weights are used as generalized coordinates. The size of the control point mesh is provided by the user and the problem of adapting the size of control point mesh to increased degrees of freedom is not addressed. Also, no examples of recovering a conic section or quadric surface with appropriate weights for exact representation without prior knowledge about the shape are shown.

5 Examples

In this section, examples on reconstructing a spline surface from range data [27] are given using both tensor-product and triangular spline representations. The image resolution is 256-by-256 pixels and 16 bit quantization is used for the depth values. The Face Mask data is used in order to show the ability to describe very complicated shapes whereas the Hand data forms a shape where the rectangular parametric space of a tensor product B-spline surface is not appropriate. Therefore, it is approximated using the trimmed surface representation. The surface approximations are refined to meet a used-defined tolerance value. The parameterization is first refined to make the error orthogonal to the the fitted surface. If the error still exceeds the tolerence value, the flexibility of the surface is increased by inserting new knots. The positions of the new control points generated by knot insertion are then modified in order to obtain the required accuracy [14, 5, 24].

The same data are also approximated using triangular Bezier patches [15, 14, 16]. The domain vertices are obtained through adaptive 2/3 Delaunay triangulation. The normal of the tangent plane at each vertex is estimated using a technique which stems from robust estimation. The intermediate control points are obtained through Hermite interpolation. Each triangle is subdivided into 3 smaller ones in order to meet the continuity requirement independently across each boundary. The original data, the approximating NURBS surfaces and the domain vertices of the Bezier triangles are depicted in Figure 5.

An example where rational splines are useful is also given using simulated Cylindrical Pin data. An example of rational surface is shown in Figure 6. Rational splines are useful in recovering conics and quadric surfaces exactly. The recovery is significantly simplified by detecting that the surface is a quadric first. Here, a superellipsoid surface fit is employed and the result strongly indicates that the part has a rotational symmetry [14]. Therefore, the weights in the spline representation can be set appropriately. The profile curve has discontinuities and multiple knots and contol points are inserted in those locations in order to introduce the discontinuity.

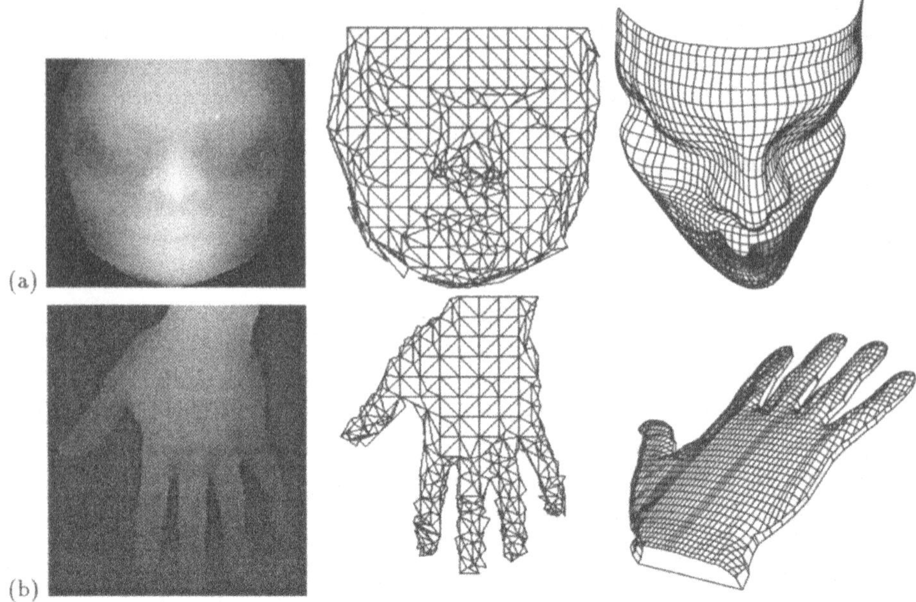

Fig. 5. The test images and the obtained triangular and NURBS approximations. Only the domain vertices are shown for the surface consisting of triangular Bezier patches: a) the Face Mask and b) the Hand are real range data. The Hand is represented using trimmed NURBS.

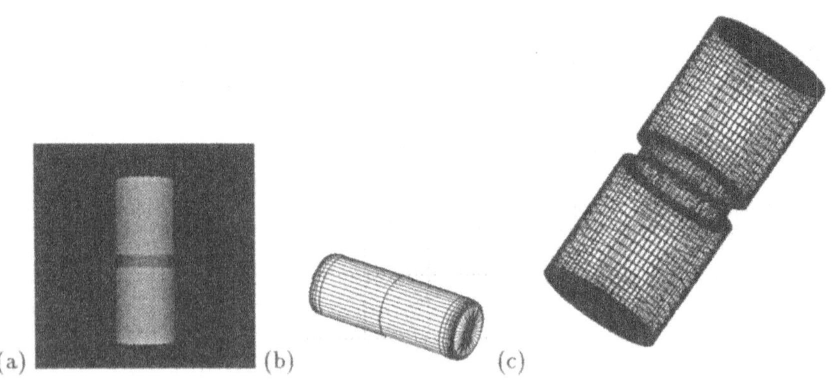

Fig. 6. The Cylindrical Pin: (a) simulated data (b) superellipsoid surface, and (c) the obtained NURBS surface.

References

1. Abdelmalek, N., "Algebraic error analysis for surface curvatures and segmentation of 3-D range images", Pattern Recognition, Vol. 23, No. 8, pp. 807-817, 1990.
2. Besl, P. "Surfaces in range image understanding", Springer-Verlag, 339 p., 1988.
3. Besl, P., McKay, N., "A Method for Registration of 3-D Shapes", IEEE Transactions on Pattern Analysis and Machine Intelligence, Vol. 14, No. 2, pp. 239-256, 1992.
4. Bolle, R., Vemuri, B. "On three-dimensional surface reconstruction methods", IEEE Transactions on Pattern Analysis and Machine Intelligence, Vol. 13, No. 1, pp. 1-13, 1991.
5. Cohen, E., Lyche, T., Riesenfeld, R., "Discrete B-Splines and Subdivision Techniques in Computer Aided Geometric Design and Computer Graphics", Computer Graphics And Image Processing, 14, pp. 87-111, 1980.
6. De Boor, C. "A practical guide to splines", Springer-Verlag, 392 p., 1978.
7. De Floriani, L., Puppo, E., "An On-Line Algorithm for Constrained Delaunay Triangulation", CVGIP: Graphical Models and Image Processing, Vol. 54, No. 3, pp. 290-300, 1992.
8. Farin, G., "Curves and Surfaces for Computer Aided Geometric Design" Academic Press, 1990.
9. Field, D., "A Generic Delaunay Triangulation Algorithm for Finite Element Meshes", Adv. Eng. Software, 1991, Vol. 13, No. 5/6 combined, pp. 263-272, 1991.
10. Gregory, J., "Geometric Continuity", Mathematical Methods in CAGD, eds. Lyche & Schumaker, Academic Press, pp. 353-371, 1989.
11. Hettmansperger, T., Sheather, S. "A cautionary note on the method of least median squares", The American Statistician, Vol. 46, No. 2, pp. 79-83, 1992.
12. Initial Graphics Exchange Standard, Version 5.1., IGES/PDES Organization, 1991.
13. Koivunen V., "A Robust Nonlinear Filter for Image Restoration", Technical Report, December 1992, To appear in IEEE Transactions on Image Processing, May, 1995.
14. Koivunen, V., Vezien, J-M., "Multiple representation approach to geometric model construction from range data", IEEE Cad-Based Vision Workshop '94, Seven Springs, PA, 1994.
15. Koivunen V., Vezien, J-M., "Machine Vision Tools for CAGD", Technical Report, September 1994, Submitted for Publication.
16. Koivunen, V. "A robust approach to enhancement of multivariate images", 1st IEEE ICIP, Nov., Austin, TX, USA, 1994.
17. Lipschutz M., "Differential Geometry", McGraw-Hill, 1969.
18. Lounsbery, M., Mann, S., DeRose, T., "Parametric Surface Interpolation", IEEE Computer Graphics and Applications, Vol. 12, No. 5, pp. 45-52, 1992.
19. Marr, D. "Vision", Freeman, New York, 397 p., 1982.
20. Metaxas, D., Terzopoulos, D. "Shape and nonrigid motion estimation through physics-based synthesis", IEEE Transactions on Pattern Analysis and Machine Intelligence, Vol. 15, No. 6, pp. 580-591, 1993.
21. Mortenson, M. "Geometric modeling", John Wiley & Sons, 763 p., 1985.
22. Naik, S., Jain, R., "Spline-based surface fitting on range images for CAD applications", IEEE Conference on Computer Vision and Pattern Recognition, pp. 249-253, 1988.
23. Nalwa, V. "Representing oriented piecewise C^2 surfaces", International Journal of Computer Vision, 3, pp. 131-153, 1989.

24. Piegl, L., "Modifying the shape of rational B-splines. Part 2: Surfaces", Computer Aided Design, Vol. 21, No. 9, pp. 538-546, 1989.
25. Piegl, L., "On NURBS: a survey", IEEE Computer Graphics and Applications, Vol. 11, No. 1, pp. 55-71, 1991.
26. Pratt, V., " Direct Least-Squares Fitting of Algebraic Surfaces" Computer Graphics 21, No. 4, pp. 145-152, 1987.
27. Rioux, M., Cornoyer, L. "The NRCC Three Dimensional Image Data Files", CNRC 29077, National Research Council of Canada, 1988.
28. Rogers, D., Fog, N., "Constrained B-spline curve and surface fitting", Computer Aided Design, Vol. 21, No. 10, pp. 641-648, 1989.
29. Schumaker, L. "Triangulations in CAGD", IEEE Computer Graphics and Applications, Vol. 13, No. 1, pp. 47-52, 1993.
30. Sinha, S., Schunck, B. "A two-stage algorithm for discontinuity preserving surface reconstruction", IEEE Transactions on Pattern Analysis and Machine Intelligence, Vol. 14, No. 1, pp. 36-55, 1992.
31. Taubin, G. "Estimation of planar curves, surfaces, and nonplanar space curves defined by implicit equations with applications to edge and range image segmentation", IEEE Transactions on Pattern Analysis and Machine Intelligence, Vol. 13, No. 11, pp. 1115-1138, 1991.
32. Terzopoulos, D., Qin, H., "Dynamic NURBS with Geometric Constraints for Interactive Sculpting", ACM Transactions on Graphics, Vol. 13, No. 2, pp. 103-136, 1994.

Triangles as a Primary Representation

Paul J. Besl

General Motors R&D Center
Warren, Michigan 48090-9055
besl@iris7.cs.gmr.com

Abstract. Triangles have often been overlooked as a serious general-purpose geometric representation for many applications owing to storage, accuracy, and combinatoric concerns. This paper advocates triangles as a very workable common representation for almost all types of 3D surface data and for almost all applications, including 3D computer vision, by examining the major concerns and other representation issues in light of practical industry experience with such models. In addition, a shape signature, or feature vector, called the "crease angle histogram" is introduced and discussed as a possible tool in geometric indexing for recognition.

1 Introduction

The processing of surface geometry is common in many endeavors, including CAD, CAM, rapid prototyping, CAE, finite element analysis, computer graphics, as well as 3D computer vision. There has often been the implicit hope that a "magic custom representation" will permit easy solutions to a given application problem [22]. The natural tendency has been for each application to focus only on its own geometric modeling needs and not on general geometric processing issues. Unfortunately, much effort is expended solving purely geometric problems that are as common as addition, multiplication, or file systems. Geometric problems are often difficult and should not have to be solved anew in each geometric application for each geometric representation. It would be beneficial to have a single practical representation that provides most solutions to most application problems in a reasonable manner.

Parametric and implicit curves and surfaces [4] have conventionally been used as primary geometric representations in many areas, including CAD/CAM applications. Just as curves are typically decomposed into many straight line segments to be drawn, surfaces are often decomposed into triangle meshes to be rendered. Prior to the 1990's, memory constraints seriously impeded the acceptance of piecewise-triangular representations as valid primary geometric entities for curved surfaces. For example, the IGES standard still does not include a piecewise-triangular, or triangle mesh, entity even though it has been revised four times since 1981.

This paper discusses some of the issues surrounding the use of triangles as a primary geometric model for multiple applications including computer vision. In the next section, the requirements of an ideal geometric representation that serves many different applications are examined. The strengths and weaknesses of triangle representations are commented on with each requirement. Since triangle representations are viewed as weak in indexing for visual recognition, the "crease angle histograms" are introduced in the subsequent section with several examples. The final section summarizes the discussion and looks at future directions for triangular geometric representations.

2 Representation Requirements

If real 3D computer vision applications can be successfully implemented without using explicit 3D models, see e.g. [23], or without using 3D data (only monocular intensity images), see e.g. [12], then this is definitely the preferable approach, all other aspects being equal. Of course, when this will be possible in relatively unconstrained environments is uncertain. In the meantime, there is much to be gained and learned by processing explicit 3D data from available sensors. Moreover, I believe it is impossible to achieve the vast potential that computers have for assisting us with the 3D world in all aspects of engineering unless computers can qualitatively and quantitatively reason about 3D geometric models and their accompanying physics easily aside from any perception issues. Henceforth, explicit 3D data and explicit 3D models will be assumed with the hope that such efforts may eventually assist in the solution of the much, much more difficult problem of working only with video information.

Although great strides have been made in computer aided design and reverse engineering from physical prototypes, it generally is not possible to simply "look at" an existing object or "think about" a non-existent object and then instantly obtain a sufficiently detailed geometric model suitable for multiple applications. Given the above assumptions about 3D data and models, and given that a significant amount of effort will go into the geometric and physical modeling of 3D objects for the foreseeable future, it makes sense to optimize that effort by attempting to insure that such models will be as compatible as possible with the widest range of engineering applications. That is, any geometric modeling effort should enable data sharing with CAD/CAM, finite element modeling, NC, rapid prototyping, point measurement devices, graphics systems, etc. In other words, there is much to be gained by tying efforts in automating geometry-based perception with other geometry-based disciplines. Hence, the listed requirements for a geometric representation suitable for computer vision will be affected by this wider view.

Any hypothetical ideal geometric representation would possess a wide variety of useful properties. The following list describes what is needed in a given area and then describes what triangle representations have to offer.

Data Acquisition: Various sensor devices output 3D coordinates [5]. It should be as simple as possible to go from a physical object to a useable 3D geometric model. An ideal geometric representation should be derivable in seconds from digitized data from 3D scanners.

Computed tomography (CT) and magnetic resonance imaging (MRI) scanners can produce triangles via discrete isosurface algorithms applied to stacks of image slices, e.g. "marching cubes" [21]. Many range imaging sensors can produce quadrilateral grids; triangles are obtained directly by splitting each quadrilateral grid element into two triangles. Other non-contact 3D scanners produce structured light scans that can be meshed with triangles by connecting nearest neighbors across adjacent scans. Unorganized point clouds can also be triangulated as in [15]. In general, obtaining a triangle mesh from raw point data is not a problem.

Owing to the standardization of tensor product NURBS surfaces (Non Uniform knot sequence Rational B-Splines), it would be greatly beneficial if an automated system could directly output NURBS surfaces [19] from raw data. However, the difficulty level is extremely high in automatically determining a useful surface patch topology where the network of fitted surfaces (1) accurately represents the sensor data without introducing artifacts and (2) maintains the desired surface, normal, and/or curvature continuity that a trained person would create if they were doing the job manually. Methods such as [3], when implemented in an industrial setting to output NURBS in IGES format, are useful to some extent, but do not solve enough of the problems. Imposing CAD surface structure where it may not have previously existed generally results in surface geometry that does not reflect the real shape as well as the smoothed raw data does.

Model Conversion: Various geometric models already exist, mostly in the CAD/CAM/CAE context, but now also in the VR (virtual reality) context. An ideal geometric representation should be derivable from existing data files in seconds.

The current *de facto* standard for the exchange of triangle geometry, the binary or ascii STL file formats (STL stands for STereo Lithography), evolved due to the popularity of rapid prototyping technology. Every major CAD/CAM system can now output their native geometry (usually based on NURBS) in a triangle form. Typically, one can select a desired threshold tolerance for the accuracy of the triangles to underlying surface. Therefore, obtaining triangle models from CAD/CAM systems is not the bottleneck it was just a few years ago.

Accuracy: An ideal geometric representation should be as accurate as required by the most demanding application, e.g. machining of dies and molds.

Triangle meshes can be made as accurate as desired. Figure 1 shows a shaded display of more than two million triangles that requires less than 3 seconds to render. The triangle representation is as accurate as the physical parts can be manufactured.

Storage Requirements: An ideal geometric representation should have reasonable storage requirements.

What is reasonable changes over time. It requires at least 4N bytes to represent N triangles in an uncompressed triangle strip format and at most 36N bytes if all triangles are isolated from one another. For the assembly shown in

Fig. 1. Outer Surfaces of a Sport Utility Vehicle. Represented by about two million triangles. Average deviation from spline surfaces: 0.025 mm. Maximum deviation from spline surfaces: 0.100 mm.

Figure 1, all the triangles could be represented in at least 8 megabytes and at most 72 megabytes depending upon the exact representation. The surface data represented in uninterpretable proprietary binary database formats requires on the same order of bytes, but the neutral IGES files for this data are much, much larger. Thus, a minimal size triangle format is a reasonable alternative.

Coarsening Process: It must be possible to simplify a detailed ideal geometric model into simpler forms based on meaningful scale parameters. This capability is necessary both to support (1) real-time rendering of complex scenes and (2) perceptual abstraction.

Triangle meshes can be reduced, decimated, re-tiled, or optimized with fewer triangles by sacrificing some accuracy in the original mesh [2] [16] [24] [27] [30]. Data smoothing is often an integral part of the process. The memory savings can be sometimes phenomenal. A fine-to-coarse scale-space of models can easily be constructed to aid in matching and indexing using reduction algorithms of this type.

Smoothing Process: High-frequency random, systematic, or quantization noise can arise as part of a geometric description, especially in geometric models created directly from raw scanner data.

Simple Laplacian smoothing of triangle meshes creates smoother meshes, but suffers from the classic shrinkage problem. Taubin [29] has developed a smoothing method that avoids shrinkage, works on arbitrary topology triangle meshes, and is fast.

Shape Modification Process: Any geometric shape should be easily modifiable. Of course, moving any given vertex in triangle mesh is easy. How the vertex neighborhood is affected is a more complicated matter. When working with spline surfaces, users and algorithms create desired changes by moving control points and/or construction lines while maintaining surface continuity. A variety of alternatives are possible in the triangle mesh case. A simple modification method is to embed the triangle geometry in a volume spline, modify the volume spline, and map the triangle geometry through the volume spline mapping. If the triangle geometry is bent too much, the mesh has to be refined via post-processing. Welch and Witkin [33] have created a design system that works directly with triangles. Halstead et al. [13] and others have methods to maintain a smooth surface based on triangles.

Refinement Process: Given a coarse model, it should be possible to refine that shape into a more detailed, smoother shape without additional information. Such refinement operations must be possible locally and globally.

Triangle meshes can be subdivided into finer meshes via simple center point interpolation based on vertex normals, mesh insertion, edge swapping with adjacent triangles to optimize triangle shape quality, and vertex smoothing. Some alternatives are discussed e.g. in [13] [17] [20].

Texturing Process: It should be easy to attach surface markings and visual texture information to an ideal geometric model since the visual world does not consist of geometry alone.

Today's graphics hardware easily allows image textures to be applied to geometry to simulate real world appearance in real-time. The geometric model requires the attachment of (s, t) image texture coordinates to each triangle vertex as well as a texture id for groups of triangles [1].

Occluding Contour Process: Any geometric representation for computer vision must support computation of occluding contours. Computing occluding contours of triangle meshes is quite easy compared to the alternative algorithms for parametric or implicit surfaces. Each edge in the indexed triangle representation is shared by a pair of triangles. Given an eye point or direction, determine if one triangle normal points toward the viewer and the other triangle normal points away. If so, add the two-point line segment for that edge to the occluding contour list of edge segments and link together. Figure 2 shows an example of an occluding contour computed from a triangular representation of a torus.

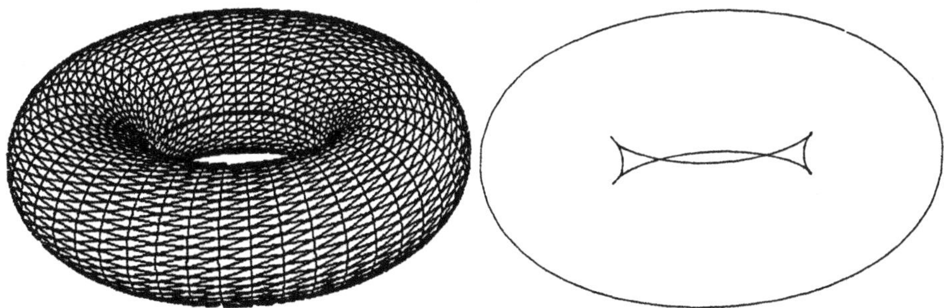

Fig. 2. A triangle model of a torus and its occluding contour.

Edge Extraction Process: Any geometric representation for computer vision must support extraction of discontinuities in surface, normal, or curvature. Depth discontinuities (jump edges) are relatively easy to detect in accurate 3D data. Such discontinuities do not exist as a part a given object model though they may occur between different regions. Normal discontinuities (crease edges) in a model can be detected by by comparing a simple dot product between adjacent triangle normals to a specified threshold. If the triangle mesh is noisy, it may be difficult to accurately determine the significant crease edges. Smoothing and triangle decimation are necessary for adequate detection. Perceptual curvature discontinuity extraction is difficult even in ideal spline-based CAD models; triangles provide no additional benefit or disadvantage in this case.

Hidden Line Drawing Process: Any geometric representation for computer vision must support creation of "hidden line removed" drawings for use in exact or approximate aspect graph methods. Such algorithms are also very useful in the CAD/CAM world for technical publications applications.

A wide variety of methods are available for such purposes and are documented in graphics texts. A simple, robust method that takes advantage of graphics hardware is to (1) access all triangle mesh models for a scene, (2) extract crease edges, occluding contours, and texture image edges in 3D, (3) do a z-buffer rendering of the triangles for the given view, (4) subsample all lines at a desired tolerance, (5) delete all hidden points that map behind the z-buffer, (6) perform chordal-deviation point reduction to get rid of any additional points introduced at the sub-sampling phase. This method yields 2D and 3D line models that are useful for approximate aspect graphs.

Surface Area, Genus, Volume, Moments, etc.: An ideal geometric representation should support computation of useful features such as surface area, genus, volume, moments.

Total surface area of triangle mesh model is obtained by simply summing the areas of the individual triangles. The accuracy of the surface area is dependent on smoothing capabilities if the data comes directly from a sensor. A closed mesh with N_T triangles permits computation of a genus feature value G, which is zero for single solid bodies and varies for multiple bodies and holes: $G = N_P + N_T - N_E - 2$ where N_E is the number of edges and N_P is the number of points in the mesh. The total volume is given by $V_{tot} = 1/3 \sum_{i=1}^{N_T} A(t_i) h(t_i, \mathbf{r_0})$ where $h(t_i, \mathbf{r}_0)$ is the signed height of the tetrahedron from the ith triangle t_i to some point $\mathbf{r}_0$ with respect the triangle's normal. Moments of inertia can be computed via similar one pass summations.

Pose Estimation: An ideal geometric representation should support computation of the 3D pose of 3D data relative to a canonical 3D object model.

Triangle mesh models are quite suitable for fast pose estimation. Besl and McKay [6] described a general representation independent method for shape matching based on an iterative closest point (ICP) algorithm. Simon et al.[26] discuss a real-time implementation of an improved ICP matching algorithm with a triangle-based model that operates at 10 Hz. Delingette et al. [11] describe the SAI (simplex angle) method of pose estimation which is also suitable for triangle mesh models. Ikeuchi and Hebert [18] discuss all classes of EGI methods (CEGI,DEGI) and the SAI method, which also are easily used with triangle models.

Matching: An ideal geometric representation should allow at least $O(\log N)$ complexity searching among N models.

Fast matching into a large database of fixed non-abstraction-oriented 3D models (e.g. more than 10,000 specific model instances) remains an elusive goal. A few years ago, it would have been a major lifetime undertaking to assemble a database of even a 1000 3D models. Today, one can order individual detailed 3D models from catalogs [32] or get CDROM's with over 500 3D models for less than one hundred dollars [28]. These commercially available databases come in a triangle mesh form using one of many possible common formats.

In the subsequent section of this paper, "crease angle histograms" are introduced. This histogram has potential has a quick discriminator among object classes, such as blocky polyhedral objects and smoothly curved shapes.

There are only a few systems worldwide that have ever made it into the class of 3D recognition systems that can be evaluated on execution speed, misses, and false alarms on a wide range of objects. Most of these systems are compatible with the geometry of triangle models.

Learning: An ideal geometric representation and its matching metrics should allow the determination of a condition where no sufficiently close match exists. Assuming the acquired data can easily be converted into the ideal geometric representation, the unknown data set should be insertable into the knowledge base of known models.

Triangle meshes at least allow the conversion of unknown data into a model by simply copying the data into the model library and recompiling whatever indexing structures would need updating.

Abstraction: Learning, matching, recognizing often imply the ability of a automated perception system to deal in abstract concepts. Geometric representation will impact the abstraction capabilities of a system. It is difficult to evaluate the limitations of triangle meshes in this context since it is a relatively undefined area. It would appear at first glance that triangle meshes are much more suitable for object shape instances rather than object shape classes, but it is not clear whether or not it is any less practical to consider building higher level constructs using triangle mesh representations than say NURBS, for example.

Articulations, Features, Deformations, Constraints: Geometric models should include directly the ability for articulated motion of component geometries, feature maintenance under object transformations, general deformation modeling from physical simulation or user interaction. In general, the representation should be able to handle arbitrary mathematical constraints.

Triangle meshes are as suitable as any other geometric representation for articulated motion or deformations. Features and constraints are harder to evaluate, but triangle meshes do not seem to be at any major disadvantage compared to alternatives.

Combine Overlapping Surfaces: When 3D surface data is acquired from an optical 3D scanner, it is generally necessary to acquire data sets that overlap in order to insure complete surface coverage. It should be easy to perform optimized partial matching of overlapping data followed by coalescing data sets into a single surface description.

Turk and Levoy [31] have demonstrated a method for zippering triangle meshes to create a single unified mesh. See also [8] [15] [27]. Triangle meshes are better suited to this sort of operation than most alternatives.

Inside, On, Outside Function: Given a point, an ideal geometric representation should provide a procedure that can *quickly* evaluate whether that point is inside the shape, outside the shape, or directly on the surface of the shape.

Implicit surface representations will always have the lead in this area. However, triangle mesh representations should not be dismissed prematurely. Efficient spatial querying methods do allow fast point-in-polyhedron and point-in-polygon tests.

Stability: Small changes in anything should yield small effects. Small perturbations to a triangle mesh yield only small effects to any of the items discussed in this list.

Invariants: Some geometric representations allow the use of invariants computed directly from function coefficients. Implicit surface representations certainly also have the lead in this area [9]. Aside from surface area, volume, moments, and genus, triangle meshes are not likely to allow the use of any remotely similar method.

Intersection and Distance Functions, Combinatorics: Given two geometric models, an ideal geometric representation should provide a procedure that can *quickly* evaluate whether those represented shapes intersect, just touch, or are spatially separated. All basic minimum distance functions should also be supported: point to surface, curve to surface, and surface to surface.

A mesh of triangles might not seem the best representation for distance or intersection operations since the complexity of the computation grows quickly with the number of triangles and/or with the required accuracy of the intersection output. As part of the SurfSeg project at the GM R&D Center, we have a working system that performs the surface-surface intersection problem to machine precision between a triangle mesh of size M and one of size N in approximately $O(M \log N)$ time without using any volume occupancy buffers. For distance computations, the complexity is approximately the same, but the minimum distance execution time increases a bit over the intersection time since larger spatial queries are usually necessary. One comparison of this code with a commercial CAD system's comparable operator showed that a 90 second task for the fast intersection operator took over 39 hours in the CAD system. For lower resolution, real-time interference checking methods are commercially available.

Arbitrary Topology: An ideal geometric representation should be able to deal with arbitrary topological arrangements of geometric features.

Triangle meshes are ideally suited to arbitrary surface topologies. The tensor product based NURBS parametric surfaces used in existing CAD/CAM systems are limiting and must often be trimmed. Maintaining continuity and smoothness over trim boundaries is quite difficult.

Drawing, Displaying, Rendering: It should be possible to interact with the geometric model in real-time on a computer screen. Indexed triangle strips are the preferred geometric representation for driving high-performance graphics displays [1] [14] [10]. There are tremendous benefits from being able to visually interact with geometry both as a system developer and a system user in all areas of geometric processing.

Sectioning, Offsetting, STL: Any primary representation should allow for offsetting and sectioning operations to support basic NC operations as well as STL file output to support rapid prototyping systems. As it turns out, some NC machining packages compute intermediate triangle meshes while generating cutter paths. In the past, saving such high resolution meshes was not practical due to the storage requirements. However, today it is, and so even NC, the most important phase of CAD/CAM after the design is defined, is amenable to triangle meshes. This can provide a major productivity advantage over touch probe copy mills when coupled with 3D laser scanner input.

3 Crease Angle Histograms

The shape of an object has unlimited degrees of freedom in any representation. How can we search for objects in a database with a fixed length key? The crease angle histogram is a simple, admittedly unproven, idea to map a shape to a k-vector in a k-D space for a k of manageable size (e.g. $7 \leq k \leq 360$). A k-D tree would be built for all objects encountered. Given 3D data from a single view, a k-vector is computed, and an appropriate range search is done on the tree using constraints from the input k-vector. This could allow for approximately $\log(N)$ search time for N objects of various shapes, including both polyhedra and curved objects. If the range search retrieves R objects with similar signatures, each object retrieved is then explicitly matched against the input 3D data from a single view using fast ICP [6] [26], (X)EGI and SAI [18], and/or any other 3D matching / pose estimation algorithm. The best explicit match is chosen.

3.1 Triangle Definitions

All triangle representations involve 3D point lists. Let the i-th point in a point list be denoted as $p_i = (x_i, y_i, z_i)$, a triplet of coordinates. Let the n-th triangle be given by $t_n = (p_i, p_j, p_k)$, a triplet of points, or equivalently by a triplet of integers $t_n = (i, j, k) = (j, k, i) = (k, i, j)$ where the integers are pointers into the point list and where invariance to cyclic reordering is indicated. Without loss of generality, we assume that all points in the point list are spatially unique. If not, the point list can be processed to insure this condition. It is often convenient to store the adjacency relationships for each triangle t as $a_n^t = (m_1, m_2, m_3)$ to allow direct indexing to the adjacent triangles of any given triangle.

The normals and the area for each triangle are needed. For each triangle $t = (\mathbf{p}_1, \mathbf{p}_2, \mathbf{p}_3)$, define two edge vectors: $\mathbf{e}_a = \mathbf{p}_2 - \mathbf{p}_1$ and $\mathbf{e}_b = \mathbf{p}_3 - \mathbf{p}_1$. The cross product of the edge vectors yields the area of the parallelogram defined by the two edges, and the triangle's area $A(t)$ is half of that: $A(t) = ||\mathbf{e}_a \times \mathbf{e}_b||/2$. The unit normal $\mathbf{n}_t$ to the triangle t is of course $\mathbf{n}(t) = \mathbf{e}_a \times \mathbf{e}_b / 2A(t)$.

A crease angle histogram (CAH) is a histogram of the crease angles at the edges between adjacent triangles $t_i = (a, b, c)$ and $t_j = (f, g, h)$. Adjacency implies that a cyclic reordering exists so that $c = g$, $b = h$, and $a \neq f$. The associated edge $\mathbf{e}_{ij}$ runs between the points $\mathbf{p}_{c=g}$ and $\mathbf{p}_{b=h}$. The crease angle θ_{ij} between triangles t_i and t_j across the edge $\mathbf{e}_{ij}$ is given by the inner product

$$\mathbf{n}_i \cdot \mathbf{n}_j = \cos(\theta_{ij}) \tag{1}$$

This angle varies between 0 and 180 degrees. However, correct maintenance of outward-pointing triangle normals indicates that there is a distinct difference between convex crease edges, which may be labeled as positive crease angles, and concave crease edges, which may be labeled negative. One way to determine the sign of the crease angle then is to compute a direction vector $\mathbf{d}_{af}$ where a and f are the unequal point vertices: $\mathbf{d}_{af} = \mathbf{p}_a - \mathbf{p}_f$. If $\mathbf{n}_i \cdot \mathbf{d}_{af} > 0$ and $\mathbf{n}_j \cdot \mathbf{d}_{af} < 0$, then $\theta_{ij} = \cos^{-1}(\mathbf{n}_i \cdot \mathbf{n}_j)$. If $\mathbf{n}_i \cdot \mathbf{d}_{af} < 0$ and $\mathbf{n}_j \cdot \mathbf{d}_{af} > 0$, then $\theta_{ij} = -\cos^{-1}(\mathbf{n}_i \cdot \mathbf{n}_j)$. If $\mathbf{n}_i \cdot \mathbf{d}_{af}$ and $\mathbf{n}_j \cdot \mathbf{d}_{af}$ have the same sign and are larger than a machine tolerance, then triangle normals have not been maintained correctly and the triangle model requires a normal vector rectification process. Figure 3 shows a 3D graphical view of the crease angle between two triangles and also shows a single triangle in a mesh.

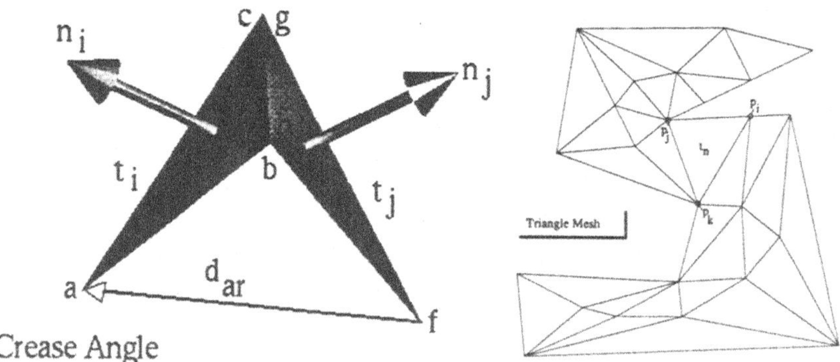

Fig. 3. The Crease Angle between two triangles and the definition of a single triangle with a mesh.

A crease angle histogram (CAH) is specified either via the number of histogram bins N_h or the histogram quantization resolution $\Delta\phi$ (in degrees) subject to the constraint:

$$(N_h - 1)\Delta\phi = 360 \text{ degrees}. \tag{2}$$

The crease angle histogram function may be defined as

$$h(k) = |\{\mathbf{e}_{ij} | \phi_k \leq \theta_{ij}(e_{ij}) < \phi_{k+1}\}| \tag{3}$$

where the outer bars indicate the set cardinality function and $\phi_k = k\Delta\phi - 180$. An *area-weighted* CAH is computed similarly except that the histogram accumulator array $h_a(k)$ is a floating point array, and each edge at the given angle accumulates $(A(t_i) + A(t_j))/2$ An *edge-length-weighted* CAH is computed by accumulating the length of the edge $||\mathbf{e}_{ij}||$ for the given angle bin. A histogram of length 361 can be regarded as high-resolution histogram, and a histogram of length 7 would be regarded as low-resolution. The goal of the crease angle histogram is to provide a simple feature vector to discriminate 3-D shapes. If only a subset of the shape is available, then $h_{subset}(k) \leq h(k)$ for each k for the whole shape. In the CAH plots of the next sections, the vertical axis is actually accumulator log $\log(1 + h(k))$ versus the crease angle ϕ_k.

Figure 4 shows the critical zones of the crease angle histogram. Typically, there is a large cluster of data located around the zero angle for nearly flat creases (i.e., no crease). Smoothly curved objects contribute to lobes at small positive angles for convex angle surface bending and at the small negative angles for concave angle surface bending. In the blocks world, large peaks occur at the convex and concave right angles, +90 and -90. A +180 count is a razor sharp surface edge, and a -180 occurs if the surface folds back on itself. The simplest

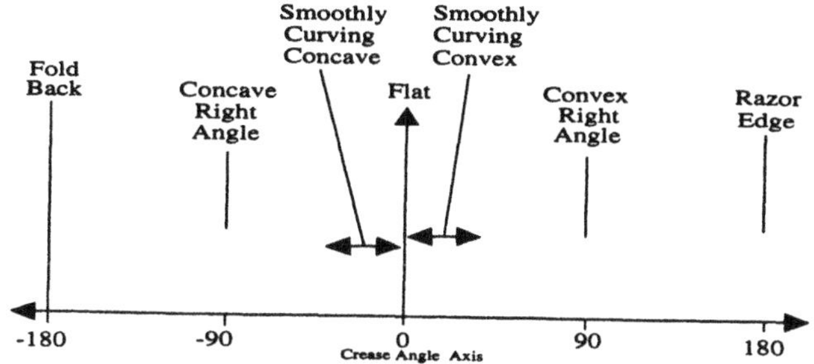

Fig. 4. The critical zones of the crease angle histogram.

CAH type signature of use might be a 7-vector where the histogram elements count angles in bins centered at -90,0,90 as well as the ranges above +90 and below -90 and the ranges between [0,+90] and [0,-90].

3.2 CAH Examples

First, several simple shapes are considered. Figure 5 shows the crease angle histograms and hidden line drawings for eight simple shapes: a block, a cylinder, a sphere, a block with channel, a "soap-shape" superquadric, two blocks glued together, a "double horn" superquadric, and a "jack-shaped" superquadric.

Working from the bottom up, we see the block CAH consists of two simple peaks: one peak at +90 degrees for the 12 edges and one peak at zero for the adjacent triangles within a face. The cylinder's creases will have angles that are zero or small and positive as well as a peak at 90 degrees. The three ideal peaks, one for flatness, one for convex curvature, and one for 90 angles, are the signature for the cylinder. An ideal cone's histogram will look very, very similar except the peak at 90 degrees should be half the size.

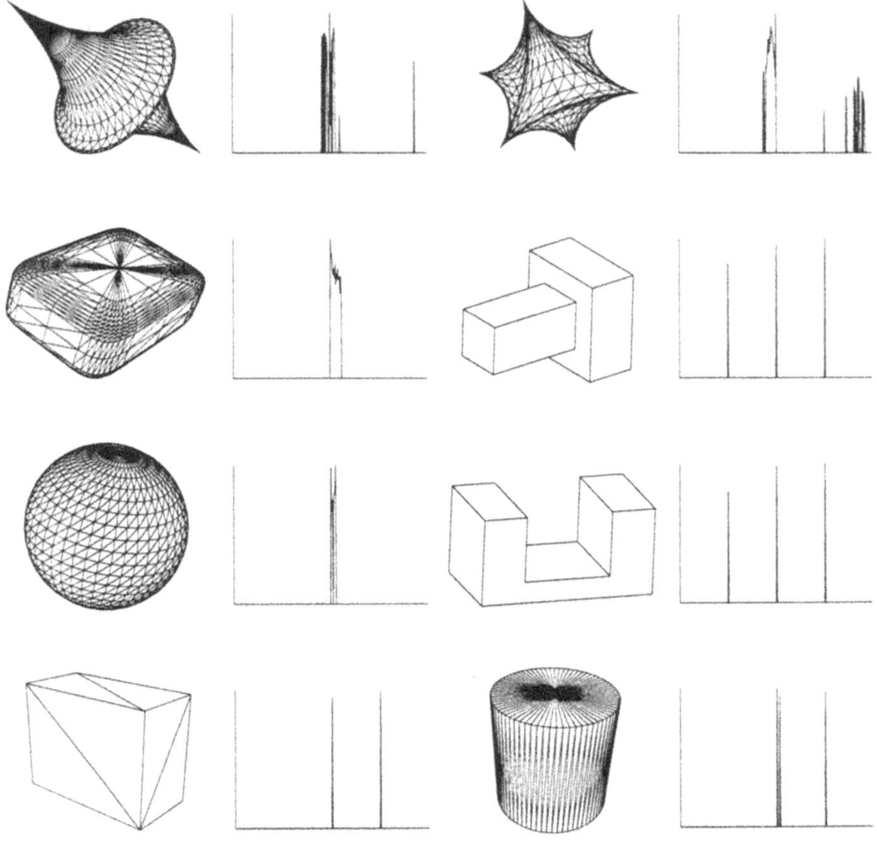

Fig. 5. Crease Angle Histograms for Simple Shapes: Double-Horn superquadric, Jack-shaped superquadric, Soap superquadric, Two Blocks Glued, Sphere, Block with Channel, Block, Cylinder.

A sphere's creases will have angles that are zero or small and positive. The split in what should be a single peak owes to the fact the most of the triangles in the model as shown come from split planar quadrilaterals. Note that the soap shape has a similar simple signature, but is more spread out. The block with channel and the two blocks glues together also have very similar looking signatures; there is a slight difference in the height of the concave -90 degree peak.

The double-horn shape is characterized by concave and convex creases with small angles and a razor sharp peak representing the cusp edge around the middle. The jack-shape superquadric is characterized by all concave creases except for twelve different razor sharp cusp edges that are spread about above the +90 line.

Next, several complex shapes are considered. Figure 6 shows the crease angle histograms and hidden line drawings for six complicated shapes: Triceratops, SittingPerson, Computer, Hammer, Quarter-Panel, Fighter-Jet. Again working

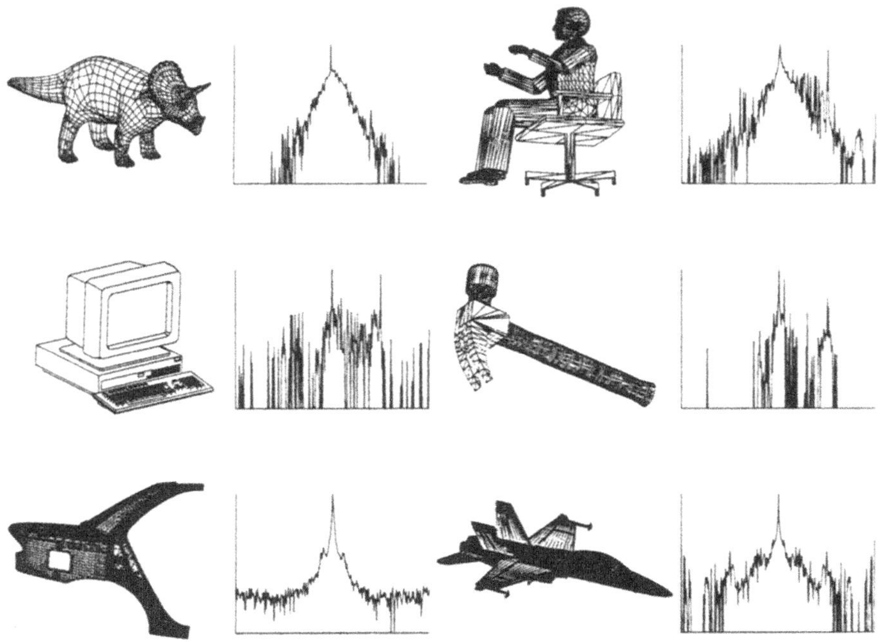

Fig. 6. Crease Angle Histograms for Complicated Shapes: Triceratops, SittingPerson, Computer, Hammer, Quarter-Panel, Fighter-Jet.

from the bottom up, we note that the quarter panel has a very full "spectrum" of creases including flanges folding back and abundant free-form curvature. The fighter jet has a very full histogram, too, but has more gaps in the above +90 and below -90 regions.

The computer signature is quite spikey owing to the number of different angles on the shape. The spread in the hammer signature is much smaller with an isolated peak for the concave crease in the claw.

The lifeform signatures are both much fatter about the 0 degree peak than any other histograms in this sample. Note the overall lack of razor sharp and fold back regions in the triceratops plot. A "superposition" principle holds for objects and signatures. If the chair were removed from the sitting man shape, the spike at 90 degrees would also be removed from the histogram.

All these histograms come from idealized object models. The stability of this shape signature is very dependent upon good triangle decimation schemes and good smoothing algorithms that do not perturb the 3D shape of the object, yet simplify it down to its major features with a minimal number of triangles. I believe the Taubin smoothing method and the Lorensen-Cline decimation method are more than adequate for this task.

The crease angle histogram is invariant to rotations, translations, and scalings. The histogram mapping is homomorphic:

$$h(\text{Object-A}) + h(\text{Object-B}) \;=\; h(\text{Object-A} + \text{Object-B}) \tag{4}$$

where object addition is object union. In addition, h(visible-surfaces) $\leq$ h(whole-object) componentwise. Note also that it is the finiteness of this description that helps to make it work. In the limit, all crease angles on a smooth surface model tend to zero. Overall, the existence of simple signatures like the CAH lends additional credibility to the notion of using triangle models even for vision purposes.

4 Conclusions

Triangle models are not new; they are polyhedral models. The basic new message is that as computer memorys, disks, processors, and graphics pipelines have improved, it has become reasonable to think about using triangles for everything including highly precise curved surfaces for manufacturing.

For industrial applications, the only competition to triangle meshes is NURBS surfaces. NURBS in fact rule the day with triangles only being used to help drive rapid prototyping devices. The best things that NURBS surfaces have going for them is that (1) they are very good for design, and (2) they have had an IGES definition since 1983. It took about 4 to 6 years for industry to accept and convert to this standard. Industry does *not* need something different from a CAD/CAM perspective today. However, a representation that allows more frequent use of data from 3D laser scanners will encourage major productivity increases, but will require some changes in existing industry processes. If dense 3D data cannot be turned into useable, high-quality NURBS surfaces automatically or semi-automatically, industry may be better off adapting design and surface editing tools and NC cutter path generation tools to accept triangle meshes as input. If computers have enough memory, disk space, and processing power to utilize the productivity advantages possible from sensor-compatible triangle meshes, then the possibilities of additional commonplace integration with 3D computer vision systems will be enhanced.

If it is true that there is no magic representation to solve the vision problem, then we will be better off taking the simplest, most flexible representation (triangles) and moving ahead with it in all areas. My experience with these representations over the last six years is that the accuracy, storage, and combinatorics of triangle meshes are not really a problem on today's computers. In fact, many operations are faster and require less disk storage than comparable

NURBS solutions in some commercial packages. When manufacturing tolerances are maintained in the triangle model, there is very little lost, but a lot is gained.

References

1. Akeley, K., "RealityEngine graphics," *Computer Graphics*, ACM Siggraph 93 Proc., Anaheim, CA, Aug. 1-6, pp. 109-116, 1993.
2. Bern, M., Eppstein, D., "Mesh generation and optimal triangulation," Tech. Report P92-00047, Xerox PARC, March 1992.
3. Besl, P. and Jain, R., "Segmentation via Variable-Order Surface Fitting," *IEEE Transactions on Pattern Analysis and Machine Intelligence,* PAMI-10, 2, March 1988.
4. Besl, P., "Geometric Modeling and Computer Vision," *Proceedings of the IEEE,* Aug. 1988, 936-957.
5. Besl, P., "Active Optical Range Imaging Sensors," in *Advances in Machine Vision: Architectures and Applications*, J. Sanz, Ed., Springer-Verlag, 1989.
6. Besl, P. and McKay, N., "A Method for the Registration of 3-D Shapes," *IEEE Transactions on Pattern Analysis and Machine Intelligence,* PAMI-10, 2, February 1992.
7. Bresenham, J., Jacobs, P., Sadler, L., Stucki, P., "Real virtuality: stereolithography - rapid prototyping in 3D," Panel Discussion, *Computer Graphics*, ACM Siggraph 93 Proc., Anaheim, CA, Aug. 1-6, pp. 377-378, 1993.
8. Chen, Y., Medioni, G., "Object modeling from multiple range images," *Image and Vision Computing*, Vol. 10, No. 3, (April 1992), pp. 145-155.
9. Cooper, D. and Keren, D., "Representing complex objects by patches of implicit polynomials and recognizing them using self and mutual invariants." Proc. Workshop on 3D Representations in Computer Vision, Spinger, NY, Dec. 5-7, 1994.
10. Deering, M., and Nelson, S., "Leo: A system for cost-effective 3d shaded graphics," *Computer Graphics*, ACM Siggraph 93 Proc., Anaheim, CA, Aug. 1-6, pp. 101-108, 1993.
11. Delingette, H., Hebert, M., Ikeuchi, K., "A spherical representation for the recognition of curved objects," Proc. ICCV, Berlin, Germany, May 1993.
12. Forsyth, D., Zisserman, A., Malik, J., "Distinctive representations for the recognition of curved surfaces using outlines and markings," Proc. Workshop on 3D Representations in Computer Vision, Spinger, NY, Dec. 5-7, 1994.
13. Halstead, M., Kass, M., DeRose,T., "Efficient, fair, interpolation using Catmull-Clark surfaces," *Computer Graphics*, ACM Siggraph 93 Proc., Anaheim, CA, Aug. 1-6, pp. 35-44, 1993.
14. Harrell, C., and Fouladi, F., "Graphics rendering architecture for a high-performance desktop workstation," *Computer Graphics*, ACM Siggraph 93 Proc., Anaheim, CA, Aug. 1-6, pp. 93-100, 1993.
15. Hoppe, H., DeRose, T., Duchamp, T., McDonald, J., Stuetzle, W., "Surface reconstruction from unorganized points," *Computer Graphics*, ACM Siggraph 92 Proc., Chicago, Ill., July 26-31, pp. 71-78, 1992.
16. Hoppe, H., DeRose, T., Duchamp, T., McDonald, J., Stuetzle, W., "Mesh optimization," *Computer Graphics*, ACM Siggraph 93 Proc., Anaheim, CA, Aug. 1-6, pp. 19-26, 1993.

17. Hoppe, H., DeRose, T., Duchamp, T., Halstead, M., Jin, H., McDonald, J., Schweitzer, J., Stuetzle, W., "Piecewise smooth surface reconstruction," *Computer Graphics*, ACM Siggraph 94 Proc., Orlando, Fla., July 24-29, pp. 295-302, 1994.
18. Ikeuchi, K. and Hebert, M., "Spherical representations: from EGI to SAI." Proc. Workshop on 3D Representations in Computer Vision, Spinger, NY, Dec. 5-7, 1994.
19. Koivunen, V. and Bajcsy, R., "Spline representations in 3D vision," Proc. Workshop on 3D Representations in Computer Vision, Spinger, NY, Dec. 5-7, 1994.
20. Loop, C., "Smooth subdivision surfaces based on triangles," Masters thesis, University of Utah, 1987.
21. Lorensen, W., and Cline, H., "Marching cubes: a high resolution 3D surface reconstruction algorithm," *Computer Graphics*, ACM Siggraph 87 Proc., pp. 163-169, 1987.
22. Mundy, J., "Object recognition: the search for representation," Proc. Workshop on 3D Representations in Computer Vision, Spinger, NY, Dec. 5-7, 1994.
23. Nayar, S. and Murase, H., "Parametric appearance representation," Proc. Workshop on 3D Representations in Computer Vision, Spinger, NY, Dec. 5-7, 1994.
24. Schroeder, W., Zarge, J., Lorensen, W., "Decimation of triangle meshes," *Computer Graphics*, ACM Siggraph 92 Proc., Chicago, Ill., July 26-31, pp. 65-70, 1994.
25. Sedgewick, R., *Algorithms in C*, Chap. 26. Addison-Wesley, Reading, Mass. 1990.
26. Simon, D., Hebert, M., Kanade, T., "Real-time 3D pose estimation using a high-speed range sensor," CMU Tech. Report RI-TR-93-24, Nov. 1993.
27. Soucy, M., Lauendeau, D., "Multi-resolution surface modeling from multiple range views," *Proc. IEEE CVPR-92*, Champaign, Ill., June 15-20, pp. 348-353, 1992.
28. Syndesis, Inc., Product Literature, Madison, Wisc., 1993.
29. Taubin, G., "Curve and surface smoothing without shrinkage," Proc. Workshop on 3D Representations in Computer Vision, Spinger, NY, Dec. 5-7, 1994.
30. Turk, G., "Retiling polygonal surfaces," *Computer Graphics*, ACM Siggraph 92 Proc., Chicago, Ill., July 26-31, pp. 55-64, 1994.
31. Turk, G. and Levoy, M., "Zippered polygon meshes from range images," *Computer Graphics*, ACM Siggraph 94 Proc., Orlando, Fla., July 24-29, pp. 311-318, 1994.
32. ViewPoint Data Labs, Inc., Product Literature, Orem, Utah, 1994.
33. Welch, W. and Witkin, A., "Free form shape design using triangulated surfaces," *Computer Graphics*, ACM Siggraph 94 Proc., Orlando, Fla., July 24-29, pp. 247-256, 1994.

Body-centered Representation And Perception

Thomas O. Binford

Robotics Laboratory
Computer Science Department
Stanford University
Stanford CA 94305

Abstract. Intuitive notions of shape, object, and part are informative for perception and modeling. Intuitive sense of shape is a volume concept, not a surface concept; shape relates surfaces typically as ratios of diameters.
We propose a natural 3-d definition based on function. Parts are volume elements derived as a visual observable from a fundamental part definition, based on C^1 continuity of generalized cylinder (GC) primitives. Parts observable from quasi-invariants from GC primitives implement a useful object-background discrimination.
The value of such a shape representation is that the concepts of the representation map one-one into nodes in a Bayesian network for perception.

1 Introduction

Imaging has growing applications in 2-d and 3-d domains, e.g. document preparation, text and image manipulation. Much image manipulation requires 3-d. In medical imaging, 3-d data are used to build 3-d models used for diagnosis and intervention. Likewise, in entertainment, 3-d surface data are used for building 3-d models. Automatic target recognition (ATR) and site surveillance require interpretation, often from 2-d projections.

The following comments are relevant for modeling and for interpretation with 3-d data and 2-d projection data. Shape representation is only a part of object representation. The approach here is 3-d shape representation in body-centered reference frames for both modeling and interpretation. We relate shape representation to human intuition about shape. We define objects and parts based on functions and observable consequences.

Mundy raises the question of a computational definition of objects and classes to support perception, and questions the importance of parts in machine perception. In this paper we argue for the importance of parts. Much of the viewpoint expressed here comes from the AI origins of computer vision with Minsky, Papert, and McCarthy; it was legitimate to ask epistemological questions with practical consequences. Much of this point of view was described in [1].

An important issue in perception is object-background discrimination. Some success has been demonstrated in object-background discrimination of wide ranges of objects in complex scenes by discrimination of generalized cylinder (GC) parts [6]. An operational definition of parts is given below in terms of

GCs and continuity. Parts are important in object-background discrimination, among other aspects of perception.

Perception relates observed imagery with underlying objects in a scene interpretation. Object representation is one of the representations needed in a perceptual system. To make a well-founded interpretation, we need representation of the scene, sensors and observation, perceptual operations, and the decision process. That is, how do we relate pixels (and image features, edges, vertices and regions) to surfaces and volumes of objects? The alternative to a fundamental model is phenomenological correspondence between image features and objects, the training set paradigm. In the Successor system we make statistical decisions based on system models of observability and explicit representations of available information about relevant elements of interpretation.

2 Surfaces

Surface representation will be used here to mean the abstract mathematical basis for approximating a surface, at one level a surface patch in differential geometry, at another level a spline. A description or a model is an approximation of a particular surface as a parameterization of some surface representation, e.g. an instantiation of a particular second order patch.

Effective geometric models of complex objects are structural, "a divide and conquer" approach, i.e. local to global. They subdivide surfaces into simple local elements, build models of simple elements, then build up complex surfaces from simple elements by a local to global process. A second "divide and conquer" process is necessary to subdivide volumes into parts and build up objects from parts.

Local representation by continuous surface patches rests on approximation to a portion of a continuous surface [3]. The simple elements are continuous surface patches. Surface patches are represented in differential geometry as second order or third order differential patches, characterized by first derivative (tangent plane), second derivative tensor (curvatures), and third derivative tensor. The parameterization may be in cartesian coordinates or so-called intrinsic form. A surface patch is a local surface description, typically diameter-limited. We are oriented to thinking about how descriptions are estimated from data. In analyzing data, the extent of validity of the surface patch approximation is initially unknown. A surface patch is an approximation to a portion of a continuous surface. In graphics and 3-d modeling, bicubic tensor product patches are common. There are some serious problems in practical manipulation with bicubic patches; e.g. intersections of two bicubics are polynomials of order 356. We don't like them much. Implicit second order and third order patches are also considered by researchers.

Differential geometry represents only continuous surfaces effectively. Where there are discontinuities, continuous surfaces give poor approximations. Because boundaries of surfaces are important, we include discontinuities by considering semi-differential geometry, piecewise differential surfaces. In a tesselation, many patches will overlap discontinuities between two or more surfaces.

Local geometry is chosen for simplicity, limited to describing simple surface patches but not adequate to approximate complex surfaces. Complex surfaces are described by a mesh of patches, compound or composite surfaces. A mesh of surface patches extends continuous local surface patches to global compound surfaces by adding continuity conditions at the seams. These are splines, i.e. piecewise continuous patches with appropriate continuity conditions. Boundary patches add surface discontinuities. Splines can be built with various bases; the most usual are parametric polynomials, e.g. bicubic splines. Among parametric polynomials, various splines differ in their boundary conditions. There are some major problems to be solved in doing this automatically with discontinuous surfaces.

The only reason that this approach works is that objects that we work with are very simple, i.e. much simpler than continuous or even analytic. Analytic is a weak constraint, too weak to be useful in practical vision. Surfaces are smooth enough that there are few enough patches to allow a decomposition into few patches relative to pixels. The approach implicit here does guarantee useful constraints where surfaces are simple enough that the constraints apply. There are no such constraints otherwise.

Some of us might say that this is shape description because it is surface description. It is generic, i.e. a very broad class of surfaces can be approximated effectively by splines. This we call a locally-generated approximation. It does include discontinuities that are very important to humans.

3 Shape

However a spline surface description is not a shape representation that corresponds to human intuition. A surface is not the level to describe parts and objects. We need another level of "divide and conquer" to define parts and objects. Intuitive concepts of of shape appear to focus on part-whole decomposition for which parts are characterized by their dimensions and proportions. It is usual to represent a volume by its boundary, its surface. The surface and volume are not equivalent. In CT or MRI images or in finite element problems in fluid flow, thermal or viscoelastic behavior, the interior is non-uniform. Representation of a volume is not equivalent to representation of its surface because there is information in a volume representation that is not in a surface representation.

A surface representation defines a parameterized point in space. We call this a one-finger representation, like running a finger over a surface. Modeling the surface shell is adequate for computer graphics of opaque objects. In finite element problems, it is often convenient to treat the volume with isoparameter surfaces rather than voxels. Iso-parametric representations are object-centered. In a similar way, generalized cylinders are object-centered. They describe relations among surfaces, e.g. certain diameters, information that is not available in surface representations. We call these representations two-finger representations that express special diameters of the object. Voronoi diagrams express similar volume representations.

There is a similar relation between curve representations and area representations in 2-d. A representation of the boundary curve of a 2-d figure corresponds to a representation of the boundary surface of a 3-d figure. A representation of the interior of a 2-d figure corresponds to a representation of the interior of a 3-d figure. An area representation of a 2-d figure adds structure that is not in a curve representation of its boundary.

From a computational point of view, the first level of importance in object representation is the part-whole decomposition. The representation of individual parts as GC primitives has second level importance. GCs have importance in that they lead to an intuitive part-whole decomposition.

Intuitive descriptions of objects give insight into effective representation. Intuitive descriptions of complex objects begin with their part-whole structure. Geometric relations between parts describes the part-whole structure. In describing parts, their proportions are descriptive, e.g. length to width. The grade school level of drawing the human body has an ellipse for the head, the body is seven heads high, arms and legs are ellipses. A building is tall and thin. These part descriptions correspond to relations between diameters of surfaces.

In estimating volume descriptions from data, the local level is diameter-limited and data-independent, based on a view-centered image-independent covering of unknown data. The global level is complexity-limited and data-dependent, potentially bounded by the data boundary.

4 Objects, Classes, and Parts

An answer to defining classes and defining objects is to find semantics for defining meaningful similarities and less meaningful differences. Much discussion of classes starts with classification of points in R^n. Discussing classification in R^n is mostly meaningless; what it ignores is the choice of the few n parameters from among thousands of measurements of voxels or pixels. What is crucial is extraction of the appropriate descriptors that define similarity within class. Mundy quotes the Ugly Duckling Theorem of Watanabe [5]. The conclusion from that theorem is that there is no way to do define classes if all groupings of attributes are equivalent.

The fundamental definition of objects is based on function. Description of object shape by GCs reflects observables from function. GCs define similarity in a way that relates to function. One of the motivations in inventing GCs was to provide a quantitative way to describe similarities of the sort: a snake is similar to a stick. This provides a mechanism for defining classes based on their GC parameterization (structure).

Why are parts important? Parts reflect an important function, i.e. motion and articulation. Articulation particularly deals with separating out invariant subsets that isolate non-invariant aspects of shape.

Hoffman and Richards propose an observational definition of parts based on successive pairs of concavities in 2-d shapes [4]. This is a 2-d criterion that proves to be unreliable for several reasons: 1) it is a curve criterion, not an area criterion; parts in 2-d are area elements; thus it does not describe parts, even in 2-d; 2) its intent is to define parts in 3-d; the criterion is 2-d; it does not extend to 3-d; its implications for 3-d are not understood.

The criterion is not effective even for 2-d figures because effective part description is area description. If we consider the 2-d example in Figure 1 that mimics veins in a leaf, it is clear that parts relate curve elements that are arbitrarily distant on the boundary although "close" in 2-d space. In the figure, pairs of concavities 1 and 2 are not successive on the boundary curve or even close. 2-d part termination does not require pairs of concavities, although this example has symmetry that does produce paired concavities. Successive concavities are not closely related to 2-d parts. The same is true for 3-d figures; parts that correspond in 3-d space may be distant on the surface.

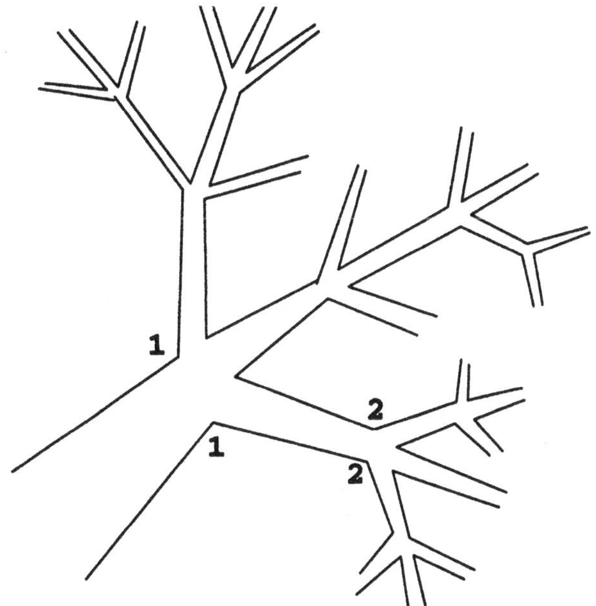

Fig. 1. 2-d Shape: Veins in a Leaf

In the 2-d figure, parts are not well defined by curve descriptions such as successive concavities. Instead, parts are determined by "opposite relations" determined by diameter relations of the 2-d figure. E.g. corresponding concavity pairs 1 and 2 are not close on the boundary of the 2-d shape.

Our alternative definition is that parts are defined by function. There are many functions of an object. One of the fundamental functions is kinematics, motion and articulation that are important in defining parts. Here we are concerned with observational consequences of function that are the basis for generating hypotheses of parts.

Our observational definition of parts is based on the observation that change of form reflects change of function. Parts are defined by continuity of GCs, i.e. continuity of volumes. Parts are volumes, not surfaces. Defining parts by continuity of surfaces would have the non-intuitive consequence that a cube has six parts; by our definition a cube has one part. A GC is defined by a cross section that is transformed as it is swept along a curve (the axis). Continuity of a GC implies continuity of the sweep and the axis. For our purpose, C^1 continuity seems appropriate.

The power of GC representation is that it recognizes and exploits continuity, i.e. smoothness, thus defining natural parts that lead to an effective part-whole description. Complex objects are decomposed into structures of parts. Complex parts are represented by complex cross sections with simple sweep functions.

5 Perception

Previous sections were done with an eye toward building models and perception or recognition. In computational geometry, representation is motivated by computational problems to implement efficient algorithms. The same is true here. We need to consider representation from the point of view of perception.

Figure 2 contrasts object-centered and view-centered approaches to recognition. View-centered approaches (aspect graph) can be summarized as: compare a set of m features from all views of all objects with all combinations of m features from an image (graph endomorphism). This approach is epistemologically weak since it has no concept of class; there is no relations among objects. It has no concept of object; there is no relation among views of an object. Every view of an object is a separate object. The approach is weak in practice because the computational complexity is prohibitive. The dominant computational complexity comes from scene complexity, i.e. the number of combinations of m features required to match from among M features in the image. The secondary complexity comes from object-view complexity, the number of object views. Our dominant objection to view-centered methods of recognition is that they accomodate very little variation in object and background. Background can change observed features of boundaries of objects. Any variation that changes observed features causes failure of view-centered methods or increases their computational complexity drastically. Segmentation errors, surface markings, articulation, variations of objects in class, obscuration all cause variation of observed features of objects that challenge view-centered methods.

Advocates of view-centered methods will argue that this is not an accurate description of view-centered matching because they include methods to restrict the number of combinations of image features and to restrict the number of object views tested. Our challenge to advocates of view-centered methods is to demonstrate well-founded, general methods rather than ad hoc methods to deal with scene complexity and object-view complexity.

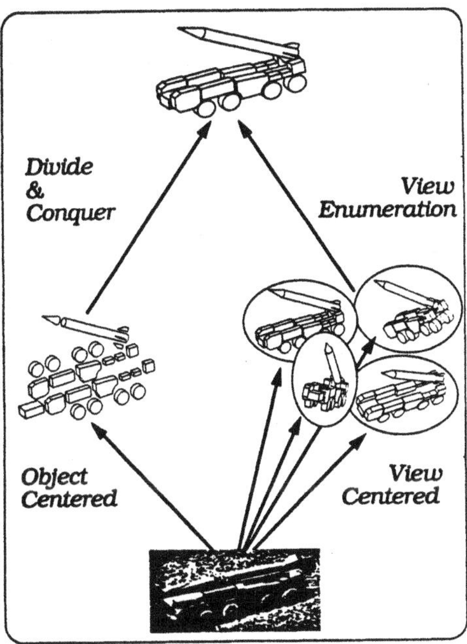

Fig. 2. Object-Centered Perception

View-centered recognition enumerates possible views and matches them against subsets of features in the image. Object-centered interpretation finds object parts (object-background discrimination), determines objects from parts, finds 3-d shapes of parts and objects then determines a subset of object models as model hypotheses.

Each combination of image features is an observed object hypothesis. In a complex image, there may be 10^{18} combinations, potential observed object hypotheses, e.g. 10,000 features taken 5 at a time. Each object view is a model hypothesis; for 10^4 objects there may be 10^6 object view hypotheses. The number of matches is the product of model hypotheses and observed object hypotheses. Our approach is to generate few observed object hypotheses, about 10^4, and few model hypotheses, about 10^2. Object-background discrimination generates object hypotheses. We have demonstrated relatively general methods for object-background discrimination and outlined the basis for extension. We have demonstrated methods for model hypothesis generation.

The prescription for object-background discrimination is: find GC parts based on quasi-invariant relations among their cross section edges and limbs.

6 Perceptual System

We introduced Bayesian networks that integrate the physics of observation with the geometry of interpretation [2]. Practical computation on Bayesian networks is an important topic since these networks are very large for non-trivial images. Dynamic instantiation and localizing network computations are crucial. Object-centered Bayesian network methods depend on avoiding combinatorics by embedding variation implicitly by using parameterized models in hierarchical network models with local hypotheses, rather than enumerating unrelated combinations with global hypotheses.

In the Successor system, there is a one-one correspondence between the geometric hierarchy and nodes in the Bayesian network. That correspondence enables automatic translation of object models to recognition networks. The implementation uses Classics, a class system based on a mathematical hierarchy.

Figure 3 illustrates components of the Successor system. Object-centered recognition in 3-d matches 3-d models with observed 3-d structures of objects from images. The goals are generic object recognition with low complexity. Object-centered recognition requires solutions to the problems of: 1. effective segmentation; 2. discriminating object from background; 3. constructing 3-d object descriptions from data; and 4. generating object hypotheses from 3-d object descriptions (indexing).

Object-centered recognition enables generic object recognition, a mechanism for perception of parts, objects, and object classes. Object-centered recognition is based on a divide-and-conquer mechanism that deals with parts of the image that are complexity-limited. Appearance-based methods are based on an enumerate-and-test paradigm.

A new Wang-Binford family of edge segmentation and linking operations produces linked edges for uniform surfaces in complex imagery. When segmentation for uniform surfaces is complete, there will remain problems of estimation of vertices and effective segmentation in texture. These same methods are important for segmentation of 3-d data. This generic segmentation is model-based in that it is based on generic models of the physics of image formation and the local geometry of surfaces (semi-differential geometry).

Discriminating object from background has been demonstrated by finding generalized cylinder (GC) parts. Spatial organization of curves, based on quasi-invariant relations for self-similarity of cross sections of generalized cylinders and for limbs, provides a solution to the problem of object background discrimination in complex imagery. Discrimination of object from background depend on definitions of part, object, and object class used in this work; they are plausible for perceptual mechanisms.

Objects are formed as joints among GC parts. Quasi-invariants enable joining of parts, and estimating shape of parts and object, i.e. estimation of 3-d object proportions from image data. Quasi-invariants enable estimation of combinations of Euclidean invariants that give quantitative shape descriptors for parts and relations between parts. Quasi-invariants are local invariants.

Scene interpretation is formulated as solution of a Bayesian network. A Bayesian network for a complex scene has a depth of about 15 levels, with a width of about 10,000 at the level of extended curves. Bayes nets implement a system of uncertain constraints. Solution of the Bayesian network fights potentially great computational complexity. Mechanisms for efficient solution include dynamic instantiation of supported hypotheses, indexing based on shape for hypothesis generation, lazy evaluation, and utility-based control of evaluation.

The natural complexity is multiplicative. Each element has variability. Jointly, they have independent variability that leads to multiplicative complexity. By dividing the complex structure into parts, the complexity can be the sum of the complexities of the parts and complexities of combinations of parts at joints between parts. If branching has fixed complexity, the net complexity is additive, i.e. logarithmic.

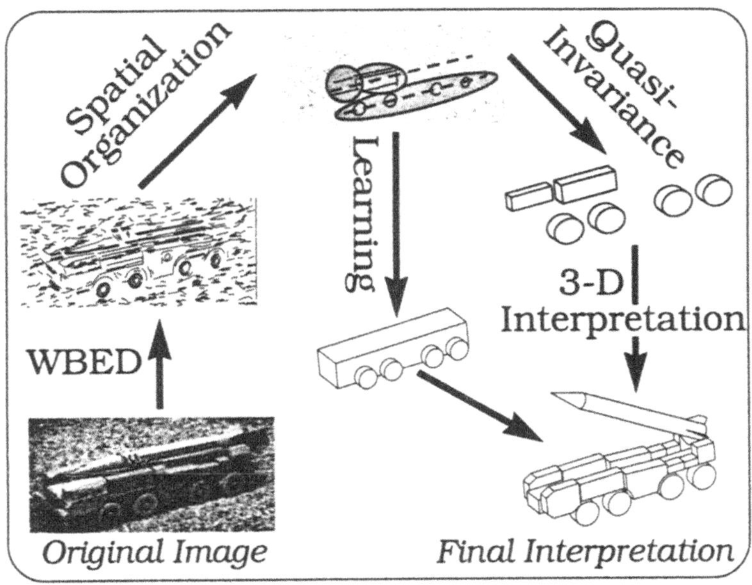

Fig. 3. The Successor System

Components of the Successor system include: the Wang-Binford edge detector that produces extended edges, spatial organization that uses relations among extended edges to generate object part hypotheses (figure-ground discrimination); quasi-invariant generation of 3-d shape hypotheses for object parts; learning to build a database of objects and expectations; and 3-d interpretation by Bayesian networks.

References

1. T.O. Binford. *Survey of Model-Based Image Analysis Systems.* Robotics Research, Vol. 1, No. 1, March 1982.
2. T.O. Binford, Tod S. Levitt, W.B.Mann. *Bayesian Inference in Model-Based Machine Vision.* Proc Workshop on Uncertainty in AI, AAAI87, 1987.
3. T.O. Binford. *Shape.* Encyclopedia of Artificial Intelligence, Second Edition, 1992, Wiley.
4. D.D. Hoffman and W. Richards. *Representing smooth plane curves for recognition.* Proc AAAI, 1982.
5. J. Mundy. *Object Recognition: The Search for Representation.* In these proceedings.
6. H. Sato and T.O. Binford. *Finding and Recovering SHGC Objects in an Edge Image.* CVGIP: Image Understanding; V 57, pp 346-358, 1993.

The Challenge of Generic Object Recognition

Mourad Zerroug and Gérard Medioni

Institute for Robotics and Intelligent Systems, University of Southern California, Los Angeles, CA 90089-0273

Abstract. We discuss the issues and challenges in the development of generic object recognition systems. We argue that high-level, volumetric part-based, descriptions are essential if we want to recognize objects which are similar but not identical to pre-stored models, under wide viewing conditions, and to automatically learn new models and add them to our knowledge base.We discuss the representation scheme and its relationships to the description extraction, recognition and learning processes. We then describe the difficulties in obtaining such descriptions from images and outline steps for robust and efficient implementations. We also demonstrate the viability of the arguments by reporting on recent progress.

1 Introduction

Recognizing 3-D objects in an image is a complex problem along multiple dimensions. One has to extract relevant "descriptions" from the image, "match" them with one or more of the objects from a previously constructed database, "learn" new objects for inclusion in the database. While these tasks are necessary, no consensus has emerged regarding the choice and level of features (2-D or 3-D), the type of indexing used, the matching strategy, and the order in which these tasks should be performed.

Most existing recognition systems use low-level representations such as points, lines and collections of these. Since these descriptions only carry minimal semantic contents, the indexing scheme cannot be very efficient, and the computational burden falls on the localization phase which uses geometric constraints. This can only succeed when the database consists of exact models. Furthermore, learning can only occur with user supervision. Instead, we argue that the *only* way to solve *all* the previous tasks is to generate and use high-level representations in terms of volumetric parts, their hierarchy and their arrangement. We further propose that part-based representations based on subsets of Generalized Cylinders constitute suitable means for shape abstraction of large classes of complex 3-D objects, and that these descriptions can indeed be generated from 2-D images under imperfect and realistic viewing conditions.

This paper discusses the challenging sub-tasks that need to be accomplished in order to build systems capable of describing complex objects from monocular intensity images, and the elements of a recognition engine to manipulate and reason about the resulting hierarchical descriptions. The challenges of extracting structured descriptions from real intensity images are to achieve robust and efficient segmentation and grouping in the presence of spurious information due to noise, surface markings and shadows, and missing information due to occlusion and feature fragmentation. The choice of the representation scheme itself is crucial because it is this scheme which provides the needed generic constraints for this process.

We organize the paper as follows. In Sect. 2, we motivate the importance of volumetric part-based representations in generic object recognition. In Sect. 3, we discuss the tasks and issues that need to be dealt with in implementing robust and efficient segmentation and recognition systems using part-based descriptions. In Sect. 4, we describe recent progress on the extraction of such descriptions from a single imperfect image. We conclude in Sect. 5 by discussing the limitations of the approach and open problems that have yet to be solved.

2 Representation Issues

2.1 Representation Level

Object recognition is an ubiquitous task in computer vision. The necessary components of an object recognition system are description, matching and learning.

The first task, *extraction of descriptions*, consists of an interpretation of the image data into meaningful entities which could correspond to objects or portions of objects. The second task, *matching*, consists of assigning an identity to the extracted descriptions, a process which involves accessing stored models and comparing them with the image descriptions. The *learning* task consists of acquiring objects not previously known to the system and describing similarities and differences with existing objects. This third component is generally ignored in many systems as it is performed by the user.

All these tasks rely strongly on the representation scheme used. It is this scheme which affects the choice of the strategies for the description process, the way models are accessed and compared, and the methods for learning. The nature of the representation also affects the complexity of each of the above tasks. Thus, choosing an adequate representation scheme is one of the most important design issues.

In restricted domains, such as a factory floor with few models given by their exact geometry, it is possible to ignore the learning task by creating and storing models off-line. The description can be in the form of low-level primitives such as points and lines. Matching proceeds by geometric verification, or alignment [8].

This approach, unfortunately, does not scale up to less restricted domains, as low-level representations may place a heavy burden on the image-model matching stage. This is because they carry minimal semantic contents and thus their discriminatory capabilities are low. This weakness does not affect the image-model matching only. In realistic viewing conditions, an image may be noisy and cluttered with objects and background. Extracting relevant descriptions based on low-level features cannot be done without exploring huge search spaces, or without making use of object models. In this case, however, image features are not meaningful unless they are successfully matched with some stored, typically exact, object model.

Motivation. There are many reasons why it is important to develop vision systems capable to derive meaningful descriptions in an image-driven fashion; i.e. without using specific objects. Such systems are necessary when a complete database is not available, or cannot reasonably be constructed. They are also important where the application requires the capability to learn new objects or simply derive descriptions for objects never seen before. These systems must be able to detect relevant objects and describe them adequately. Consider, for example, the image of Fig.1. In this case, we would like to develop systems capable of detecting objects from the clutter and even classify them as "desk", "book", "monitor", etc., without having previously exposed the system to these specific objects (in fact, it is unlikely that all readers of this paper have seen the particular objects of the scene, though no one would have difficulty interpreting it).

Fig. 1. An image of a cluttered scene. We can readily "recognize" many objects, even if never seen before.

The Problem is not Decomposable. To solve this problem clearly requires generic capabilities to identify features of interest, group the ones projecting from the same object, and derive relevant descriptions for them, while at the same time filtering out false interpretations. This high-level reasoning process has been demonstrated largely on "perfect" data, as presented in [1, 6, 11, 15], for example. Unfortunately, this knowledge does not transfer to real images as it bypasses the central difficulty: description and segmentation are inter-dependent and cannot be performed in sequence. Our belief is that generic capabilities and imperfect images are inseparable, except in restricted scenarios.

The Need for High-Level Representations. In order to achieve these capabilities, it is essential that high-level representations be used. These high-level representations would provide elaborate information about the shape (geometry) of an object, or a meaningful portion of it, such as how it is composed, how it changes in some directions, etc. They also provide means to decide which subsets of the image constitute relevant objects, of course, with respect to the representation scheme itself. Thus, high-level representations are useful both to capture elaborate shape properties and to extract relevant descriptions from the image.

2.2 Choice of High-Level Representations

As with any other problem, a solution to the problem of extracting generic descriptions requires an adequate representation scheme. Such a scheme becomes itself a *generic* model from which description methods are derived. There is no universal representation scheme which can deal with arbitrary objects. Consequently, it is useful to consider *classes* of objects. Part of addressing the above challenge is to use generic shape representations which capture a large class of objects.

Desirable Properties. The desirable characteristics of a representation scheme for the purposes of generic shape analysis and recognition are well known [12, 15]. They include *local support*, *stability*, *discriminatory* power and *unambiguity*. These characteristics are desirable both to represent objects in the object (or representation) space and to recover descriptions from an image (or set of images) under realistic imaging effects. Below, we focus on some of their aspects only, and the discussion deliberately supports the above characteristics for both purposes (representation and recovery of descriptions) simultaneously.

The stability of a representation is with respect to changes in the scene properties including small irregularities on an object's surface, sensor noise, illumination changes and, especially, changes in viewpoint. We would like the descriptions recovered for the same object to be "similar", even if its appearances are substantially different. This suggests that the shape representation scheme itself provides viewpoint tolerant image features from which final descriptions are built.

The search for such features has triggered much research in the community. An instance of which is the use of geometric invariants to characterize point-sets and certain curves [13]. Another instance is the derivation of higher-level viewpoint- tolerant properties of generic shape representations schemes, such as generalized cylinders [3]. To use point-sets requires visibility and selection of key points on an object. Even if certain invariant signatures for a number of points exist, they may not be stable or discriminatory enough. Building higher-level structures is necessary for stability and expressiveness. This constrains us to use a high-level representation scheme from which such structures can be identified in the image.

Besides the above characteristics, one would certainly wish to use a representation which allows the control of what is represented in an object. For each task, and at each of its steps, not all what can be represented in the object is needed. Rather, some tasks may require only coarse-level shape information whereas others some finer details. Representing an object's shape hierarchically is a natural way to control the representation scale.

Part-Based Representations. The complexity of viewed objects forces us to represent them in a way which makes their interpretation tractable. Objects are often made up of a number of components. This is the case for both man-made and natural objects. This component (part-based) aspect stems from functional factors, as well as manufacturing constraints (a keyboard is not made all at once).

For some objects, a part-based representation may be difficult (such as bushes, terrain, face features), but for many others it is evident. For example, we have no hesitation in describing a teapot as consisting of a main body, a spout, a handle and a lid; similarly for an airplane, a keyboard, most mammals, etc. For such objects, an explicit representation in terms of parts is advantageous. It allows us to capture their rich structure which, as discussed above, is important for the image interpretation process. It also allows us to represent them even if some components are missing. A part-based representation is also useful for reasoning about the function of an object [18, 21].

Such a representation is by design hierarchical, where parts can themselves be decomposed into sub-components, etc. This makes it convenient to control the description scale during image interpretation and during object recognition. Thus, it becomes possible to judge the similarity of objects in a coarse-to-fine fashion, a highly useful capability, especially in object learning. There exist, of course, many difficulties in generating such descriptions, as indicated later.

Volumes vs. Surfaces. The two main representations for capturing the structured nature of an object are surfaces and volumes. Volumetric representations provide the correct level of abstraction for viewpoint tolerant descriptions. In fact, volumes lead themselves more naturally than surfaces to abstract an object's shape. For example, a stick figure using only the axes of generalized cylinders is often a good abstraction of the structure of a complex object (a horse for instance; see Fig.2). Abstraction from a surface-based representation is unclear in this case. Surface attributes can be easily determined from a volumetric description, whereas the reverse is not as direct. The differences between two objects' shapes may not be directly accessible at a symbolic level when using surface-based descriptions. In fact, surface-based representations of otherwise "similar" objects can be drastically different. A volumetric representation provides

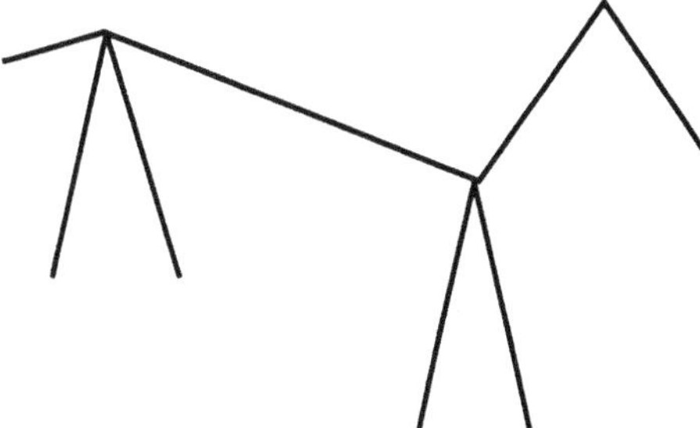

Fig. 2. Shape abstraction is more natural when volumetric part-based descriptions are used (here the parts are represented by their major axes).

means to analyze shape similarities (or differences) along several criteria (for each of the intrinsic functions of the volumetric description). By decoupling the representation into a set of attributes, it becomes easier to compare objects and decide in which ways they are similar or different. Consider the two objects of Fig.3. Despite their surface differences, we are still able to say that they have a similar structure, thanks to the volumetric parts composing them. Furthermore, the corresponding parts of the objects have quite different surface representations, a cylindrical and a piecewise polyhedral one for the "heads", for example. But we are still able to judge their similarities (sizes along the parts axes and cross-sections), and differences (circular vs. polygonal cross-sections). This type of analysis is very desirable for recognition and even learning. In this case, the representation is symbolic and discriminative.

Fig. 3. Comparison of shapes is easier when volumetric representations are used.

We find *generalized cylinders* (GCs) to be particularly suited in that they have many of the above-mentioned characteristics and they capture a large number of both natural and man-made objects. They have been used in many occasions, in past and on-going work, for the purpose of addressing one or more of the steps needed for generic object recognition [5, 14, 15, 17, 20, 22, 24]. However, without any restrictions, GCs are too general to be practical. Rather, *subclasses* of GCs are needed which are defined well enough to be practical yet general enough to capture a large set of objects. Examples of such subclasses of GCs are *straight homogeneous generalized cylinders* (SHGCs) [17, 20, 24], *planar right constant generalized cylinders* (PRCGCs) [22], and *circular planar right generalized cylinders* (circular PRGCs) [23].

Segmentation into Parts. An issue that naturally arises at this point is the *definition of a part.* One could justifiably claim that the decomposition into parts is not a property of the object, but of our perception of it and, therefore, there might not be a rigorous way to define parts. However, in the case of GCs, a definition which was adopted since early work [15], consists of delimiting parts at discontinuities of any of the intrinsic functions, the axis, the sweep or the cross-section. An example is shown in Fig.4. In this example, the segmentation of the "handle" (left-most part) is based on the discontinuity of its axis with the other axes (Fig.4 middle), whereas the segmentation between the top part and the central one is based on the (first order) discontinuity of the sweep function (Fig.4 right), their axes being continuously connected. This definition can be used both to partition a 3-D model and to segment the image of an object. The discontinuities in the intrinsic functions create discontinuities in the image, such as in symmetry relationships, and thus provide generic criteria by which to decompose shape. This captures, for example, the natural decomposition of parts at articulation joints where

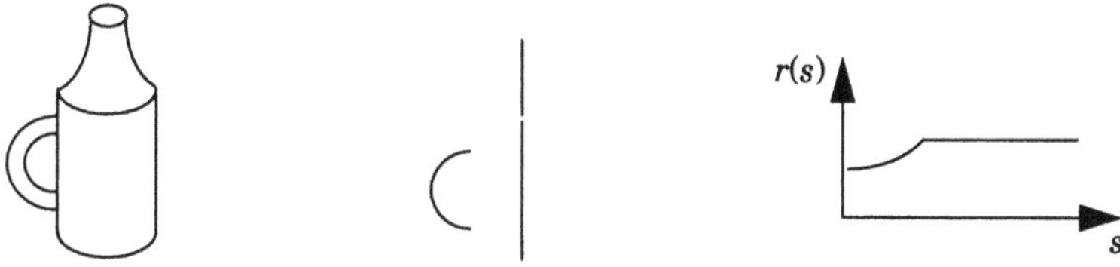

Fig. 4. Segmentation into parts can be based on discontinuities in the intrinsic functions, such as axis discontinuity for the handle (middle), and first order discontinuity of the sweep function (right).

due to the relative motion of parts, discontinuities in their shape are produced (in this case, axis orientation discontinuities, as between the upper and lower legs of a human).

Another issue is given a certain definition of a part, in terms of 3-D properties and image observables, to what extent can it faithfully capture the shape properties of an object. It is true that part models, such as GCs, are idealized in that they typically consist of smooth intrinsic functions and thus may not capture all shape details of an object. However, such part models, in the worst case, can still encode the coarse structure of the object's shape. This structure is often powerful enough to differentiate between objects and is thus very useful in recognition.

Psychological Evidence. Part-based representations, based on volumetric primitives, appear to be psychologically plausible. Studies of human perception and recognition of 3-D objects from their line drawings show a strong influence of the structured nature of objects on the perception performance of subjects [2]. The crux of the findings is the fact that simple geometric primitives (geons, analogous to simple GCs) and their interactions provide substantial viewpoint invariant object shape information. The geons are volumetric by nature and their perception has been found to be stable under rotations in depth, contour deletions and even occlusion.

This theory does not account for textured or highly irregular objects. Nonetheless it accounts for a large set of man-made and even natural ones (such as animals). It also accounts for new objects, not known a priori to the observer.

3 Implementation Issues

In such an approach to generic object recognition, several tasks need to be solved. They can be viewed as forming two functional modules: one for computing high-level generic part-based descriptions from the image, the other for recognizing the detected objects (descriptions) from the database. This approach has a long history in computer vision [3, 5, 12, 15, 16, 24]. Although they have been studied by many, the main criticism part-based descriptions have received is the difficulty of recovering them from a real image; i.e. solving the segmentation/description problem. Despite this difficulty, notable progress is being made. We discuss below what we view as essential tasks in segmentation and recognition of part-based descriptions and also report on relevant recent progress.

3.1 Computing Part-Based Descriptions

Deriving generic descriptions from an imperfect image (or images), in a data-driven fashion, is a challenging problem. In fact, it is one of the most challenging problems in computer vision. A promising methodology is to adequately implement principles of perceptual grouping. But the difficulties lie in the derivation of grouping (segmentation) and description methods which work not only on imperfect images but for large classes of objects as well.

Organization. We view the description module as consisting of a bottom-up process. However, this process makes use of expectations about image properties, which can be thought of as top-down generic knowledge.

An important aspect of this process is its organization. We do not believe it can be solved at a single feature level, for example by finding the best edge detector or boundary grouping method. There is a direct link between the hierarchical nature of the representations and the organization of the description process. This latter should proceed in successive stages where features are detected then grouped to form higher-level descriptions. The feature groups become themselves features of a higher level type which in turn can be grouped, and so on. Each level increases the scope of the interpretation process by building a geometric context based on previous feature groups. Thus, feedback loops can be used from higher levels to lower ones in order to refine the features or their groupings based on the information gathered from the increased scope.

Figure 5 illustrates this organization. The description module starts from an intensity image and ends at a level where structured representations in terms of parts (generalized cylinders) and their relationships are obtained. This representation includes information about the 3-D shape of each complex object so detected.

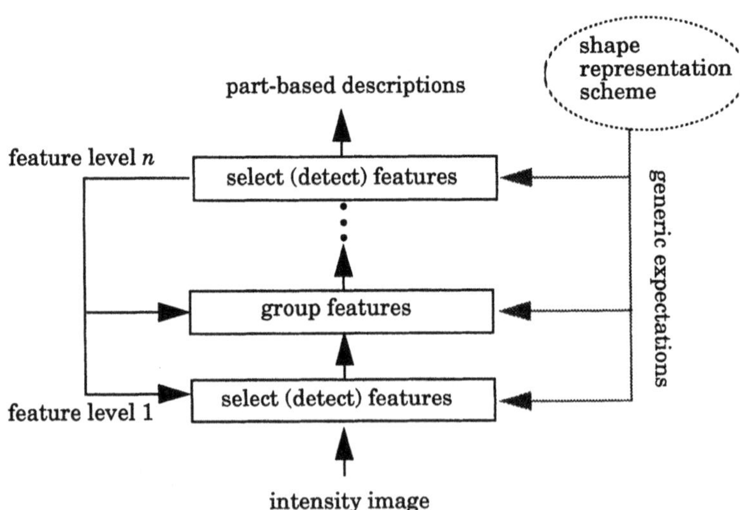

Fig. 5. Hierarchical organization of the description extraction process.

Tasks. The tasks that need to be performed at each feature level are:

- extraction of features
- grouping of features
- selection (and refinement) of feature groups

Each of the above tasks constitutes a real challenge in itself. Important issues here are which features to extract, what criteria to use in grouping them into meaningful clusters and how to know which ones are indeed meaningful and which are not.

The role of the generic shape representation scheme in this process is crucial. Knowledge of this scheme allows the derivation of generic constraints (in the form of expectations) in most of the above tasks, especially as we go higher in the hierarchy, to overcome noise and other image imperfections. Part of these constraints are image properties that any element of the class of objects captured by the shape scheme must satisfy if it is observed. In using these properties to extract object descriptions, the extraction process takes explicit account of the shape representation (the desired output of the system) including its dimensionality. In the case of 3-D representations, the expectations are in the form of *projective properties* which bear the 3-D attributes of the shape of the object. Figure 6 illustrates this view. The "3-D object" projects onto the "image object" (a set of features to be extracted). The 3-D representation scheme (based on generalized cylinders) projects onto a set of "projective properties". The segmentation and description process makes use of the projective properties to extract relevant descriptions from the image features (finds image features which satisfy these properties). This results in image (projective) descriptions from which 3-D attributes of the detected object's shape are inferred.

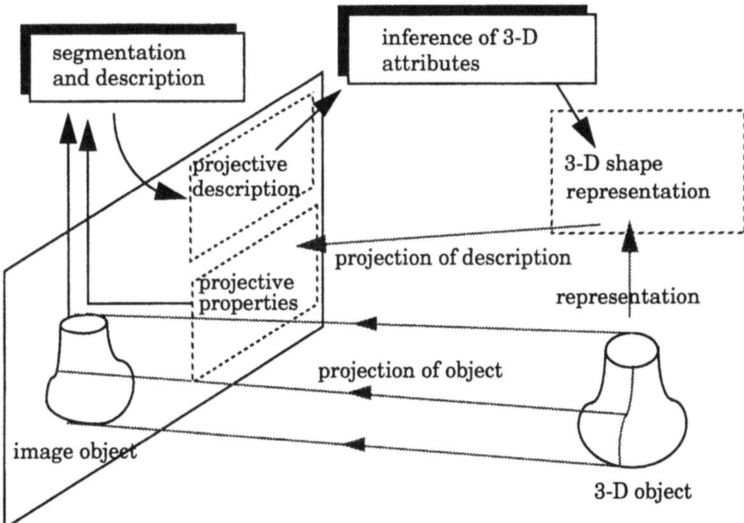

Fig. 6. Illustration of the description extraction process.

A Hierarchy. The typical hierarchy used in past and current work for the description process includes:

- boundaries
- surface patches
- volumetric parts
- compound objects

The tasks at these levels are all inter-related as we believe a robust system should perform all levels of grouping in a concurrent manner. Unfortunately, in most systems the process is linear with little (or no) feedback, making higher levels sensitive to errors at lower levels.

For example, no working (robust and efficient) boundary grouping system has been demonstrated yet. Most of the work has either outlined some of the grouping criteria and/or implemented them in some prototype system demonstrated on few (real) images (e.g. [10, 9]).

Some progress has been made in going beyond "signal" edges, detecting perceptual edges by enforcing continuity and co-curvilinearity generic constraints. While they significantly improve the detection of contours, they cannot be expected to "solve" the contour extraction problem, as they miss the importance of higher- level structures in the segmentation (and description) process. Thus, we believe it is important not to be over-ambitious in the design of "continuity" criteria solely based on boundary information. The more promising direction is to exploit feedback from higher-level processes.

Once partial contours are extracted, the next step is to infer surface patches from them. The most powerful constraint is that of symmetry. An important issue is which symmetry relationships to use. To serve the right purpose, they should have the characteristics outlined for the shape representation scheme. Symmetries which are chosen intuitively may not serve the purpose as their information contents, about the presence of relevant scene objects, is unknown. Rather, they should be viewpoint tolerant properties of the shape representation scheme itself. One would also expect these properties to be stable not only to imperfections in object shape but to approximations of the viewing geometry as well.

There have been several efforts on this issue, most notably on the derivation and use of invariant and quasi-invariant properties of certain generalized cylinders. For example, a perspective invariant property of SHGCs derived by Ponce et al. [17] is the intersection of tangent lines at corresponding (symmetric) points of the outline is a straight line which is the projection of the SHGC axis. Quasi-invariants are looser versions of invariants which have a small range of variation over a large portion of the viewing sphere [4]. For example, a quasi-invariant property of circular PRGCs derived by Zerroug and Nevatia [23] is that the segments joining the projection of co-cross-sectional points of the outline and the locus of their mid-points are "very close" to forming a right ribbon under "most viewing directions". Other symmetry relationships used are parallel and skew symmetries [22].

These properties apply to certain subclasses of GCs and more research is needed to derive other (more general) properties. For this, we view quasi-invariance as a central element because invariant properties are rare or so general as to be useless.

The inferred patches need to be pieced together to form volumetric part hypotheses. This process involves the merging of surface fragments of the same part surface together as well as different surfaces of a part into a single part hypothesis. The grouping criteria must be rigorous and derived from the properties of the shape scheme.

Similarly, the object-level grouping should also exploit the projective properties of the shape representation scheme for grouping different parts which are likely to belong to the same object into a single object hypothesis and inferring the 3-D shape attributes (qualitative or quantitative) of each object.

Summary. In summary, knowledge of the shape representation scheme determines the capabilities of the system, but also its limitations. The system is not capable to deal with objects which cannot properly be represented by the chosen scheme. The capabilities considered in this discussion are of interest because the scope of part-based representations using Generalized Cylinders is large. They are also of interest because the conditions they address, imperfect viewing conditions, are realistic. A central, challenging, issue is of course how to achieve all the above tasks in an efficient and robust manner. Despite past and current achievements on the problem, much still remains to be done in order to build robust and efficient systems for extracting part-based descriptions.

3.2 Recognition Using Part-Based Descriptions

Once "good" descriptions (i.e. part-based, volumetric, hierarchical) are extracted from the image, we need to define a recognition engine to manipulate them. Each new object needs to be stored into the database in such a way that it can be retrieved efficiently when a candidate description is proposed for recognition. We therefore identify two issues:

- the organization of the database (indexing based on high-level repre- sentations), and
- the retrieval methodology (matching high-level representations)

One possible solution to these problems is to organize the database hierarchically also, using the structure of the descriptions and the local description of parts. This allows the database to evolve dynamically with minimum cost to update the index. This is schematically illustrated in Fig.7.

The retrieval step needs to proceed from partial matches to take into account possible occlusions and uncertainty. A partial match, based on the connection structure and the aggregation of dissimilarities between parts, is computed incrementally level by level between the new shape and the possible candidates. The combination of the incremental process with the hierarchical indexing effectively decreases the number of shapes processed at each step, therefore dramatically reducing the average complexity of the retrieval. The selected retrieved shape(s) are used to give a classification for the new shape.

We have implemented the core of this recognition engine [7] for 2-D shapes descriptions as produced in [19]. Experimental validation indeed confirms that retrieval

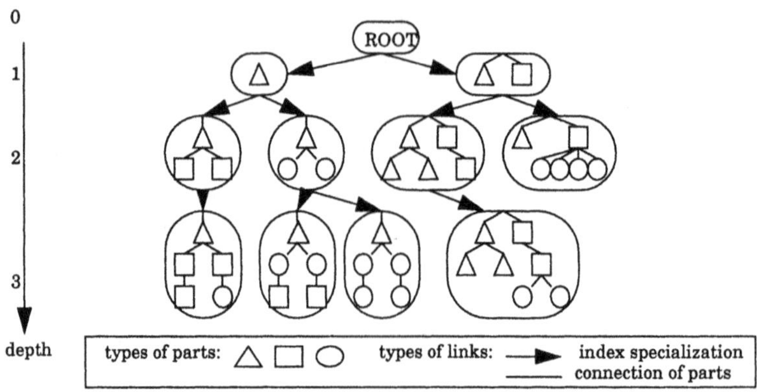

Fig. 7. Hierarchical organization of the database for dynamic and efficient access.

time is nearly independent of the number of objects stored in the database, even with thousands of entries.

It should be noted that the described approach does not obviate the need for geometric reasoning and pose estimation, but rather proposes to invoke them once the object has been recognized. The purpose is then to estimate pose, but also to emphasize differences with existing stored models.

4 Recent Progress

In [24] the approach described in Sect. 3.1 has been adopted for extracting structured descriptions based on certain sub-classes of generalized cylinders: SHGCs, PRCGCs and circular PRGCs. The method addresses the hierarchy of features discussed in that section. To give an example of the working of the method, a sample image is shown in Fig.8 together with some of the intermediate steps, shown in subsequent figures, leading to the final volumetric part-based descriptions.

At the *curve level* grouping, relatively short gaps are bridged between linked sequences of edges (Fig.8 right) extracted from the input intensity image (Fig.8 left). At the *surface patch level*, the detection of surface patches uses a set of generic invariant and quasi-invariant properties of the contours of SHGCs, PRCGCs and circular PRGCs. Figure 9 (left) shows sample surface patches hypothesized as being fragments of SHGC surfaces. The basis for hypothesizing these latter are invariant properties of SHGCs relating (hypothesized) cross-sections and the side boundaries (outlines of the surface). These properties allow the determination of corresponding points of the side boundaries (projections of points lying on the same 3-D cross- sections), based on the visible cross-section, and of the projection of the straight SHGC axis in the image. Simple tests for generating SHGC surface patches, thus, include the collinearity of points which are the intersections of lines joining symmetric points on different cross-sections along the area between two side boundaries [25]. Notice that this process results in both correct and false hypotheses since no notion of part is yet formed at this stage and due to errors and thresholds in judging the collinearity of points. Figure 9 (right) shows PRCGC and circular PRGC surface patches hypotheses formed based on other (quasi)invariant properties of curved-axis parts [25].

Fig. 8. A sample image of a complex curved object to be segmented in the presence of noise, clutter and partial occlusion.

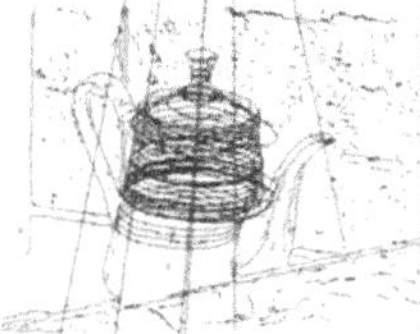

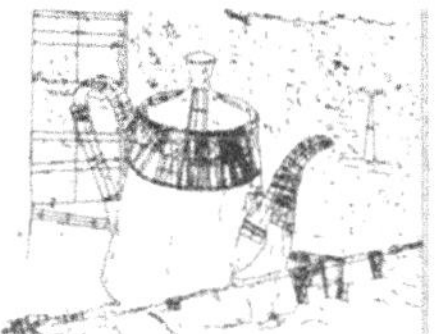

Fig. 9. GC surface patches hypothesized from the boundaries of Fig.8 based on the invariant and quasi-invariant properties of SHGCs (left), PRCGCs and circular PRGCs (right)

The grouping of surface patches uses compatibility criteria based on the similarity of their image descriptions. For example, SHGC surface patches which have collinear axes are good candidates to be merged into a single part hypothesis (see the top and bottom halves of the teapot lid in Fig.10). The (grouped) part hypotheses are subsequently verified for global consistency including valid closure at their ends.

Verified parts are then analyzed at the object level for possible interactions which might suggest the existence of joints between them. The detection of joints is based on generic structural properties consisting of a catalog of lawful junctions and their relationships. Connected parts are used to hypothesize compound objects, graph structures whose nodes are parts and whose arcs are the joints between parts. The compound object resulting from this process on the previous example is shown in Fig.11.

This implementation does not exploit feedback loops but uses a least commitment strategy at each level, leaving critical verification tests to higher-levels. The implemented system serves as an illustration of the current performance, and does not claim to solve all problems.

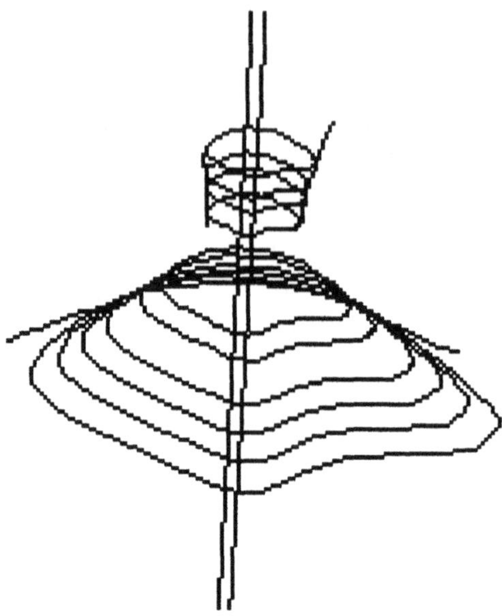

Fig. 10. Example of surface patch level grouping: two SHGC surface patches having "collinear" axes determined from the properties of SHGCs.

5 Limitations and Open Problems

Any shape representation scheme has its limitations. Part-based representations using generalized cylinders are suited to capture the coarse structure of objects made up of basic primitives (a very large set). However, they are not suited to represent certain objects which are better described by statistical features, for instance rocks. Even for objects which can be approximated by generalized cylinders, the resulting descriptions may not be sufficient to capture fine aspects of the shape. Thus, generalized cylinders need to be complemented with other representation schemes which are better suited to capture these details.

The actual extraction of part-based descriptions from realistic images in unrestricted environments still remains a largely unsolved problem. Some of the issues that need to be solved have been mentioned in this paper. They include the use of non-linear hierarchies (with feedback between description steps as well as partial image-model matches), and the use of rigorous methods.These should contribute to robustness as well as efficiency.

Most work, so far, has relied on a single cue to perform segmentation into part-based descriptions, typically boundaries due to their geometric aspect. Despite the powerful nature of boundaries alone, we believe it is also important to exploit several cues in order to achieve robust and efficient methods. For example, intensity information might be used together with edges in (rapidly) localizing features of interest in the image. The use of other cues will undoubtedly be accompanied with additional assumptions. In order to be realistic, these assumptions should be generic, not specific to particular reflection models or combinations of viewpoint and illumination for example. There is

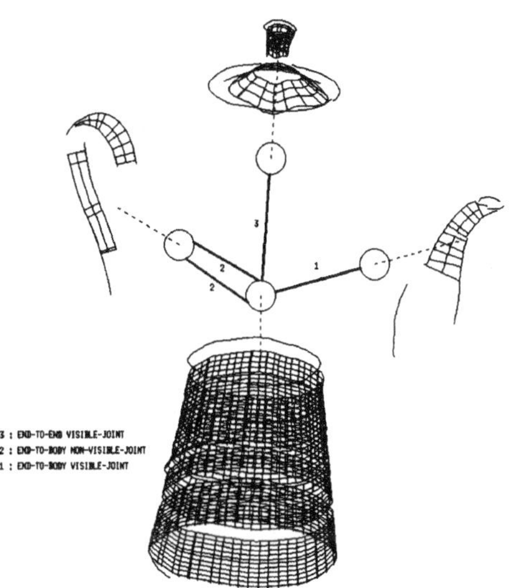

Fig. 11. Resulting part-based description in terms of subclasses of GCs from the image of Fig.8.

much to explore on how to best exploit the image(s) contents for generic extraction of object descriptions.

The recognition issues need to be explored in much greater detail. While everyone agrees that high-level descriptions reduce the complexity of the indexing and matching steps, the methods need not be ad-hoc. More work is needed beyond the largely intuitive methods proposed so far. For example, it is important to derive optimal ways to use the hierarchical descriptions in indexing and matching which account for possible segmentation or description errors. These tasks should selectively use the rich descriptions in order to achieve optimal performance in each situation. The indexing mechanisms and the matching strategies need to be considered as serious challenges, especially if their results are to be fed back to the segmentation process.

References

1. Bergevin, R., Levine, M.D.: Generic Object Recognition: Building and Matching Coarse Descriptions from Line Drawings. IEEE Transactions PAMI **15** (1993) 19–36
2. Biederman, I.: Recognition by Components: A Theory of Human Image Understanding. Psychological Review, **94** (1987) 115–147
3. Binford, T.O.: Visual Perception by Computer. IEEE Conference on Systems and Controls (1971), Miami.
4. Binford, T.O, Levitt, T.S. and Mann, W.B.: Bayesian Inference in Model-Based Machine Vision. Proceedings of AAAI Uncertainty Workshop (1987)
5. Brooks, R.A.: Model-Based Three Dimensional Interpretation of Two Dimensional Images. IEEE Transactions PAMI **5** (1983) 140–150
6. Clowes, M.B.: On Seeing Things. Artificial Intelligence **2** (1971) 79–116
7. Francois, A., Médioni, G.: Hierarchical Indexing for Generic Shape Recognition. USC-IRIS Technical Report, August 1994

8. Grimson, W.E.L.: *Object Recognition by Computer: The Role of Geometric Constraints.* MIT Press, Cambridge, MA 1990
9. Guy, G., Médioni, G.: Inferring global perceptual contours from local features. Proceedings of the CVPR (1993)
10. Lowe, D.G.: *Perceptual Organization and Visual Recognition.* Kluwer Academic Publishers, Hingham, MA. 1985
11. Mackworth, A.K.: Intrepreting Pictures of Polyhedral Scenes. Artificial Intelligence **4** (1973) 121–137
12. Marr, D.: *Vision* W.H. Freeman and Co. Publishers, 1981
13. Mundy, J.L., Zisserman, A. editors.: *Geometric Invariance in Computer Vision.* MIT Press, 1992
14. Mundy, J.L., Huang, C., Liu, J., Hoffman, W., Forsyth, D.A., Rothwell, C.A., Zisserman, A., Utcke, S., and Bournez, O.: MORSE: A 3-D object recognition system based on geometric invariants. Proceedings of Image Understanding Workshop (1994) 1393–1402
15. Nevatia, R. and Binford, T.O.: Description and Recognition of Complex Curved Objects. Artificial Intelligence **8** (1977) 77–98
16. Pentland, A.: Recognition by Parts. Proceedings of ICCV (1987) 612–620
17. Ponce, J., Chelberg, D. and Mann, W.B.: Invariant Properties of Straight Homogeneous Generalized Cylinders and their Contours. IEEE Transactions PAMI **11** (1989) 951–966
18. Rivlin, E., Dickinson, S.J. and Rosenfeld, A.: Recognition by Functional Parts. Proceedings of Image Understanding Workshop (1994) 1531–1539
19. Rom, H. and Médioni, G.: Hierarchical Part Decomposition and Axial Shape Description. IEEE Transactions PAMI, **13** (1993) 973–981
20. Sato, H. and Binford, T.O.: Finding and Recovering SHGC Objects in an Edge Image. Computer Vision Graphics and Image Processing **57** (1993) 346–356
21. Stark, L. and Bowyer, K.: Achieving generalized object recognition through reasoning about association of function to structure. IEEE Transactions PAMI **13** (1991) 1097–1104
22. Ulupinar, F. and Nevatia, R.: Recovering Shape from Contour for Constant Cross Section Generalized Cylinders Proceedings of CVPR (1991) 674–676
23. Zerroug, M. and Nevatia, R.: Quasi-invariant Properties and 3-D Shape Recovery of Non-Straight, Non-Constant Generalized Cylinders. In Proceedings of CVPR (1993) 96–103
24. Zerroug, M. and Nevatia, R.: From an Intensity Image to 3-D Segmented Descriptions. Proceedings of the ICPR (1994)
25. Zerroug, M. and Nevatia, R.: Using invariance and quasi-invariance for the segmentation and recovery of curved objects. Applications of Geometric Invariance in Computer Vision, LNCS **825**, Springer-Verlag (1994)

A Physics-Based Framework for Segmentation, Shape and Motion Estimation

Dimitri Metaxas

Department of Computer and Information Science
University of Pennsylvania, Philadelphia, PA 19104-6389

Abstract. This paper summarizes our research efforts towards the development of a physics-based modeling framework that addresses the difficult problems of segmentation, shape and motion estimation in a uniform way. The framework is based on the sophisticated integration of mathematical techniques from geometry, physics and mechanics, with special emphasis on the design of algorithms with close to real-time performance. We demonstrate the usefulness of this framework in experiments involving image and range data, as well as in biomedical applications.

1 Introduction

We present our recent research efforts towards the development of new model representations for shape and motion estimation and the development of techniques for simultaneous segmentation and shape estimation. These techniques stem from a careful and sophisticated integration of mathematical methods from geometry, physics and mechanics with special emphasis on the design of algorithms with close to real-time performance. While most of the existing research in the above areas has resulted in separate techniques for each of the above problems, our integrated physics-based segmentation, shape and motion estimation framework supports several shape representation schemes. The choice of shape representation depends on the given task.

To address the problem of shape representation and estimation, we have developed a framework [22, 14] which addresses in a rigorous way several challenging problems related to the analysis of 3D shape and nonrigid motion. The framework demonstrates vision algorithms for reconstructing and tracking complex objects through forces from noise-corrupted visual data. A methodology based on Lagrangian mechanics was proposed for combining conventional geometric primitives, such as splines and superquadrics, and transforming them into dynamic models whose behaviors are governed by simulated forces, torques, mass distributions, elastic strain energies, and other physical quantities. As a result, a new class of models has been introduced that can simultaneously abstract the low complexity global shapes and motions of the salient parts of natural objects, while precisely capturing the local details of their true shapes and motions. Finally, in order to account formally for noise and uncertainty in visual data, we demonstrate how the developed dynamic models can be used as system models in a continuous Extended Kalman filtering framework.

One of the major challenges in computer vision is the ability to represent complex models with a minimum number of parameters. In that respect we have recently proposed

[3, 17] to extend our above developed shape representation paradigm, which is based on global and local parameters. First, we used axial shape blending [3] to increase the vocabulary of objects that we can represent using a small number of global shape parameters. For examples, using such shape techniques we can combine spherical and cylindrical primitives into a single shape and we can achieve the transition of shape from a sphere to a torus. This initial axial blending of primitives was subsequently extended [4] to a new theory for adaptive blending of primitives, where parts of different primitives are selected and "glued" together. Based on this new theory we were able to compactly represent complex shapes such as a cup starting from a sphere. Furthermore, this new theory allows topological changes in the shape of an initial primitive (e.g., the formation of holes), thus significantly increasing the shape coverage of our deformable primitives. Second, we developed [17, 19] a new class of deformable models whose global parameters are functions. These parameters are able to represent the local shape variations based on intuitive global parameter functions, instead of local deformations. We used such models in biomedical applications to estimate the shape and motion of the heart's left ventricle.

Our shape representations offer a very powerful way for estimating the shape and motion of pre-segmented data. However, shape estimation and segmentation, traditionally studied in isolation, are definitely interconnected. We have explored two ways for integrating shape estimation and segmentation. The first is based on the integration of qualitative and quantitative shape estimation techniques [7, 15, 16], while the second is based on the use of motion [9, 10] to segment the shape and estimate the motion of articulated objects using single or multiple cameras.

In the following sections we elaborate more on the above techniques and we present several experiments.

2 Deformable Models: Geometry and Dynamics

We have developed [22, 14] a physics-based framework which provides deformable models with broad geometric coverage along with robust techniques for inferring shape and motion from noise-corrupted data.

In this framework, the positions of points on the model relative to an inertial frame of reference Φ in space are given by $\mathbf{x}(\mathrm{u}, t) = (x_1(\mathrm{u}, t), x_2(\mathrm{u}, t), x_3(\mathrm{u}, t))^T$, where ^{T} denotes transposition. We set up a noninertial, model-centered reference frame ϕ and express the position function as $\mathbf{x} = \mathbf{c} + \mathbf{Rp}$, where $\mathbf{c}(t)$ is the origin of ϕ at the center of the model and the rotation matrix $\mathbf{R}(t)$ gives the orientation of ϕ relative to Φ. Thus, $\mathbf{p}(\mathrm{u}, t)$ gives the positions of points on the model relative to the model frame. We further express $\mathbf{p} = \mathbf{s} + \mathbf{d}$ as the sum of a global reference shape $\mathbf{s}(\mathrm{u}, t)$ and a local displacement $\mathbf{d}(\mathrm{u}, t)$.

We define the global reference shape as $\mathbf{s} = \mathbf{T}(\mathbf{e}(\mathrm{u}; a_0, a_1, \ldots); b_0, b_1, \ldots)$. Here, a geometric primitive $\mathbf{e}$, defined parametrically in u and parameterized by the variables a_i, is subjected to the *global deformation* $\mathbf{T}$ which depends on the parameters b_i. Examples of global deformations are the parameterized bending, tapering and twisting deformations.

Through the application of Lagrangian mechanics, we have developed a method [14] to convert systematically the geometric parameters of the solid primitive, the global (parameterized) and local (free-form) deformation parameters, and the six degrees of freedom of rigid-body motion into generalized coordinates or dynamic degrees of freedom.

In the following sections we first present our new shape representation and estimation techniques and then we present our efforts for simultaneous segmentation and shape estimation.

3 New Models for Shape Abstraction Based on Shape Blending

In order to represent complex shapes (e.g., cup, light bulb, donut) with a small number of parameters we have developed a new theory of shape representation based on shape blending. We use shape blending to increase the vocabulary of objects that we can represent using a small number of global shape parameters. Based on this new theory we can estimate the shapes of complex objects starting from a sphere, undergoing topological changes in its shape if necessary (e.g., formation of a hole). In this way we can achieve the shape transition of a sphere to a torus. We first experimented [3] with the use of axial blending between parameterized primitives formulated as follows:

$$\mathbf{s} = \mathbf{s}_1 \alpha(u) + \mathbf{s}_2(1 - \alpha(u)), \tag{1}$$

where $\mathbf{s}_1, \mathbf{s}_2$ are two parameterized primitives (e.g., a sphere, a torus), $\alpha(u)$ is a blending function, and $\mathbf{s}$ is the resulting global shape.

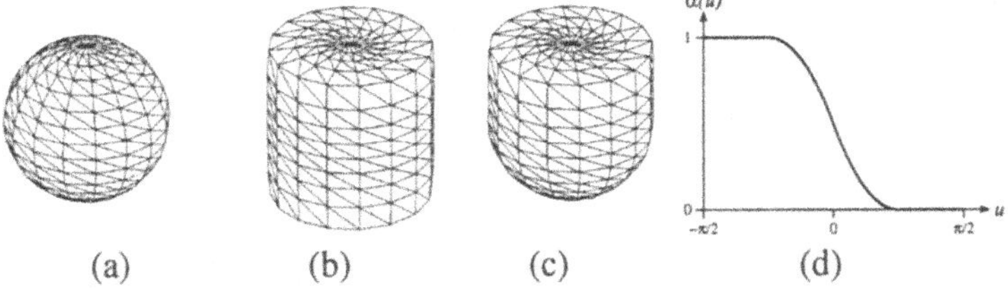

Fig. 1. (a) Shape $\mathbf{s}_1$ (b) Shape $\mathbf{s}_2$ (c) Blended shape $\mathbf{s}$ (d) Blending function $\alpha(u)$

Fig. 1(c) shows the result of blending the shapes shown in Fig. 1(a) and (b). The blending function used to blend the shapes is shown in Fig. 1(d). The blending is performed along u, which corresponds to the z-axis in these shapes (from pole to pole). Notice how the "top" of $\mathbf{s}$ looks like $\mathbf{s}_2$ since $\alpha(\frac{\pi}{2}) = 0$, and how the "bottom" of $\mathbf{s}$ looks like $\mathbf{s}_1$ since $\alpha(-\frac{\pi}{2}) = 1$.

Blended shapes of that type offer great promise in terms of the variety of objects they can represent with only a few parameters. Apart from representing shape compactly it also offers shape abstraction since the shape components (e.g., sphere, cylinder) are integrated into the representation. Furthermore, they offer the capability of genus change (e.g., a sphere turns into a torus through appropriate blending), which was not possible with any of the previous parameterized models used in computer vision.

Fig. 2 shows the fitting of torus data. The initial range data is shown in (a), and the initialization is shown in (b). The rough fit after step 1 is shown in (c). The hole is "pinched" closed in (d) after step 2, and the genus changes to 1. The hole is opened (step 3) in (e) and (f) (which are the same object from different viewpoints). A final fit using all data (step 4) is shown in (g). Notice how the blending function (f) has $\alpha(-\frac{\pi}{2}) = \alpha(\frac{\pi}{2}) = 1$, since the hole is present.

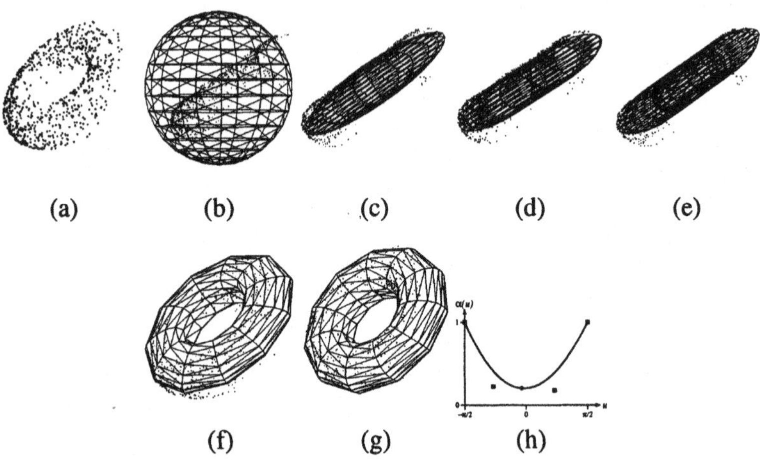

Fig. 2. Fitting of torus data with genus change in (d)

Based on the above success of axial blending, we have recently developed a theory of shape blending which allows the adaptive blending of shape primitives in complex ways (not just axially) to be able to recover non-trivial shapes such as a mug. As opposed to other shape representation schemes, the recovered shapes are represented with a small number of parameters (compactness) [4]. It also achieves shape abstraction by representing the extracted shape as a graph of components (nodes in the graph) where geometric information is also stored (as links between graph nodes).

Fig. 3 shows the fitting of a mug using this new theory. The blending region which corresponds to the mug handle forms in (d), and after rough fitting is shown in (e). After further fitting of the handle in (f), a hole blend is added in (g). After rough fitting and hole opening (h), the final fit is obtained (i). A *symbolic* description of a mug is shown in (j).

4 Deformable Models with Parameter Functions

Our goal in introducing these models was to be able to capture the local shape variation over time of complex objects with a few parameters. We therefore developed a new class of deformable models [17, 18, 19] whose global shape parameters are functions.

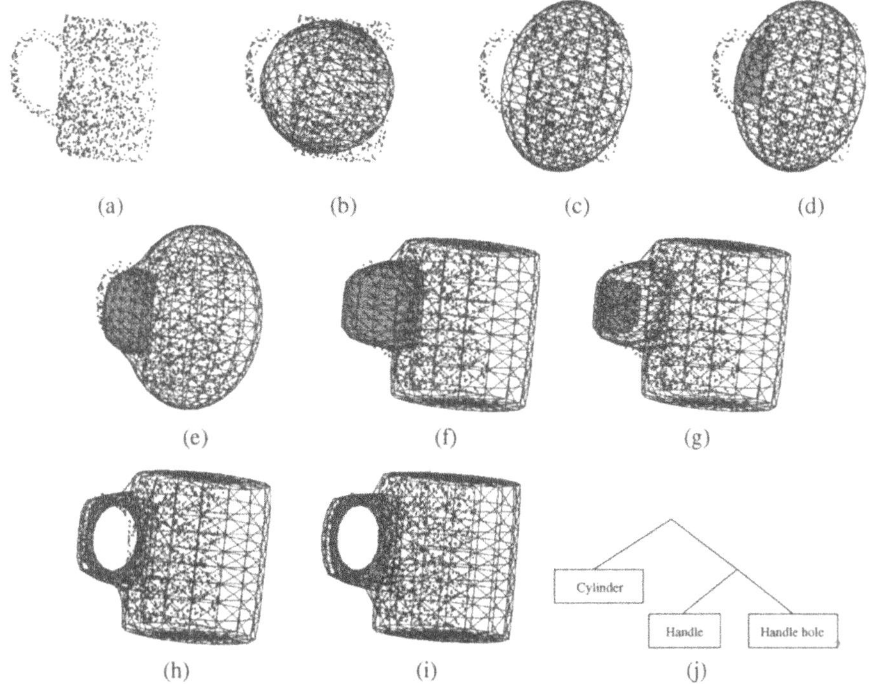

Fig. 3. Fitting of a mug (a) range data (b) initial model (c-j) fitting (k) final fit of mug (l) symbolic representation

We have so far used these models for biomedical applications, even though they can be used in many other applications.

In these new models, the global parameters $\mathbf{q}_s = (a_0(\mathrm{u}), a_1(\mathrm{u}), .., b_0(\mathrm{u}), b_1(\mathrm{u}), b_1(\mathrm{u}), ..)^T$ are functions of u, instead of being constants [14]. This definition allows us to generalize definitions of primitives (e.g., superquadrics, cubes) and parameterized deformations (e.g., twisting) as will be shown in the following example. For our applications, we will assume that $a_i(\mathrm{u}) = a_i(u), b_i(\mathrm{u}) = b_i(u)$, where the material coordinate u corresponds to the longest axis of the deformable model.

Our technique for creating primitives with parameter functions can be applied to any parametric primitive, by replacing its constant parameters with differentiable parameter functions. For our applications, we transform an ellipsoid primitive [14] to a primitive with parameter functions. The definition of such a generalized primitive $\mathbf{e} = (e_1, e_2, e_3)^T$ is given as follows:

$$\mathbf{e} = (e_1, e_2, e_3)^T, e_1 = a_0 a_1(u) \ \cos u \ \cos v, \\ e_2 = a_0 a_2(u) \ \cos u \ \sin v, \ e_3 = a_0 a_3(u) \ \sin u, \tag{2}$$

where $-\pi/2 \leq u \leq \pi/2$, $-\pi \leq v < \pi$. Here, $a_0 \geq 0$ is a scale parameter, and $0 \leq a_1(u), a_2(u), a_3(u) \leq 1$, are aspect ratio parameter functions. We can also define an

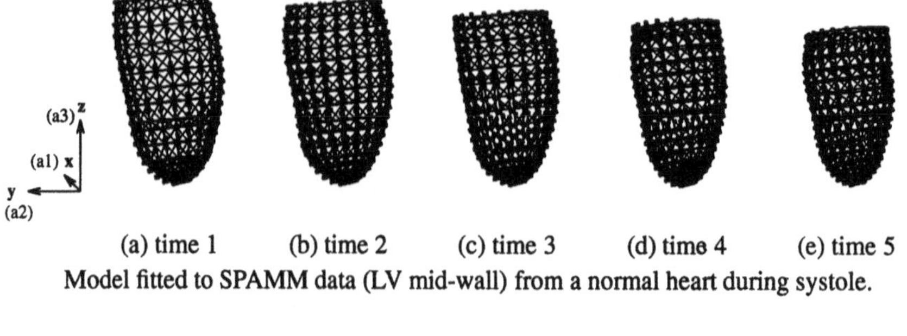

(a) time 1 (b) time 2 (c) time 3 (d) time 4 (e) time 5

Model fitted to SPAMM data (LV mid-wall) from a normal heart during systole.

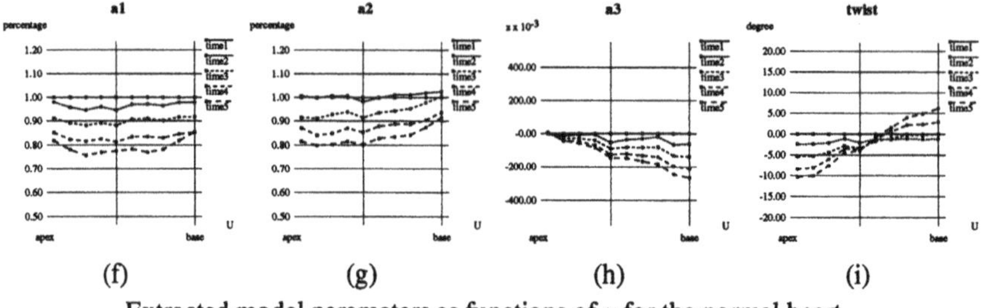

(f) (g) (h) (i)

Extracted model parameters as functions of u for the normal heart.

Fig. 4. Normal Heart

open parameterized primitive given by the above definition by restricting the ranges of the u and v parameters to a subset of the above definition.

Our formulation of deformations with continuous parameter functions is general and can be applied to any underlying shape $\mathbf{e}$. For the left ventricle's (LV) mid-wall modeling applications shown here, given the primitive $\mathbf{e}$, we first apply a twisting deformation along the principal axis 3 (or model axis z), and then apply axis-offset deformations which allow the axis to be non-straight in the x and y directions [17].

The choice of the parameter functions depends on the application. For the LV mid-wall shape and motion recovery, we assume that those parameter functions are piecewise linear along u, so we do not impose any other constraints on the LV shape.

We applied our technique to SPAMM data sequences from two normal hearts and two abnormal hearts with hypertrophic cardiomyopathy. The data obtained from the Department of Radiology at the University of Pennsylvania, were collected during the LV systole over five intervals. The SPAMM technique provides data throughout the heart wall. However, since our modeling technique is surface based, we chose to fit the LV mid-wall motion since this is most accurately defined by the SPAMM imaging technique. Young *et al.* [24] developed a technique based on snakes [11] to extract 3-D coordinates of SPAMM data from the LV mid-wall. In each case we fit our deformable models to the LV mid-wall data points, and plot the time and space varying parameter functions to extract conclusions.

Fig. 4 shows model fitting results to image data from a normal heart taken over five time sequences during systole (from end-diastole, $t = 1$, to end-systole, $t = 5$). We can

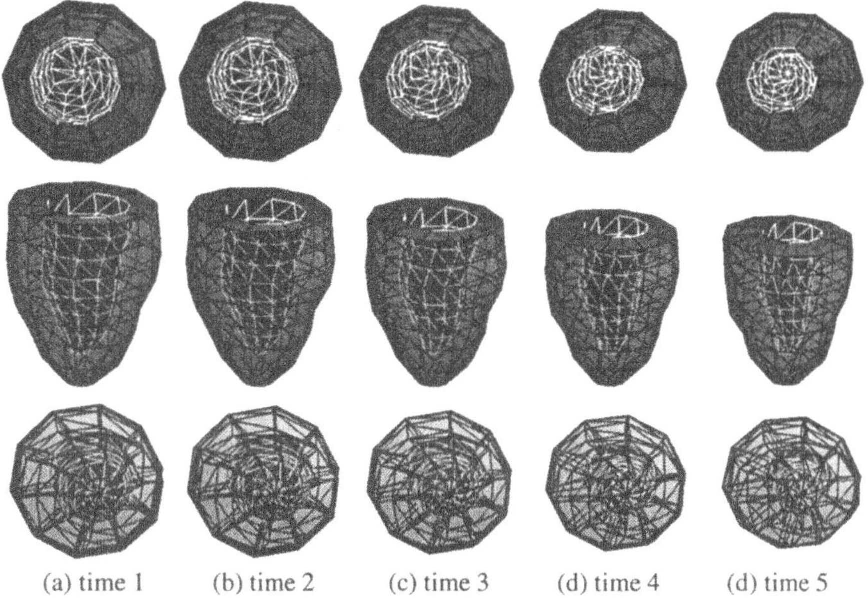

Fig. 5. Fitted models during systole

easily observe the contracting motion as well as the twisting motion of the model. The vertical lines of the mesh at the front and the back cross each other progressively due to its twisting motion.

In Figs. 4(f-i) we plot the extracted model parameter functions over the five time frames for the normal heart. Fig. 4(f) and Fig. 4(g) show the plots of the model's parameter functions $a_1(u)$ and $a_2(u)$, which are associated with its length in the x and y directions, respectively. For each frame we plot the ratio of each parameter function during frame $t = 2 \ldots 5$, with respect to its value at the initial frame ($t = 1$). Fig. 4(h) shows plots of the displacement of the length along the z direction. Fig. 4(i) shows plots of the model's twisting parameter function $\tau(u)$.

From these graphs, we can quantify the motion and shape changes of the LV during its systole (see [18] for more details). Using this technique, we have also reported in [18] shape and motion estimation results for abnormal hearts.

We have recently developed a volumetric model of the LV [19] which allows the estimation of the shape and motion of the inner and outer walls of the left ventricle as well as of any location within the walls. Although we get the temporal correspondence of individual data points from original MRI-SPAMM images, we are missing a "through-plane" motion due to the fact that the image plane is 2-D. In order to capture the missing dimension, we scan the heart in two orthogonal views (short-axis and long-axis views). Our volumetric model is designed to combine such 2 orthogonal sets of time-varying 2D tagged data points.

Fig. 5 shows fitting results of a volumetric model to SPAMM data over 5 time

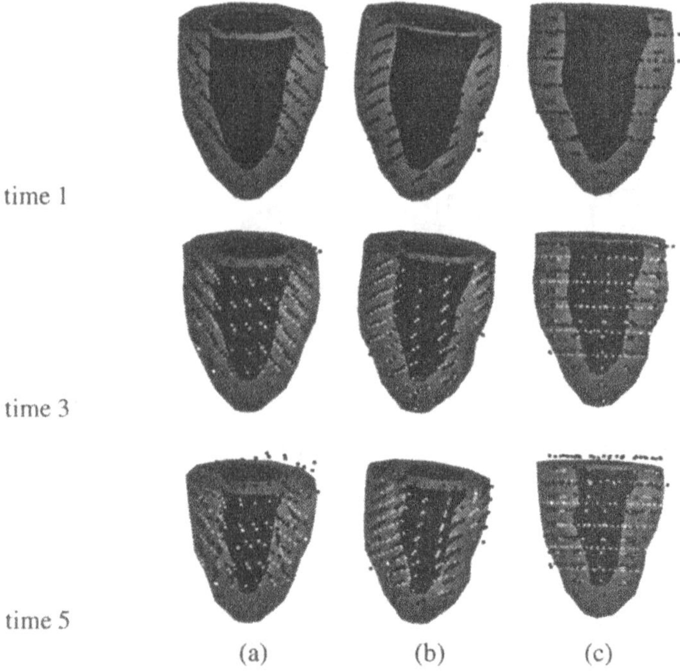

Fig. 6. Fitted Models during systole (SPAMM data and material points)

frames. The top row shows a view from the base of the LV of the fitted model. The twisting of the inner wall (shown in white) is obvious. The middle row shows a side view of the model, while the last row is similar to the first row and shows a view of the model from the apex. We can easily observe the longitudinal contraction as well as the radial contraction.

Figs. 6 (top,middle,bottom) show the fitted model superimposed to the SPAMM data at times $t = 1, 3, 5$, respectively. Columns (a),(b),(c) show the model with short axis SPAMM datapoints, long axis SPAMM datapoints and all the SPAMM datapoints, respectively. Given the method of capturing SPAMM datapoints, a SPAMM datapoint is the projection, on the image plane the SPAMM datapoint lies, of the corresponding model's material point. SPAMM datapoints are denoted with black dots, while the corresponding model material points are denoted with white dots.

5 Motion-based Segmentation, Shape and Motion Estimation

We will first give a brief presentation of our novel, active, integrated approach to identify reliably the parts of a moving articulated object and estimate their shape and motion [9]. This technique can be used in case of "chain-like" articulated objects such as robot arms and human limbs.

Instead of estimating the shape and motion of complex objects under the assumption of prior segmentation, our technique allows active, simultaneous segmentation and fitting. To identify the location of the articulation of the objects, we use a sequence of images which contain different postures of the moving object. When we observe an articulated object in a posture where the articulations are not detectable, we assume initially that the object consists of a single part, we fit a deformable model to the given time-varying data and we recover the relevant model parameters. As the object moves and attains new postures, we decide if and when to replace the initial model with two new models. This decision is based on the error of fit, the rate of change and magnitude of the bending deformation, and the continuity of the given data within the bending region.

When the criteria associated with the measures mentioned above are satisfied, we decompose (segment) the model into two models. We identify the data that correspond to the fixed, bending and relocation zones of the initial model based on the estimated bending parameters and the image projection assumptions (orthographic or perspective). We then initialize the two models to the data that correspond to the fixed and relocation zone, respectively. The data that correspond to the bending region of the initial model are marked as *orphan datapoints* since we are uncertain as to which of the two new models they should be assigned. This is necessary since we do not know in advance the extent of each of the two models. Our goal then is to fit the two new models to the given data. Furthermore, we would like them to fit in a way that allows partial overlap between the two parts.

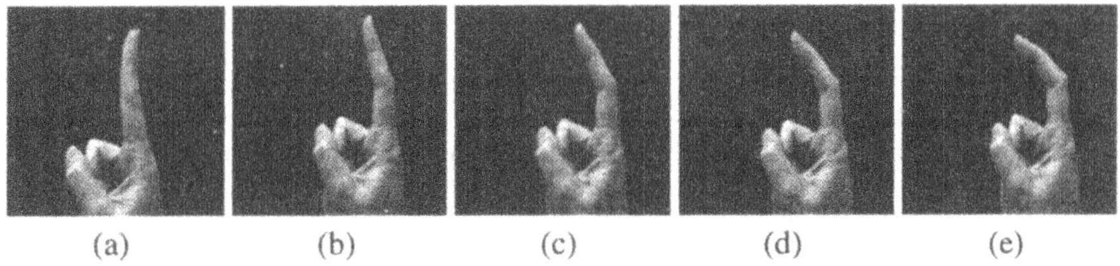

(a) (b) (c) (d) (e)

Fig. 7. Segmentation, shape and motion estimation of a human finger. A sample of the image sequence.

Unlike conventional geometric models, deformable models adapt their shapes and move in response to simulated forces as if they were made of non-rigid materials such as rubber. Since we know to which model the data in the fixed and relocation zones belong, we use our previously developed algorithm for assigning forces from datapoints to points on the model. To assign forces from the orphan data to the two models we use a novel algorithm that allows the weighted assignment of a given orphan datapoint to both deformable models [9]. We compute these weights, whose sum is always equal to one, by minimizing an appropriately selected energy expression. Once we compute all the forces from the datapoints to the two models we estimate the shape and motion of the two new models. An important property of our new force assignment algorithm

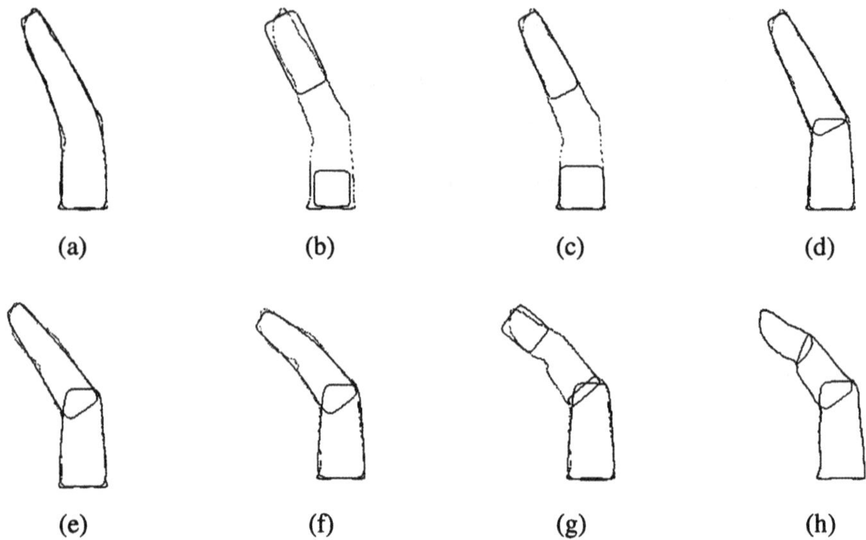

Fig. 8. Segmentation, shape and motion estimation of a human finger.

is that it allows partial overlap between the two models at a joint, whose location we can determine. Once the joint has been identified we initiate a point-to-point constraint. Having determined the location of the joints and using a Kalman Filter to predict the location of each part, we can successfully address the case where parts of the articulated object are occluded during the motion.

The above segmentation, shape and motion estimation algorithm which is applicable to outlines of "chain-like" structures also allows the determination of the joint locations as was demonstrated in [9]. We will now present some results to demonstrate the technique.

In the following experiment, we use image data obtained from a human finger moving in a plane (Figs. 7(a-e)). Fig. 8(a) shows the model fitted to the image frame where the partitioning criteria are satisfied and the hypothesis that the object is comprised from two parts is generated. Figs. 8(b-d) demonstrate the fitting of the two new models to the image data. Fig. 8(b) shows the initialization of the new models, Fig. 8(c) shows an intermediate step in the fitting process, while Fig. 8(d) shows the final, fitted models. The overlap between the two models allows us to compute robustly the location of the joint over several frames and place a point-to-point constraint between the two models. Fig. 8(e) shows the models fitted to a new frame, while Fig. 8(f) shows the models fitted to the frame where the partitioning criteria are satisfied for the upper model and the hypothesis that the upper model should be replaced by two new models is generated. Fig. 8(g) shows the initialization of the two new models based on our technique, while Fig. 8(h) shows all three models fitted to the given data.

We have recently extended this technique to segment and estimate the 3D shape of the parts of a whole human body from multiple views [10]. This technique is based on the observation that the geometry of the silhouette or apparent contour of a human body

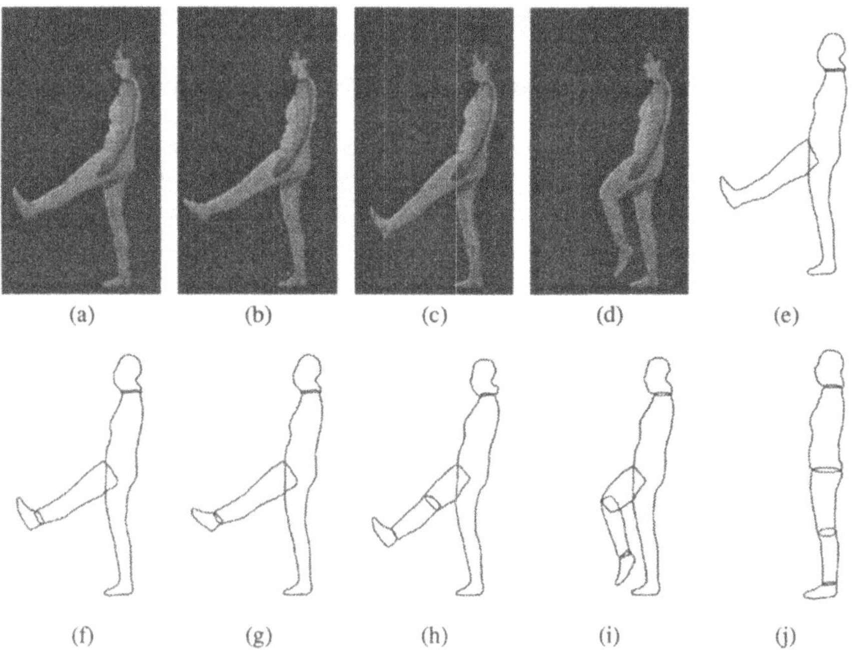

Fig. 9. 2D Human Part Identification, Shape and Motion Estimation.

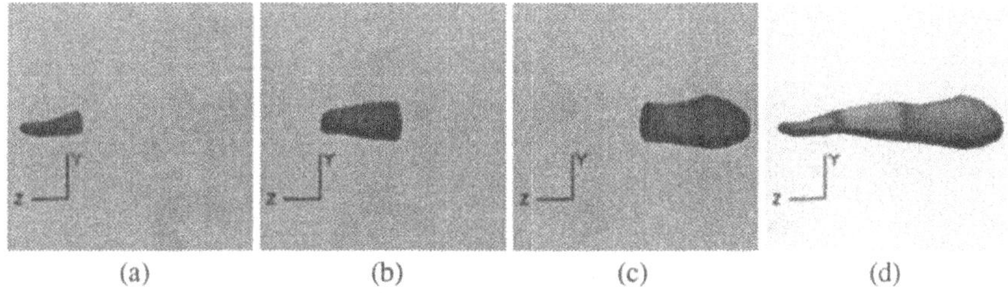

Fig. 10. 3D Human Part Shape Estimation.

and its deformation under motion provides rich information for inferring its shape and structure. Our novel algorithm simultaneously fits data from 2D human body silhouettes and segments them to the different body parts. It then combines multiple views to estimate the 3D shape of the parts.

We first present selected results from processing an image sequence captured from an actor moving according to a protocol of movements. Figs. 9(a-d) show four frames from the motion of lower body extremities, and Figs. 9(e-j) depict the parts recovered. In the absence of other assumptions, multiple cameras are necessary to overcome the problems associated with occlusion and manage the complexity of a human's motions. We have developed a new method for selectively integrating multiple views of a human body in motion. In our experiments we have chosen to integrate images taken from three cameras placed orthogonally. Figs. 10(a-c) show the 3D shape of the parts identified for the left arm of the actor and Fig. 10(d) shows the spatial layout of the parts.

6 Integration of Aspect-based and Deformable Model Techniques

To overcome the reliance of our physics-based framework [14] to prior data segmentation, we proposed a two-step shape recovery process recently [7] that first recovers the qualitative shape of an object in terms of its parts [5, 6]. If detailed shape or localization is needed to manipulate the object, for example, we then use knowledge of a part's qualitative shape and its orientation (encoded by its aspect) to provide strong constraints in fitting a deformable model to the part. Furthermore, since the qualitative shape recovery technique supports occlusion through a hierarchical aspect representation, it can selectively pass to the model fitting stage only those contours belonging to the object.

The correspondence between image faces and model surfaces encoded in the recovered qualitative primitive can be exploited to provide strong constraints on the initial placement of the deformable model. Subsequently, this correspondence has allowed us first to extend our previously developed technique for deformable model fitting, which was limited to orthographic projection of occluding boundaries [22], to the case of more general objects with internal surface discontinuities, under perspective and stereo projections.

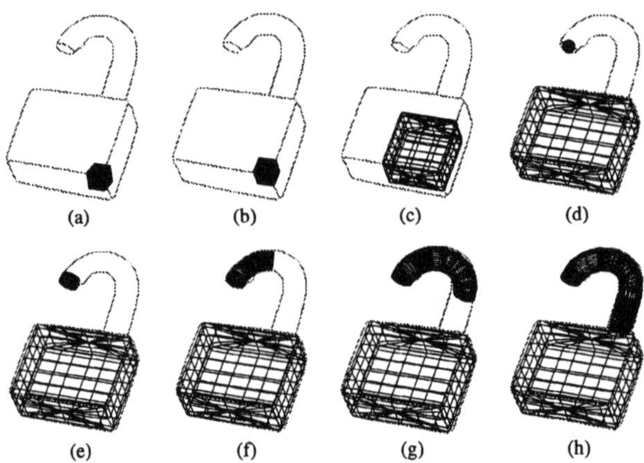

Fig. 11. Two models fitted to the image of a lock.

Significant extensions to the above work have resulted in the development of techniques for the shape estimation of objects in complex scenes with significant occlusion by using multiple views [16], as well as for the robust tracking of multiple objects in the presence of occlusion and noise [15]. An important aspect of the work on multiple view integration and in stereo tracking is that correspondence at the pixel level is not needed. We only need to know the correspondence of lines (not points on corresponding lines) in subsequent images. Furthermore, a tracked line does not have to be present in all images since forces from all the visible lines result in the estimation of the shape and motion of the tracked object.

The first experiment shown in Fig. 11 demonstrates the use of our framework to estimate object shapes under perspective projection from a single image of a lock. Using the qualitative shape recovery process OPTICA [5, 14] a box like primitive and a bent cylinder primitive are recovered. Figs. 11(a-d) and (e-h) show the quantitative fitting stages of the 2 models to the lower and upper parts of the lock, respectively.

In the second experiment, we consider a sequence of stereo images of a scene containing multiple objects, including a two-part object. Fig. 12 shows the initial stereo images of the scene. The cameras are rotated around the scene at a constant rate. Fig. 13(a) shows the initialized models recovered using the above described technique. Fig. 13(b) shows image potentials at an instant which the aspects of some parts have changed and some parts have become partially occluded. Each object is successfully tracked under these circumstances. Figs. 13(c-f) show the individual part models overlaid on the potentials in Fig. 13(b).

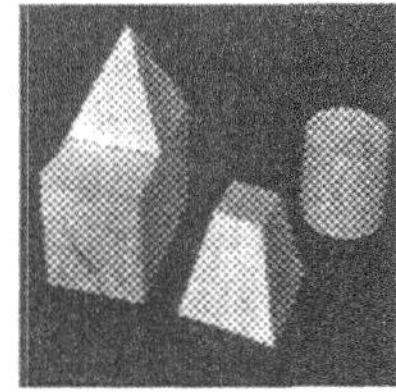

Fig. 12. Initial pair of stereo images of a multi-object scene.

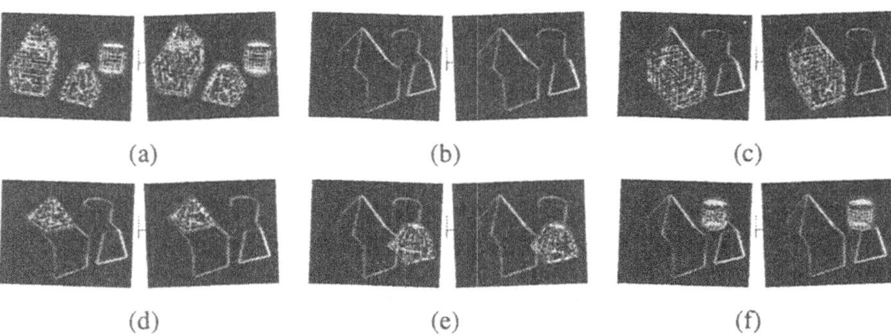

Fig. 13. Tracking multiple objects in a sequence of stereo images (a) initialized models, (b) image potentials of an inter-mediate frame (both occlusions and visual events have occurred) (c-f) each object part correctly tracked with part models overlaid on the image potentials in (b). Note that the active model nodes are highlighted.

7 Conclusions

In the past few years we have developed a physics-based framework for the segmentation, shape and motion estimation of complex objects. The framework has allowed us to solve aspects of the shape and motion estimation problems that could not be addressed previously. Furthermore, as opposed to other disparate efforts, the framework provides a unified approach to the above problems. The framework supports multiple shape representation schemes since the determination of the correct shape representation depends on the given task. Our shape representations offer both shape compaction and abstraction and can therefore be potentially useful for recognition. In the future, we plan to expand and integrate further the above methods into a single system that can address more complex problems, including shape-based recognition.

References

1. I. Biederman. "Recognition by Components: A theory of human image understanding", *Psychological Rev.*, 94(2), pp. 115–147, 1987.
2. M. Chan and D. Metaxas. Physics-Based Object Pose and Shape Estimation from Multiple Views. Procs. 12th International Conference on Pattern Recognition, October 94, Jerusalem, Israel.
3. D. DeCarlo and D. Metaxas. Blended Deformable Models. Proc. Computer Vision and Pattern Recognition Conference, pp. 566-572, Seattle, June 1994.
4. D. DeCarlo and D. Metaxas. Adaptive Deformable Model Evolution Using Blending. In Procs. IEEE Fifth International Conference on Computer Vision, MIT, Cambridge, June 1995, to appear.
5. S. J. Dickinson, A. P. Pentland, and A. Rosenfeld. 3-D Shape Recovery using Distributed Aspect Matching. *IEEE Trans. Pattern Analysis and Machine Intelligence, special issue on Interpretation of 3-D Scenes*, 14(2):174–198, February 1992.
6. S. J. Dickinson, A. P. Pentland, and A. Rosenfeld. From Volumes to Views: An Approach to 3-D Object Recognition. *Computer Vision Graphics and Image Processing:Image Understanding,special issue on CAD-based vision*, 55(2), March 1992.
7. S. Dickinson and D. Metaxas. "Integrating Qualitative and Quantitative Shape Recovery," International Journal of Computer Vision, 13(3), pp. 1-20, 1994.
8. C.M. Hoffmann. *Geometric and Solid Modeling*. Morgan-Kaufmann, Palo Alto, 1989.
9. I. A. Kakadiaris, D. Metaxas, and R. Bajcsy. Active part-decomposition, shape and motion estimation of articulated objects: A physics-based approach. In *IEEE Computer Society Conference on Computer Vision and Pattern Recognition*, pp. 980–984, Seattle,WA, June 21-23 1994.
10. I. A. Kakadiaris and D. Metaxas. Active Motion-Based Segmentation of Human Body Outlines. In Procs. IEEE Fifth International Conference on Computer Vision, MIT, Cambridge, June 1995, to appear.
11. M. Kass, A. Witkin, D. Terzopoulos, "Snakes: Active Contour Models", International Journal of Computer Vision, 1(4), pp. 321-331, 1988.
12. E. Koh, D. Metaxas and N. Badler. Hierarchical Shape Representation Using Locally Adaptive Finite Elements. Proc. European Conference on Computer Vision, pp. 441-446, Stochholm, Sweden, May 1994.
13. S. Kurakake and R. Nevatia, "Description and Tracking of Moving Articulated Objects", *Proceedings of the 11th International Conference on Pattern Recognition*, pp. 491-495, 1992.

14. D. Metaxas and D. Terzopoulos. Shape and Nonrigid Motion Estimation Through Physics-Based Synthesis. *IEEE Trans. Pattern Analysis and Machine Intelligence*, 15(6), pp. 569-579, June, 1993.
15. M. Chan, D. Metaxas and S. Dickinson. "Physics-based tracking of 3D objects in 2D image sequences". *Proceedings of the 1994 IEEE 12th International Conference on Pattern Recognition*, Vol. I, pp. 432-436, Jerusalem, Israel, October 1994.
16. M. Chan and D. Metaxas. "Physics-based object pose and shape estimation from multiple views". *Proceedings of the 1994 IEEE 12th International Conference on Pattern Recognition*, Vol. I, pp. 326-330, Jerusalem, Israel, October 1994.
17. J. Park, D. Metaxas and A. Young. Deformable Models With Parameter Functions: Application to Heart-Wall Modeling. Proc. Computer Vision and Pattern Recognition Conference, pp. 437-442, Seattle, June 1994.
18. J. Park, D. Metaxas, A. Young and L. Axel. Model-based Analysis of Cardiac Motion from Tagged MRI Data. Proc. Seventh Annual IEEE Symposium on Computer-Based Medical Systems, pp. 40-45, Winston-Salem, North Carolina, June 1994.
19. J. Park, D. Metaxas and L. Axel. Volumetric Deformable Models with Parameter Functions: A New Approach to the 3D Motion Analysis of the LV from MRI-SPAMM. In Procs. IEEE Fifth International Conference on Computer Vision, MIT, Cambridge , June 1995, to appear.
20. A. Pentland, "Automatic Extraction of Deformable Part Models", *International Journal of Computer Vision*, 4:107–126, 1990.
21. F. Solina and R. Bajcsy. Recovery of Parametric Models from Range Images: The Case for Superquadrics with Global Deformations. *IEEE Trans. Pattern Analysis and Machine Intelligence*, 12(2):131–146, 1990.
22. D. Terzopoulos and D. Metaxas. Dynamic 3D Models with Local and Global Deformations: Deformable Superquadrics. *IEEE Trans. Pattern Analysis and Machine Intelligence*, 13(7):703–714, 1991.
23. D. Terzopoulos, A. Witkin, and M. Kass. "Constraints on Deformable Models: Recovering 3D Shape and Nonrigid motion", *Artificial Intelligence*, 36(1), pp. 91–123, 1988.
24. A. A. Young, D. Kraitchman and L. Axel. "Deformable Models for Tagged MR Images: Reconstruction of Two and Three Dimensional Heart Wall Motion". Procs. IEEE Workshop on Biomedical Image Analysis, Seattle, WA, June, 1994.
25. O. Zienkiewicz. *The Finite Element Method*. McGraw-Hill, 1977.

Modal Represenations

Alex P. Pentland[1] and Stan Sclaroff[2]

[1] The Media Laboratory, Massachussetts Institute of Technology
20 Ames Street, Cambridge, MA 02139, USA
[2] Department of Computer Science, Boston University,
111 Cummington Street, Boston, MA 02215, USA

Abstract. Modal representations are a family of physically-motivated deformable shape models for machine vision, robotics, and simulation. These models employ the eigenvectors of the object's physical model to obtain canonical description of shape in terms of deformation from a prototype. This results in efficient and reliable methods for recovering, recognizing, and tracking 3-D solid models from 2-D and 3-D measurements. Modes also provide an economical, low-complexity method for computing complex, multi-body physical simulations. This chapter provides an overview of the modal shape representation, and then describes recent advances in the formulation and use of modal representations for solving problems in correspondence, alignment, object recognition, and categorization. In contrast to previous techniques, which required correspondence to be computed with an initial shape, the new formulation allows for a shape's modes to be computed directly from the object prototype or from available image information.

1 Introduction

All shapes can be represented as deformations from a standard or prototypical shape; it is thought that this is one plausible mechanism for human perception [2, 16, 17, 23, 30]. This basic premise has served as inspiration for many of the deformable shape representations for machine vision, robotics, and simulation. A key issue in using such representations is how to describe features, contours, surfaces, and volumes so that they can be recognized and matched from view to view. The primary difficulties are that object descriptions are sensitive to noise, that an object can be nonrigid, and that an object's appearance deforms as the viewing geometry changes. These problems have motivated the use of physically-motivated, deformable models [5, 7, 13, 19, 23, 28, 36, 39, 40], to exploit *a priori* knowledge about physics to interpolate and smooth raw data, based on prior assumptions about noise, material properties, *etc.*

Physically-motivated models do not by themselves provide a method for computing reliable canonical shape descriptions, or for establishing correspondence between data sets. To address this problem, we proposed using the *modal representation*, in which shapes are represented as modal deformations from some prototype object [24, 28]. Modes are the eigenvectors of the prototype's stiffness matrix, and make it possible to obtain a robust, frequency-ordered shape description. Moreover, these eigenvectors provide an intuitive method for shape description because they correspond to the object's generalized axes of symmetry. By representing objects in terms of modal deformations we developed robust methods for 3-D shape modeling, object recognition, and 3-D tracking utilizing point, contour, 3-D, and optical flow data [24, 26, 28].

However, this method still did not address the problem of determining correspondence between sets of data, or between data and models. This was because every object had to be described as deformations from a *single* prototype object. This implicitly imposes an *a priori* parameterization upon the sensor data, and therefore implicitly determines the correspondences between the data and the prototype. We have therefore generalized our earlier method by obtaining the modal shape invariants directly from the sensor data. This allows us to compute robust, canonical descriptions for recognition *and* to solve correspondence problems for data of any dimensionality [32, 33].

In the rest of this chapter, we will first give a brief review of related work in the area of eigen-representations. We will then review the modal formulation and present selected results in the original formulation for 3-D shape recognition and tracking. The original method will be illustrated using contour, grey-level, and volumetric input data. Software implementing this method, together with technical reports and papers, are available by anonymous FTP from whitechapel.media.mit.edu in `pub/modal.tar.Z` and `pub/tech-reports`.

We will then describe our new formulation, *modal matching*, and review results obtained using this new technique. For convenience the method will be illustrated only on 2D problems, using both gray-scale and point feature input data. The extension to data of other dimensionality is described in a technical report [33]. To illustrate the use of this method for object recognition and category classification, we will present an example of recognizing and categorizing images of hand tools.

2 Related Work: Eigen-representations

In the last few years there has been a revival of interest in eigendecomposition-based pattern recognition methods, due to the surprisingly good results that have been obtained by combining these methods with modern machine vision representations. Using these approaches researchers have built systems that perform stable, interactive-time recognition of faces [42], cars [20], and biological structures [5, 18, 25], and allowed interactive-time tracking of complex and deformable objects [3, 9, 26, 41].

Typically, these methods employ eigendecompositions like the modal transform or any of a family methods descended from the Karhunen-Loève transform. Some are feature-based *eigenshapes* [1, 9, 31, 32, 34, 35], others are physically-based *eigenwarps* [3, 5, 21, 25, 28, 32], and still others are based on (preprocessed) image intensity information, *eigenpictures* [12, 15, 20, 27, 41, 42].

In these methods, image or shape information is decomposed into an ordered basis of orthogonal principal components. As a result, the less critical and often noisy high-order components can be discarded in order to obtain overconstrained, canonical descriptions. This allows for the selection of only the most important components to be used for efficient data reduction, real-time recognition and navigation, and robust reconstruction. Most importantly, the orthogonality of eigen-representations ensures that the recovered descriptions will be unique, thus making recognition problems tractable.

The modal methods described here utilize the eigenvectors of a finite element model shape representation, and are therefore most closely related to eigenshapes and eigenwarps. At the core of all of these techniques is a positive definite matrix that describes the connectedness between features. By finding the eigenvectors of this matrix, we can obtain a new, generalized coordinate system for describing the location of feature points. The modal representation is reviewed in the next section.

3 The Modal Representation

A shape's modal representation is based on the eigenvectors of its physical model. This physical model is formulated using the finite element method. The major advantage of the finite element method is that it uses the Galerkin method of surface interpolation. This provides an analytic characterization of shape and elastic properties over the whole surface, rather than just at the nodes (nodes are typically the spring attachment points). In Galerkin's method, we set up a system of polynomial shape functions that relate the displacement of a single point to the relative displacements of all the other nodes of an object. By using these functions, we can calculate the deformations which spread uniformly over the body as a function of its constitutive parameters.

In general, the polynomial shape function for each element is written in vector form as:

$$\mathbf{u}(\mathbf{x}) = \mathbf{H}(\mathbf{x})\mathbf{U} \tag{1}$$

where $\mathbf{H}$ is the interpolation matrix, $\mathbf{x}$ is the local coordinate of a point in the element where we want to know the displacement, and $\mathbf{U}$ denotes a vector of displacement components at each element node.

For most applications it is necessary to calculate the strain due to deformation. Strain ϵ is defined as the ratio of displacement to the actual length. The polynomial shape functions can be used to calculate the strains (ϵ) over the body provided the displacements at the node points are known:

$$\epsilon(\mathbf{x}) = \mathbf{B}(\mathbf{x})\mathbf{U} \tag{2}$$

where $\mathbf{B}$ is the strain displacement matrix. The rows of $\mathbf{B}$ are obtained by appropriately differentiating and combining rows of the element interpolation matrix $\mathbf{H}$. Solution to the problem of deforming an elastic body to match the set of feature points requires solving the *dynamic equilibrium equation*:

$$\mathbf{M}\ddot{\mathbf{U}} + \mathbf{K}\mathbf{U} = \mathbf{R}, \tag{3}$$

where $\mathbf{R}$ is the load vector whose entries are the spring forces between each feature point and the body surface, and where $\mathbf{M}$ and $\mathbf{K}$ are the element mass and stiffness matrices, respectively.

Both the mass and stiffness matrices are computed directly:

$$\mathbf{M} = \int_V \rho \mathbf{H}^T \mathbf{H} dV \text{ and } \mathbf{K} = \int_V \mathbf{B}^T \mathbf{C} \mathbf{B} dV, \tag{4}$$

where ρ is the mass density, and $\mathbf{C}$ is the *material matrix* which expresses the material's particular stress-strain law.

This system of equations can be decoupled by posing the equations in a basis defined by the **M**-orthogonalized eigenvectors of **K**. These eigenvectors and values are the solution (ϕ_i, ω_i^2) to the following generalized eigenvalue problem:

$$\mathbf{K}\phi_i = \omega_i^2 \mathbf{M}\phi_i. \tag{5}$$

The vector ϕ_i is called the ith *mode shape vector* and ω_i is the corresponding frequency of vibration. Each mode shape vector describes how each node is displaced by the ith vibration mode.

The mode shape vectors ϕ_i are **M**-orthonormal, this means that

$$\Phi^T \mathbf{K} \Phi = \Omega^2 \text{ and } \Phi^T \mathbf{M} \Phi = \mathbf{I}. \tag{6}$$

where the ϕ_i are columns in the transform Φ, and ω_i^2 are the elements of the diagonal matrix Ω^2. This generalized coordinate transform Φ is then used to transform between nodal point displacements $\mathbf{U}$ and decoupled modal displacements $\tilde{\mathbf{U}}$, $\mathbf{U} = \Phi\tilde{\mathbf{U}}$. We can now rewrite Equation 3 in terms of these generalized or modal displacements, obtaining a decoupled system of equations:

$$\ddot{\tilde{\mathbf{U}}} + \Omega^2 \tilde{\mathbf{U}} = \Phi^T \mathbf{R}, \tag{7}$$

allowing for closed-form solution to the equilibrium problem [28]. Given this equilibrium solution in the two images, point correspondences can be obtained directly.

By discarding high frequency modes the amount of computation required can be minimized without significantly altering correspondence accuracy. Moreover, such a set of modal amplitudes provide a robust, canonical description of shape in terms of deformations applied to the original elastic body. This allows them to be used directly for object recognition [28].

4 Results

4.1 Fitting

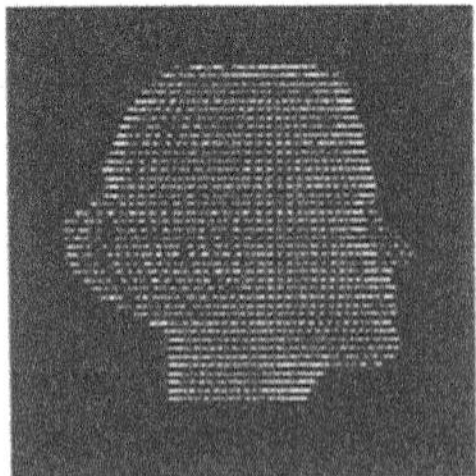

Fig. 1. Fitting laser rangefinder data of a human face as described in [28].

As described in Pentland and Sclaroff [28], the modal representation is effective for fitting models to 3-D contour, range, and surface data, and also for fitting models to

2-D image data. The left-hand image of Figure 1 shows an example using 360° laser rangefinder data of a human head. There are about 2500 data points. Using only the low-frequency 30 modes a qualitative shape model can be recovered, shown in the middle image. If more detail is desired, it is possible to either add more modes, or to use a displacement map, as shown in the right-hand image of Figure 1.

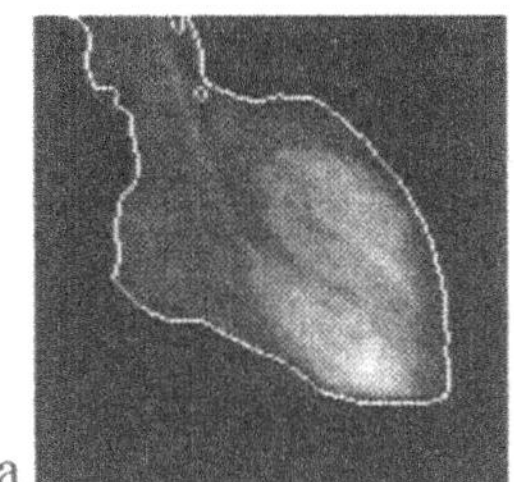

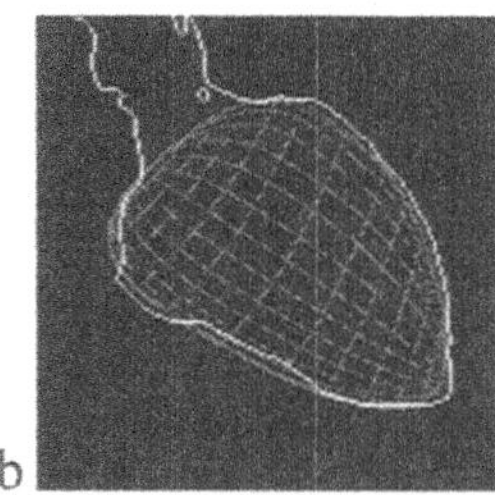

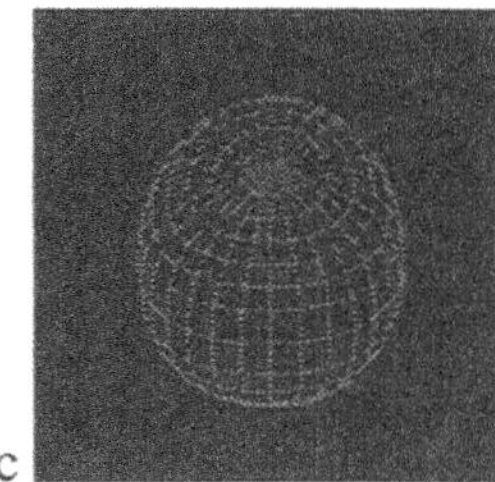

Fig. 2. An example showing the use of a 2-D image contour to recover a 3-D deformable solid model. The original image and contour are shown in (a). The model is recovered from the contour as shown in (b). An orthogonal side view is shown in (c).

Figure 2(a) shows one frame from a sequence of X-ray images, with the zero-crossing edge contours overlayed. From these contours the 3-D shape was estimated. The resulting shape is shown in Figure 2(b) as a 3-D wireframe overlayed on the original X-ray data. Figure 2(c) shows a side view.

4.2 Tracking

As first shown in Pentland and Horowitz [26], by correctly setting the material parameters of a physically-based deformable model one can obtain a simple type of Kalman filter. This can allow stable and accurate tracking of non-rigid objects.

Figure 3 shows an example of recovering non-rigid motion from contour information. The 3-D shape and motion of the heart ventricle was tracked over time using the contour information shown in the top row of boxes of Figure 3. These shape estimates were then integrated by making the physical model behave as a Kalman filter [26].

Figure 4 illustrates a more complex example of tracking rigid and non-rigid motion. This figure shows three frames from a twelve image sequence of a well-known tin woodsman caught in the act of jumping. Linear springs were employed to enforce the joint constraints[26].

4.3 Medical application

By using a principal components analysis, modes that are particular to a shape class can be learned from a training set [9, 31]. This in some ways avoids the problems associated with setting *a priori* material properties for the model. As described by Martin, Pentland, and Kikinis [18], these learned modes can then be used in 3-D fitting, recognition, and medical diagnosis.

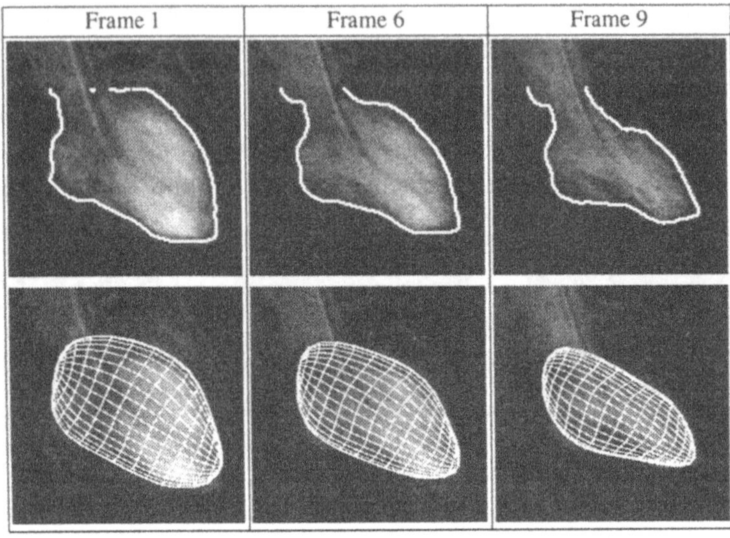

Fig. 3. A heart's nonrigid motion as it was recovered from contours taken from a motion sequence as described in [26].

Fig. 4. Three frames from an image sequence showing tracking of a jumping man.

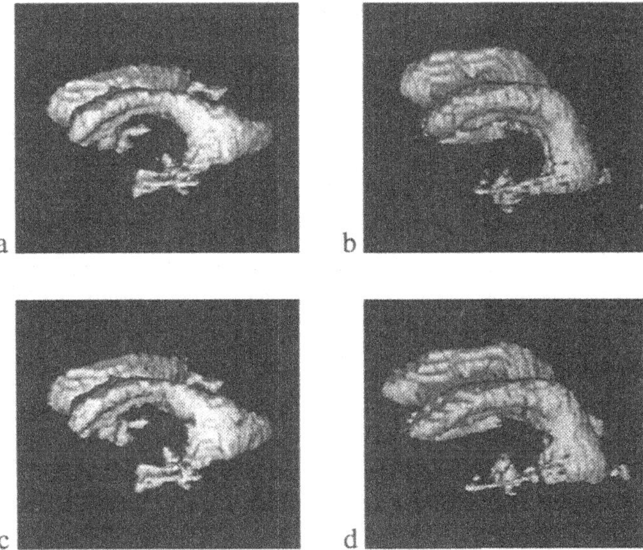

Fig. 5. Using modal models to account for normal variation in head shape. The top row shows reconstructions of the ventricular system of the human brain created from MR images: (a) Alzheimer's disease causes an enlargement of ventricles, and (b) normal-pressure hydrocephalus causes an even greater enlargement of the ventricles, along with an overall "puffy" appearance. The bottom row (c,d) shows each of the ventricles normalized for cranium shape by warping each patient's cranium to a reference cranium using a modal model as described in [18].

Various neurological disorders affect the gross anatomical shape of different brain structures, as illustrated in Figure 5. Because of both genetic and environmental factors, however, biological structures have a large normal range of variation. The onset of a disease introduces further changes in morphology, with each particular disease causing its own type of shape changes and range of variation in these changes. Therefore to properly study the pathological deformations, we must first account for the large normal variation in biological shape.

For structures inside the cranium, a significant part of this normal shape variability is due to the different head shapes seen across individuals. These macroscopic shape effects interfere with the analysis of local shape deformations caused by disease.

We use modal modeling to separate out normal and disease-process deformations, by creating a physical model that describes the macroscopic effects caused by different head shapes. This allows us to accurately warp the cranial contents so that we obtain inter-patient shape differences that are largely independent of head shape. Using modes derived from a training set, it was possible to obtain more accurate fitting classification of Alzheimer's, normal-pressure hydrocephalus, and normal patients.

5 A New Method: Modal Matching

This above method still does not address the problem of determining correspondence between sets of data, or between data and models. This is because every object has to be described as deformations from a *single* prototype object. This implicitly imposes an *a priori* parameterization upon the sensor data, and therefore implicitly determines the correspondences between data and the prototype. We would like to avoid this as much as is possible, by letting the data determine the parameterization in a natural manner.

To solve this problem, we introduced a new finite element formulation that uses Gaussian basis functions as Galerkin interpolants[32]. As a result, we can use the data itself to define the deformable object, by building stiffness and mass matrices that use the positions of image feature points as the finite element nodes. This allows us to compute robust, canonical descriptions for recognition *and* to solve correspondence problems[33].

5.1 Determining correspondence and similarity

Correspondence is often defined as the problem of matching points on two objects (or two views of one object), and has proven to be a very difficult problem. If, however, the points were given in body-centered coordinates rather than Cartesian coordinates, the problem would be nearly trivial. That is, it is easy to match up the bottom-left points on two objects, once given the up-down and left-right axes of the two objects. This suggests that the best way to match two sets of feature points is to first compute a body-centered coordinate frame for each object. The modes of an object form such a coordinate frame.

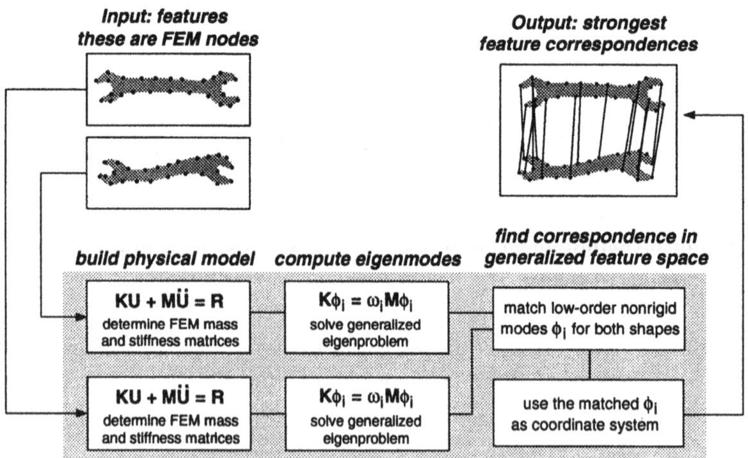

Fig. 6. Modal matching system diagram.

A flow-chart of our method is shown in Figure 6. For each image we start with feature point locations, which are used as nodes in building a finite element model of the shape. This approach avoids imposing an a priori parameterization on the data. A Gaussian is centered at each node, and used as the Galerkin interpolants in constructing the mass and stiffness matrices [32]. The use of Galerkin interpolants reduces the effects of missing or mislocated features.

We then compute the modes Φ of this model. Modes can be thought of as the shape's generalized symmetry axes; the form an orthogonal object-centered coordinate system for describing feature locations. That is, each feature point location can be uniquely described in terms of *how it projects onto each eigenvector*, i.e., how it participates in each deformation mode. The transform between Cartesian feature locations (x, y) and modal feature locations (u, v) is accomplished by using the eigenvectors Φ as a coordinate basis, i.e.,

$$\Phi = \left[\phi_1 \mid \ldots \mid \phi_{2m}\right] = \begin{bmatrix} \mathbf{u}_1^T \\ \vdots \\ \mathbf{u}_m^T \\ \mathbf{v}_1^T \\ \vdots \\ \mathbf{v}_m^T \end{bmatrix} \tag{8}$$

where m is the number of nodes used to build the finite element model. The column vector ϕ_i is called the ith *mode shape*, and describes the modal displacement (u, v) at each feature point due to the ith mode, while the row vector $\mathbf{u}_i$ and $\mathbf{v}_i$ are called the ith *generalized feature vectors*, and together describe the feature's location in the modal coordinate system.

Normally only the n lowest-order modes are used in forming this coordinate system, so that (1) we can compare objects with differing numbers of feature points, and (2) the feature point descriptions are insensitive to noise. Depending upon the demands of the application, we can also selectively ignore rigid-body modes, or low-order projective-like modes, or modes that are primarily local. Consequently, we can describe, track, and compare nonrigid objects in a very flexible and general manner.

Given a body-centered coordinate frame for each of two objects, Φ_1, Φ_2, we would now like to establish point correspondences between the two objects. The first step is to match the two sets of eigenmodes, thus establishing an equivalence between the coordinate frames for each object. This is accomplished by determining, for each eigenvector $\phi_{1,i}$ of object one, the eigenvector $\phi_{2,j}$ of object two that has the most similar direction [33]. The important idea here is that the low-order vibration modes computed for two similar objects will be very similar — even in the presence of affine deformation, nonrigid deformation, local shape perturbation, or noise.

Point correspondences are then determined by comparing the two groups of features in the eigenmode coordinate system. That is, modal signatures $(\mathbf{u}_1, \mathbf{v}_1)$ for the first data set are compared with those of the second set $(\mathbf{u}_2, \mathbf{v}_2)$. The points that have the most similar and unambiguous coordinates are then matched, with the remaining correspondences determined by using the physical model as a smoothness constraint [32, 33].

Finally, given correspondences between many of the feature points on two objects, we can track their motion and measure their difference in shape, as in the original modal method. Typically this is accomplished by measuring the modal *strain energy* required to warp the feature points of one object into alignment with those of the other object:

$$E_I = \frac{1}{2}\tilde{\mathbf{U}}^T \Omega^2 \tilde{\mathbf{U}}. \tag{9}$$

Since the modal framework decomposes deformations into an orthogonal set, we can selectively measure rigid-body, projective-like, or local deformations. For additional detail see Sclaroff and Pentland [32].

6 Recognition Experiments

This new formulation enables us both to automatically establish feature correspondences *and* to describe the amounts and types of deformations that take the features from one position to the other. This is demonstrated in Figure 7, where modal strain energy is used to compare three different prototype tools: a wrench, hammer, and crescent wrench. In each case, silhouettes were first extracted and thinned from each tool image, and then the strongest corresponding contour points were found via modal matching. Mode amplitudes for the first 22 modes were recovered and used to warp each prototype onto the other tools. The modal strain energy that results from deforming the prototype to each tool is shown below each image.

Figure 7(a) depicts the use of modal strain energy in comparing a prototype wrench with thirteen other hand tools. As this figure shows, the shapes most similar to the wrench prototype are those other two-ended wrenches with approximately straight handles. Next most similar are closed-ended and bent wrenches, and most dissimilar are hammers and single-ended wrenches. Note that the matching is orientation and scale invariant (modulo limits imposed by pixel resolution).

Figure 7(b) continues this example using a hammer as the prototype. When the hammer prototype is used, the most similar shapes found are three other images of the same hammer, taken with different viewpoints and illumination. The next most similar shapes are a variety of other hammers. The least similar shapes are a set of wrenches.

The fact that the similarity measure produced by the system corresponds to functionally-similar shapes is important. It allows us to recognize the most similar wrench or hammer from among a group of tools, even if there is no tool that is an exact match. Moreover, if for some reason the most-similar tool can't be used, we can then find the next-most-similar tool, and the next, and so on. We can find (in order of similarity) all the tools that are likely to be from the same *category*.

7 Modal Representations, Estimation, and Learning

As has been pointed out by Boult [6], there has been a well understood link between energy formulations and statistical estimation for over 25 years now. Splines were perhaps some of the first "physically-based" models employed in statistical estimation [14]; they are particularly well-suited to modeling data sampled from a Markov Random Field (MRF), where Gaussian noise has been added [4, 10]. The same principles hold

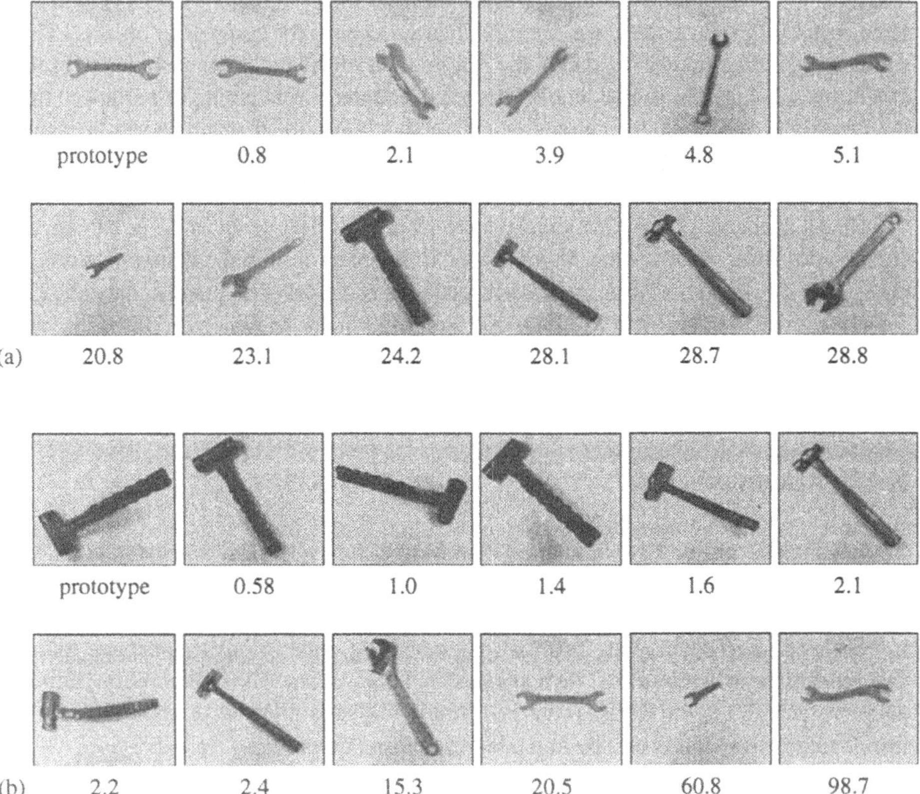

Fig. 7. Using modal strain energy to compare a prototype wrench (a) and a prototype hammer (b) with different hand tools. Silhouettes were first extracted from each tool image, and then the strongest corresponding contour points were found. Mode amplitudes for the first 22 modes were recovered and used to warp the prototype onto the other tools. In (a) the modal strain energy that results from deforming the prototype wrench to each tool is shown below each image. Similarly, in (b), the strain is measured for aligning the hammer prototype. As can be seen, strain energy provides an good measure for similarity.

true for regularization [11, 29, 38], where the energies of a physical model can be related directly with measurement and prior probabilities used in Bayesian estimation [37].

In modal models, as in regularization, a direct connection can be drawn between the uncertainty measures familiar in statistical estimation and the energy formulation of our physical model. In statistical estimation, we need to combine knowledge about the uncertainty of our measurement data with confidence in our prior model. In our energy-based paradigm, the confidence in the prior model is expressed both as the model's initial resting shape (the mean) and the local material stiffnesses of the model (the covariance).

If we assume independent distribution of measurement data, the stiffness matrix and inverse covariance matrix are equivalent [8, 18]. Similarly, there is an inverse relationship between the eigenmode frequencies and the principal variances (modulo a scale factor). In this chapter, we use modal strain energy for comparing objects. This strain energy is computed by taking the dot product of the modal amplitudes with the modal frequencies. The modal amplitudes are decoupled, each having a "variance" that is inversely proportional to the mode's eigenvalue. Our modal strain energy measure therefore yields a Mahalonobis distance metric that can be directly related to multi-dimensional Gaussian probability distributions.

Thus it is plausible to use principal components analysis to learn an shape category's eigenmodes from training data. Using a modal model as an initial estimate, we would then iteratively learn the "true" modes via recursive principal components analysis [22]. As a result, we would obtain a regularized learning scheme in which the initial covariance matrix, Ω^{-2} is iteratively updated to incorporate the observed modal parameter covariances.

8 Conclusion

The advantages afforded by our method stem from the use of the finite element technique of Galerkin surface approximation to avoid sampling problems and to incorporate outside information such as feature connectivity and distinctiveness. This formulation has allowed us to develop an information-preserving, rotation-invariant shape matrix that models the distribution of "virtual mass" within the data. The combination of finite element techniques and a mass matrix formulation have allowed us to avoid setting initial parameters, and to handle much larger deformations.

It is important to emphasize that the transformation to modal space not only allows for automatically establishing correspondence between clouds of feature points; the same modes (and the underlying FEM model) can then be used to describe the deformations that take the features from one position to the other. The amount of deformation required to align the two feature clouds can be used for shape comparison and description, and to warp the original images for alignment and sensor fusion. The power of this method lies primarily in it's ability to unify the correspondence and comparison tasks within one representation.

Finally, we note that the descriptions computed are canonical, and vary smoothly even for very large deformations. This allows them to be used directly for object recognition as illustrated by the airplane and hand-tool examples in the previous section. Finally, because the deformation comparisons are physically-based, we can determine whether or not two shapes are related by a simple physical deformation. This has allowed us to identify shapes that appear to be members of the same category.

References

1. A. Baumberg and D. Hogg. Learning flexible models from image sequences. In *Proc. European Conference on Computer Vision*, pages 299–308, Stockholm, Sweden, May 1994.
2. I. Biederman. Human Image Understanding: Recent Research and a Theory. *Computer Vision, Graphics and Image Processing*, 32(1):29–73, 1985.

3. A. Blake, R. Curwen, and A. Zisserman. A framework for spatiotemporal control in the tracking of visual contours. *International Journal of Computer Vision*, 11(2):127–146, 1993.
4. A. Blake and A. Zisserman. *Visual Reconstruction*. M.I.T. Press, 1987.
5. F. Bookstein. Principal warps: Thin-plate splines and the decomposition of deformations. *IEEE Trans. Pattern Analysis and Machine Intelligence*, 11(6):567–585, June 1989.
6. T. Boult, S. Fenster, and T. O'Donnell. Physics in a fantasy world vs. robust statistical estimation. In *Proc. NSF Worksthop on 3D Object Recognition*, New York, NY, November 1994.
7. I. Cohen, N. Ayache, and P. Sulger. Tracking points on deformable objects. In *Proc. European Conference on Computer Vision*, Santa Margherita Ligure, Italy, May 1992.
8. T. Cootes. Combining point distribution models with shape models based on finite element analysis. In *Proc. British Machine Vision Conference*, 1994.
9. T. Cootes, D. Cooper, C. Taylor, and J. Graham. Trainable method of parametric shape description. *Image and Vision Computing*, 10(5):289–294, June 1992.
10. S. Geman and D. Geman. Stochastic relaxation, Gibbs distribution, and Bayesian restoration of images. *IEEE Trans. Pattern Analysis and Machine Intelligence*, 6(11), November 1984.
11. W. Grimson. An Implementation of a Computational Theory for Visual Surface Interpolation. *Computer Vision, Graphics, and Image Processing*, 22:39–69, 1983.
12. P. W. Hallinan. A Low-Dimensional Representation of Human Faces For Arbitrary Lighting Conditions. Technical Report 93-6, Harvard Robotics Lab, Cambridge, MA, December 1993.
13. M. Kass, A. Witkin, and D. Terzopoulos. Snakes: Active contour models. *International Journal of Computer Vision*, 1:321–331, 1987.
14. G. Kimeldorf and G. Wahba. A correspondence between Bayesian estimation and on stochastic processes and smoothing by splines. *Annals of Mathematical Statistics*, 41(2):495–502, February 1970.
15. M. Kirby and L. Sirovich. Application of the Karhunen-Loeve procedure for the characterization of human faces. *IEEE Trans. Pattern Analysis and Machine Intelligence*, 12(1):103–108, 1990.
16. M. Leyton. Perceptual organization as nested control. *Biological Cybernetics*, 51:141–153, 1984.
17. D. Marr and K. Nishihara. Representation and Recognition of the Spatial Organization of Three-dimensional Shapes. In *Proc. of the Royal Society - London B*, volume 200, 1978.
18. J. Martin, A. Pentland, and R. Kikinis. Shape analysis of brain structures using physical and experimental modes. In *Proc. Computer Vision and Pattern Recognition*, June 1994.
19. T. McInerney and D. Terzopoulos. A finite element model for 3-D shape reconstruction and nonrigid motion tracking. In *Proc. Fourth International Conference on Computer Vision*, May 1993.
20. H. Murase and S. Nayar. Learning and Recognition of 3D Objects from Appearance. In *Proc. of IEEE Workshop on Qualitative Vision*, pages 39–50, New York, NY, June 1993.
21. C. Nastar and N. Ayache. Fast segmentation, tracking, and analysis of deformable objects. In *Proc. Fourth International Conference on Computer Vision*, pages 275–279, 1993.
22. E. Oja and J. Karhunen. Nonlinear PCA: Algorithms and Applications. Technical Report A18, Helsinki University of Technology, Laboratory of Computer and Information Sciences, SF-02150 Espoo, Finland, 1993.
23. A. Pentland. Perceptual organization and representation of natural form. *Artificial Intelligence*, 28(3):293–331, 1986.
24. A. Pentland. Automatic extraction of deformable part models. *International Journal of Computer Vision*, 4(2):107–126, March 1990.
25. A. Pentland. Computational Complexity Versus Virtual Worlds. *Computer Graphics*, 24(2):185–192, 1990.

26. A. Pentland and B. Horowitz. Recovery of non-rigid motion and structure. *IEEE Trans. Pattern Analysis and Machine Intelligence*, 13(7):730–742, July 1991.
27. A. Pentland, B. Moghaddam, T. Starner, O. Oliyide, and M. Turk. View-based and modular eigenspaces for face recognition. In *Proc. Computer Vision and Pattern Recognition*, pages 84–91, 1994.
28. A. Pentland and S. Sclaroff. Closed-form solutions for physically-based shape modeling and recognition. *IEEE Trans. Pattern Analysis and Machine Intelligence*, 13(7):715–729, July 1991.
29. T. Poggio, V. Torre, and C. Koch. Computational Vision and Regularization Theory. *Nature*, 317:314–319, September 1985.
30. E. Rosch. Cognitive representations of semantic categories. *Journal of Experimental Psychology: General*, 104:193–233, 1975.
31. A. Samal and P. Iyengar. Natural shape detection based on principle components analysis. *SPIE Journal of Electronic Imaging*, 2(3):253–263, July 1993.
32. S. Sclaroff and A. Pentland. A modal framework for correspondence and recognition. In *Proc. Fourth International Conference on Computer Vision*, pages 308–313, May 1993.
33. S. Sclaroff and A. Pentland. Modal Matching for Correspondence and Recognition. Technical Report Vision and Modeling TR-201, MIT, May 1993.
34. G. Scott and H. Longuet-Higgins. An algorithm for associating the features of two images. In *Proc. Royal Society of London*, number 244 in B, pages 21–26, 1991.
35. L. Shapiro. Towards a vision-based motion framework. Technical report, Oxford University, 1991.
36. L. Staib and J. Duncan. Parametrically deformable contour models. In *Proc. Computer Vision and Pattern Recognition*, pages 98–103, 1989.
37. R. Szeliski. *Bayesian Modeling of Uncertainty in Low-Level Vision*. Kluwer Academic Publishers, 1989.
38. D. Terzopoulos. The Computation of Visible Surface Representations. *IEEE Trans. Pattern Analysis and Machine Intelligence*, 10(4):417–438, July 1988.
39. D. Terzopoulos and D. Metaxas. Dynamic 3-D models with local and global deformations: Deformable superquadrics. *IEEE Trans. Pattern Analysis and Machine Intelligence*, 13(7):703–714, July 1991.
40. D. Terzopoulos, A. Witkin, and M. Kass. Constraints on deformable models: Recovering 3D shape and nonrigid motion. *Artificial Intelligence*, 36:91–123, 1988.
41. C. Thorpe. Machine learning and human interface for the CMU navlab. In *Proc.. Computer Vision for Space Applications*, Juan-les-Pins, France, September 1993.
42. M. Turk and A. Pentland. Eigenfaces for recognition. *Journal of Cognitive Neuroscience*, 3(1):71–86, 1991.

Time Representation of Deformations : Combining Vibration Modes and Fourier Analysis

Chahab Nastar[1]* and Nicholas Ayache[2]

[1] Massachusetts Institute of Technology, The Media Laboratory, 20 Ames street, Cambridge, MA 02139, USA
[2] INRIA Sophia-Antipolis, B.P. 93, 06902 Sophia-Antipolis Cédex, France

Abstract. We present a method for analysis of nonrigid motion in time sequences of volume images (4D data). In this method nonrigid motion of the deforming object contour is dynamically approximated by a deformable surface. In order to reduce the number of parameters describing the deformation, we make use of modal analysis which provides a spatial smoothing of the surface, and Fourier analysis on time signals of the main deformation spectrum components, which provides a temporal smoothing. Thus, a complex dynamic deformation is represented by very few parameters : the main excited spatial modes and the main Fourier harmonics. Therefore, 4D data can be analyzed and compressed in a very efficient way. The power of the approach is illustrated by results on a 4D scan heart data.

1 Introduction

For about a decade, computer vision researchers have shown interest in nonrigid motion analysis using physically-based deformable models [11, 5, 4].

Among these works, *modal analysis* has the advantage of being a frequency-based technique in which nonrigid motion is expressed in the free vibrations basis (modes) of the deformable object [9, 7, 6, 8].

More recently, many researchers have focused on time evolution of deformable models in order to analyze time sequences of images [10, 3, 1].

In this paper, we propose a unified approach for nonrigid motion estimation from time sequences of multidimensional images by taking into account both the spatial and the temporal frequencies of a deformable model.

Our method takes place in three stages :

1. Recover the deformation field between each pair of succesive 3D images,
2. Express the modal amplitudes of the deformation at each time t,
3. Express Fourier còefficients of the time-depending modal amplitudes.

Then it is possible to compress the information contained in the modal amplitudes, either by skipping some spatial modes and/or by skipping some temporal Fourier harmonics.

* This work was done while the author was at INRIA Rocquencourt, B.P. 105, 78153 Le Chesnay Cédex, France.

The basic justification of the approach relies in the following : for most structures that do not deform too chaotically, the main excited modes are low-frequency ones ; on the top of that, for heart data, modal amplitudes as a function of time are periodic and sine-like.

For stages 1 and 2, it is possible to constrain the modes beforehand, by letting a deformable model evolve along low-order modes. The advantage is of course a reduced numerical compexity and a low-order smoothing of the deformation allowing its robust recovery.

Note that stage 1 can be performed not only by our deformable model, but also by optical flow or any other technique providing the motion field

Canine 4D heart data from the Dynamic Spatial Reconstructor (DSR, a high speed X-ray CT scanner) was used as an input to our method. It consists of 18 volume (or 3D) images during a single cardiac cycle. Each volume image has a spatial resolution of $98 \times 100 \times 110$. In figure 1 one can observe this data partially. There is dye injected into the left ventricle, which shows up clearly. Note the difficulty of interpretation of such complex and huge data by physicians. The challenge is to analyze such tremendous amount of information (19,404,000 bytes, with each voxel being coded on 1 byte) and supply physicians with quantitative results, and a few parameters describing the motion. Our method has important implications in automatic diagnosis of heart diseases and in 4D data compression.

2 An elastically deformable model

A discrete mass spring mesh of N nodes can be elastically deformed in images to track the contour of the deforming object (e.g. the left ventricle) [7].

More precisely, in 3D, the deformation of the system is governed by the $3N$-dimensional differential matrix equation [2] :

$$\mathbf{M}\ddot{\mathbf{U}} + \mathbf{C}\dot{\mathbf{U}} + \mathbf{K}\mathbf{U} = \mathbf{F}(t) \qquad (1)$$

where $\mathbf{U}$ is a vector storing nodal displacements $\mathbf{M}$, $\mathbf{C}$ and $\mathbf{K}$ are respectively the mass, damping and stiffness matrices of the system, and $\mathbf{F}$ is the image force which has the system attracted by image edges. Equation (1) is the finite element formulation of the deformation process. It yields an estimation of the displacement field $\mathbf{U}$ between each pair of successive images.

3 Vibration modes of the model

Modal analysis is a standard engineering technique allowing more effective computations and a closed-form solution of the deformation process [2]. It was first introduced in computer vision by Pentland's team [9] ; we have recently suggested a novel and powerful approach to modal analysis of deformable surfaces, by extracting the analitical expression of modes for certain surface topologies [6]. Let us expose briefly the modal analysis technique.

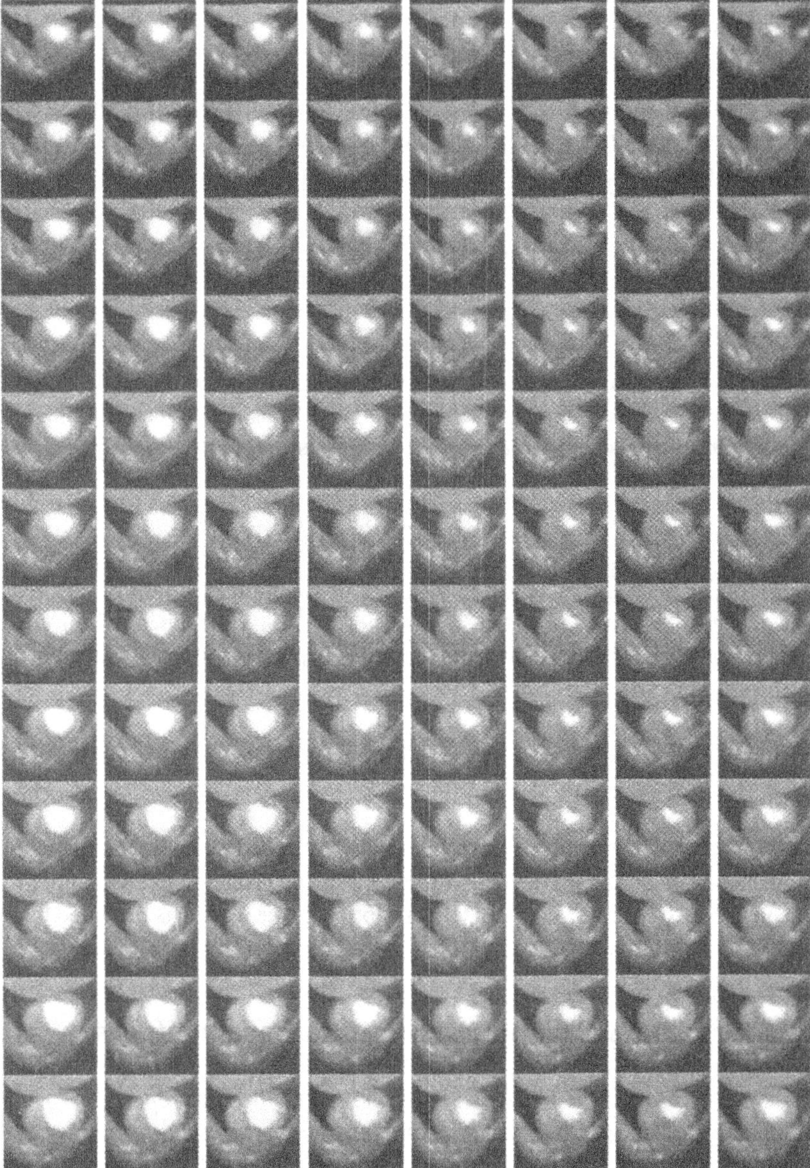

Fig. 1. 4D data. Each image is in the xy plane. z coordinate is along the vertical axis, t coordinate is along the horizontal axis.

Instead of solving directly the equilibrium equation (1), one can transform it by a change of basis :

$$\mathbf{U} = \mathbf{P}\tilde{\mathbf{U}} \tag{2}$$

where $\mathbf{P}$ is the square nonsingular transformation matrix of order $3N$ to be determined, and $\tilde{\mathbf{U}}$ is referred to as the *generalized displacements* vector. One effective way of choosing $\mathbf{P}$ is setting it to $\boldsymbol{\Phi}$, a matrix whose entries are the eigenvectors of the generalized eigenproblem :

$$\mathbf{K}\phi = \omega^2\mathbf{M}\phi \tag{3}$$

$$\mathbf{U}(t) = \boldsymbol{\Phi}\tilde{\mathbf{U}} = \sum_{i=1}^{3N} \tilde{u}_i(t)\phi_i \tag{4}$$

Equation (4) is referred to as the modal superposition equation. ϕ_i is the ith mode, $\tilde{u}_i$ its *amplitude*, and ω_i its frequency. The new modal basis simultaneously diagonalizes $\mathbf{M}$ and $\mathbf{K}$, and provided that matrix $\tilde{\mathbf{C}} = \boldsymbol{\Phi}^T\mathbf{C}\boldsymbol{\Phi}$ is diagonal as well, the governing matrix-form equations decouple into $3N$ scalar equations :

$$\ddot{\tilde{u}}_i(t) + \tilde{c}_i\dot{\tilde{u}}_i(t) + \omega_i^2\tilde{u}_i(t) = \tilde{f}_i(t) \tag{5}$$

The amplitudes $(\tilde{u}_i(t))_{i=1,\dots,3N}$, are obtained by solving these equations at time t, and the displacement of the structure's nodes is obtained by the modal superposition equation.

In practice, we wish to approximate nodal displacements by picking up p low-frequency modes, where $p \ll 3N$.

$$\mathbf{U}(t) \approx \sum_{i=1}^{p} \tilde{u}_i(t)\phi_i \tag{6}$$

The $(\phi_i)_{i=1,\dots,p}$ are the *reduced modal basis* of the system. This is the major interest of modal analysis : it allows a closed-form solution by selecting a few number of low-frequency modes. Therefore a compact description of the motion is provided by *spatial smoothing.*

4 Performing FFT on time signals of modal amplitudes

In [8], we introduced the *deformation spectrum* and studied its properties.

The deformation spectrum of a motion is the graph representing the modal amplitudes as a function of mode rank[3] : $\tilde{u}_i(t) = f(i)$. The deformation spectrum is initially drawn for a deformation occuring between two image frames : it describes which modes are excited, and how, in order to deform the object in the first image into the object in the second image. Thus a single deformation spectrum gives quite a *static* information about the spatial frequencies of the motion, whereas the time parameter t is not really taken into account.

However, time evolution is really what we are interested in. A deformation process can be much better described and interpreted if we have a time sequence of images.

[3] Note that we truncate the spectra to the p lowest frequency modes.

Therefore, we can draw T consecutive deformation spectra for $T+1$ frames of images showing the time evolution of the deformation process. For a chosen mode i, we now show interest in the time signals : $\tilde{u}_i(t) = g(t)$ for different values of i.

Let us track the moving ventricle through all image frames. We wish to find a minimum number of paramters describing this sequence. Can we compress the 4D data information for its further analysis, storage or transmission ?

Note $(P_0, \ldots, P_T)$ the positions of this surface during the $T+1$ frames of the cardiac cycle. $3N \times (T+1)$ parameters describe the 4D data (i.e. 345,600 parameters with $N = 6400$ and $T = 17$).

Let us now track the ventricle in the reduced modal space. We obtain T truncated spectra that store the low-frequency modal amplitudes through the cycle. 4D data is now stored in $3N + T \times p$ parameters (i.e. 20,985 parameters with $p = 105$).

We wish to discard as many modes as possible. Therefore we define a criterion for keeping only the most excited modes among the p low-frequency modes. We compute :

$$L^2(i) = \int_0^T [\tilde{u}_i(t)]^2 dt$$

It appears that high values of L^2 are clustered ; it is then easy to automatically select the first cluster that corresponds to the q modes of highest L^2. In this example $q = 9$.

We now can describe the motion by $3N + T \times q$ (i.e. 19,353) parameters.

Let us draw the time evolution of the q selected modal amplitudes. They are represented by groups of three in figure 2.

These curves have globally a single period sine shape, which is an expectable result for low-frequency modes during a cardiac cycle. Their shape is quite like the ventricle volume curve as a function of time during a cardiac cycle. The shape of these curves encourage us to perform a Fast Fourier Transform : most probably only a few Fourier harmonics will describe the time evolution of the curves.

In order to illustrate this point, let us choose one among these $q = 9$ spectral components (or modal amplitudes). Figure 3 shows the time evolution of this particular amplitude. In figure 4 are displayed the real and imaginary parts of the corresponding Fourier Transform ; we observe that indeed only low-frequency Fourier harmonics are excited. Therefore we keep the harmonics of rank 0, 1 and 2 of the Fourier spectra (thus, with symmetry considerations, a total of $H = 5$ harmonics, see figure 4) ; we then reconstruct the time signal in figure 5. In this way, we couple modal analysis and Fourier analysis : the number of the parameters allowing a total reconstruction of the motion is now $3N + H \times q$. Note that among these parameters, only $H \times q$ i.e. 45 parameters describe the motion itself, this very low number being quite constant for different smooth nonrigid motions that we would want to describe.

As for motion compression, instead of keeping a whole 3D image at every instant (that makes $t_x\ t_y\ t_z = 1,078,000$ parameters per instant) we now keep only $H \times q/(T+1) = 2.5$ parameters per instant. This compression of 4D data is impressive : it means that, provided that we keep the first shape, we are able to synthetize the motion with 2.5 parameters per instant. This 4D data is then described by $3N + H \times q = 19,245$ parameters instead of $(T+1) \times t_x\ t_y\ t_z = 19,404,000$, which makes a compression of more than 10^3, for a cardiac cycle whose duration is approximately 1 second.

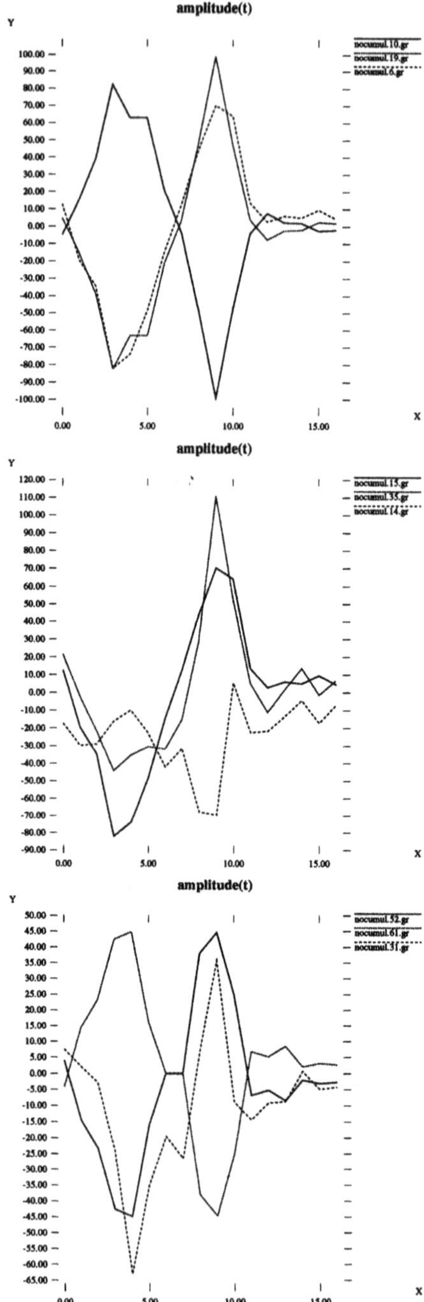

Fig. 2. Time evolution of the $q = 9$ main modal amplitudes during a cardiac cycle

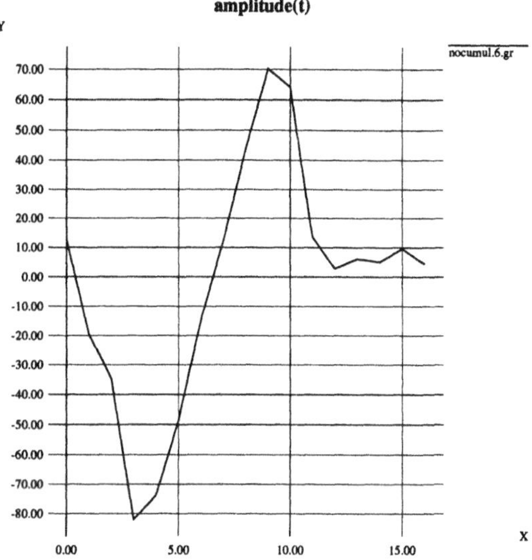

Fig. 3. Time evolution of one of the main modal amplitudes

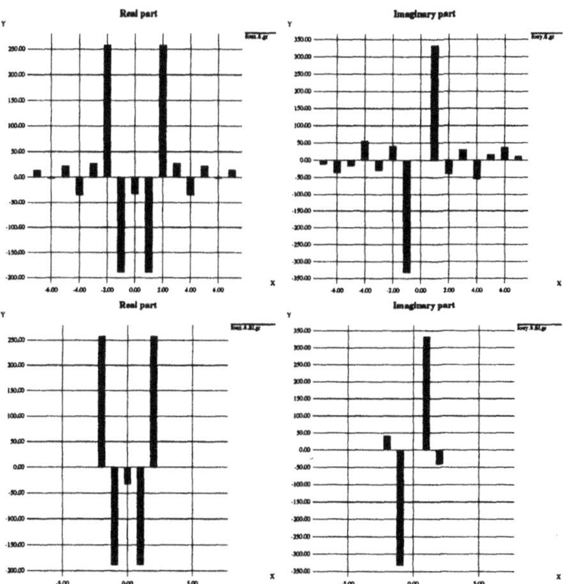

Fig. 4. Fourier and Truncated Fourier spectra of the modal amplitude displayed in figure 3

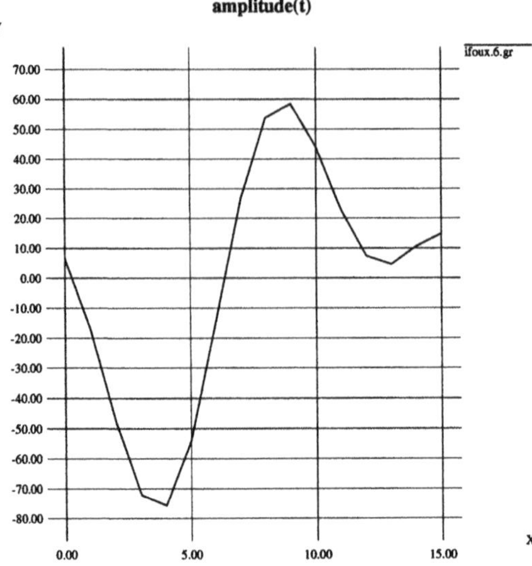

Fig. 5. Reconstruction of the time evolution of the modal amplitude in figure 3 from truncated Fourier spectra

On the top of the compression ability, the synthetic information that our space-time processing provides allows comparison of nonrigid motion by comparing very few parameters, and is an improvement of what we propose in [8]. The method is indeed useful not only for compression, but also for analysis of dynamic motion for diagnosis purposes.

In figure 9, we show the evolution of the left ventricle during a cardiac cycle. In each figure, we observe : left, the evolution of the ventricle in the real space (equation 1) ; center, the computation with $q = 9$ modes (spatial smoothing) ; right, space-time processing with 9 modes and 5 Fourier harmonics.

Finally, figure 6 displays the mean Euclidean error between the real-space mesh and respectively the spatial and spatio-temoporal mesh. Note that this error (in voxels) is extremely low. The position of the two peaks indicate that the error increases between diastole and systole, where the rate of shape change is also maximum.

5 Conclusion

We introduce space-time analysis of nonrigid motion from 4D data by performing a modal analysis on a deformable surface (spatial smoothing), and then by recovering the time evolution of the main modal amplitudes by Fourier analysis (temporal smoothing). This combination method provides a very nice description of 4D data by very few parameters. It has important implications on medical nonrigid motion analysis (automatic diagnosis...), and also provides a tremendous compression of multidimensional nonrigid motion.

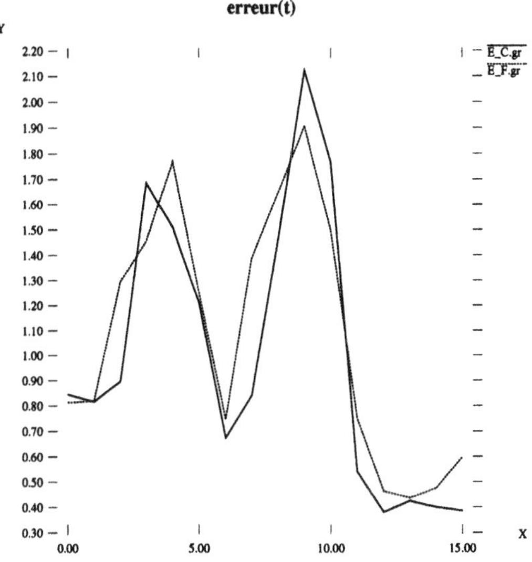

Fig. 6. Time evolution of the error, for modal and modal plus Fourier recovery.

Acknowledgements

We wish to thank Dr. Richard Robb at Biomedical Imaging Resource, Mayo Foundation/Clinic, for providing the DSR data. This work was supported in part by a grant from Digital Equipment Corporation.

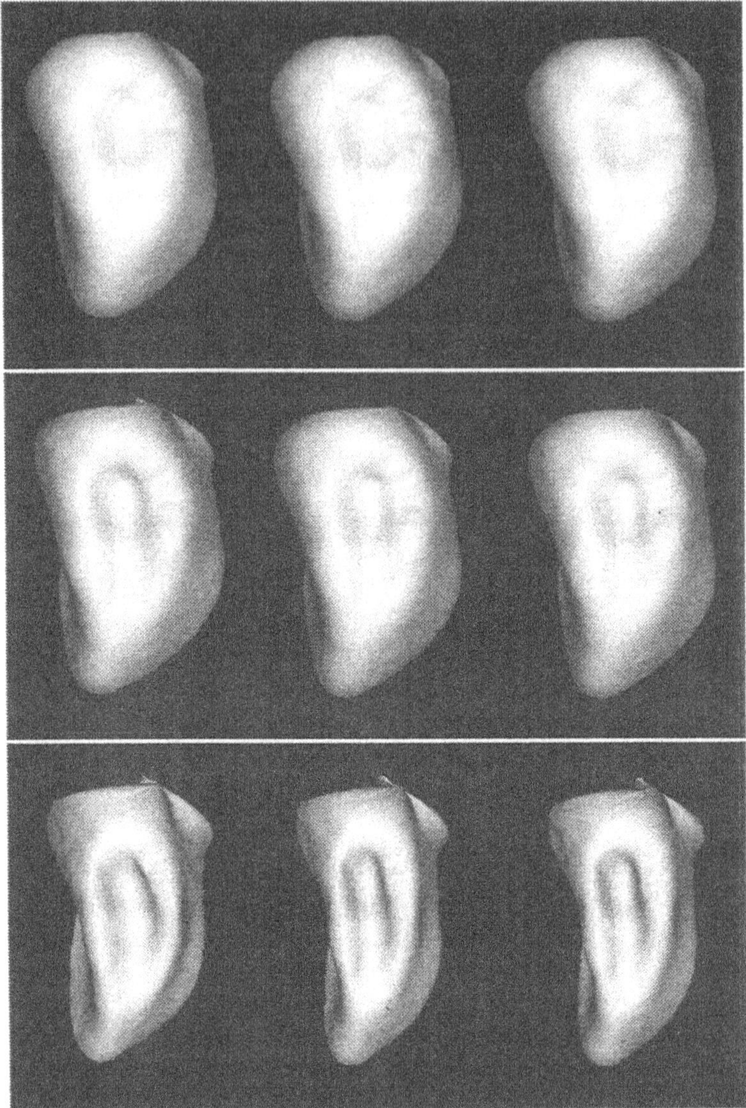

Fig. 7. Space-time evolution of the left ventricle over the 1st third of a cardiac cycle. Left : finite element formulation. Middle : Modal recovery. Right : Modal plus Fourier recovery.

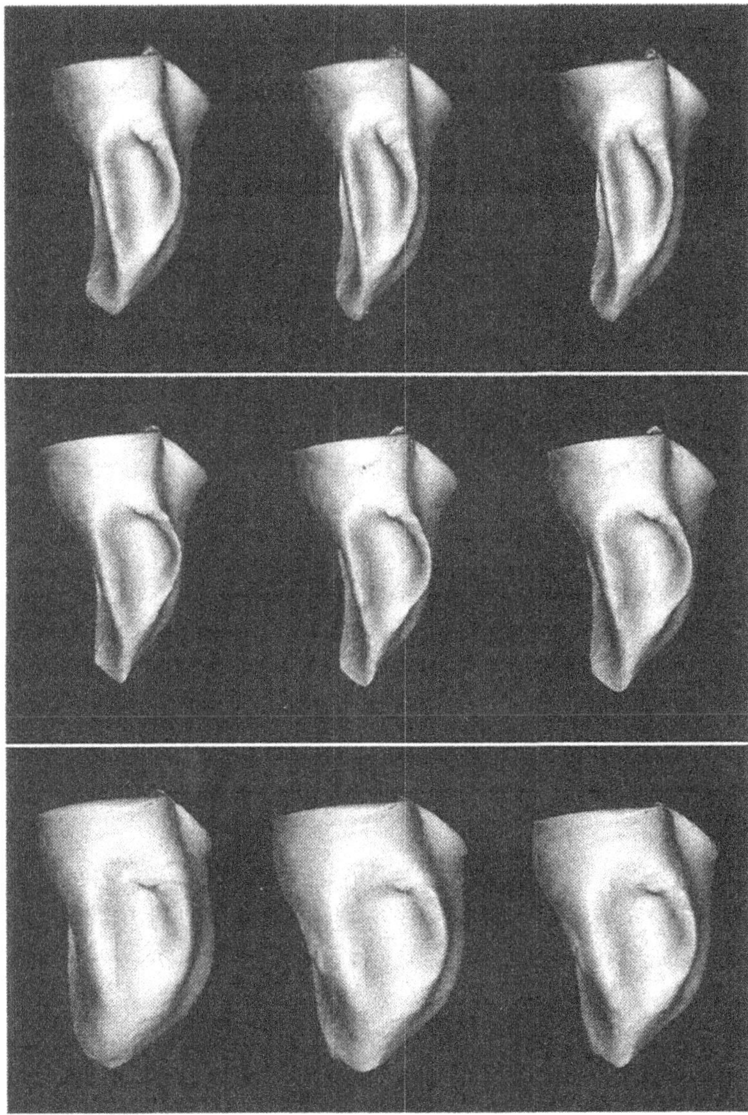

Fig. 8. Space-time evolution of the left ventricle over the 2nd third of a cardiac cycle. Left : finite element formulation. Middle : Modal recovery. Right : Modal plus Fourier recovery.

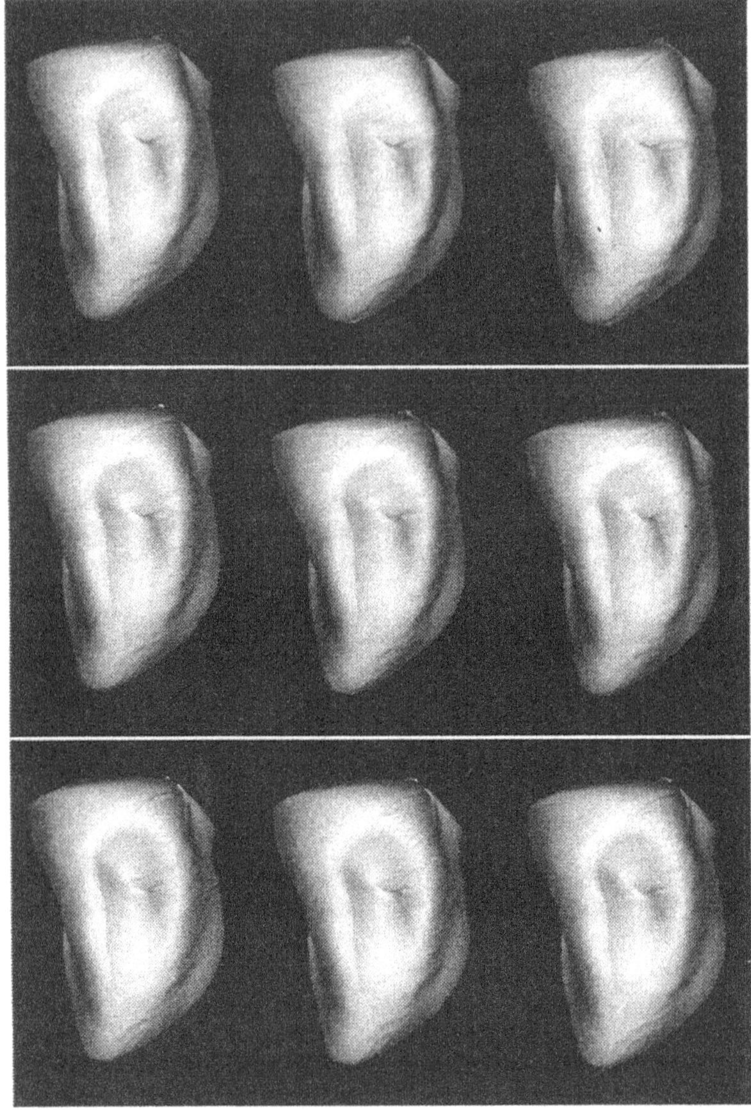

Fig. 9. Space-time evolution of the left ventricle over the 3rd third of a cardiac cycle. Left : finite element formulation. Middle : Modal recovery. Right : Modal plus Fourier recovery.

References

1. Chris A. Bartels, Alan C. Bovik, and Chris E. Griffin. Spatio-temporal tracking of material shape change via multi-dimensional splines. In *Proceedings of the IEEE Workshop on Biomedical Image Analysis*, Seattle, June 1994.
2. Klaus-Jurgen Bathe. *Finite Element Procedures in Engineering Analysis*. Prentice-Hall, 1982.
3. John C. McEachen, Arye Nehorai, and James S. Duncan. A sequential filter for temporal analysis of cardiac motion. In *Proceedings of the IEEE Workshop on Biomedical Image Analysis*, Seattle, June 1994.
4. T. McInerney and D. Terzopoulos. A finite element model for 3-D shape reconstruction and nonrigid motion tracking. In *IEEE Proceedings of the Fourth International Conference on Computer Vision*, pages 518–523, Berlin, June 1993. IEEE.
5. D. Metaxas and D. Terzopoulos. Shape and non-rigid motion estimation through physics-based synthesis. *IEEE Transactions on Pattern Analysis and Machine Intelligence*, 15(6):580–591, 1993.
6. Chahab Nastar. Vibration modes for nonrigid motion analysis in 3D images. In *Proceedings of the Third European Conference on Computer Vision (ECCV '94)*, Stockholm, May 1994.
7. Chahab Nastar and Nicholas Ayache. Fast segmentation, tracking, and analysis of deformable objects. In *Proceedings of the Fourth International Conference on Computer Vision (ICCV '93)*, Berlin, May 1993.
8. Chahab Nastar and Nicholas Ayache. Classification of nonrigid motion in 3D images using physics-based vibration analysis. In *Proceedings of the IEEE Workshop on Biomedical Image Analysis*, Seattle, June 1994.
9. Alex Pentland and Stan Sclaroff. Closed-form solutions for physically based shape modelling and recognition. *IEEE Transactions on Pattern Analysis and Machine Intelligence*, PAMI-13(7):715–729, July 1991.
10. P. Shi, A. Amini, G. Robinson, A. Sinusas, C. T. Constable, and J. Duncan. Shape-based 4d left ventricular myocardial funtion analysis. In *Proceedings of the IEEE Workshop on Biomedical Image Analysis*, Seattle, June 1994.
11. Demetri Terzopoulos, Andrew Witkin, and Michael Kass. Constraints on deformable models: recovering 3-D shape and nonrigid motion. *AI Journal*, 36:91–123, 1988.

Physics in a Fantasy World vs Robust Statistical Estimation

Terrance E. Boult[1]* Samuel D. Fenster[2] and Thomas O'Donnell[2]

[1] tboult@eecs.lehigh.edu
EECS Dept., 304 Packard Laboratory, Lehigh University,
19 Memorial Drive West, Bethlehem, PA 18015, USA

[2] [odonnell | fenster]@cs.columbia.edu
Dept. of Computer Science, Columbia Univ., NYC NY 10027, USA.

Abstract. Deformable models in the "physically-based" paradigm are almost always formulated in an ad-hoc fashion, not related to physical reality – they apply the equations on physics in a fantasy world. This paper discusses some of the drawbacks of this approach. Still these techniques have shown themselves to be useful, so there must be something here. This paper reinterprets these "physics-based" techniques by putting them into a framework of robust statistics. We use this framework to analyze the problems and ad-hoc solutions found in common physically-based formulations. These include incorrect prior shape models; bad relative weights of various energies; and the two-stage approach to minimization (adjusting global, then local shape parameters). We examine the statistical implications of common deformable object formulations. In our reformulation, the units are meaningful, training data plays a fundamental role, different kinds of information may be fused, and certainties can be reported for the segmentation results. The robust aspects of the reformulation are necessary to combat interference from the necessarily large amount of unmodeled image information.

1 Introduction

Deformable models have proven to be a useful tool for finding shapes in image data, for addressing both segmentation and model acquistion. This paper examines two views of the same processes: the first is the "physics-based" analogy; the second is an interpretation in terms of robust statistical estimation. The physical analogy, however, is an analogy to a "fantasy world" where data actually exerts force on the model and materials have elastic properties unrelated to the actual object, e.g. one can have a model deforming to describe a bone, despite the fact that the bone is very rigid!

While the physical analogy may help some understand the fitting process and be useful for graphics designers, we believe it leads to many problems for vision/recover systems and may hamper future developments. This paper examines the assumptions that are usually made when formulating a deformable model, and the issues that are

* This work supported in part by ARPA Contract DACA-76-92-C-007 and by NSF PYI award #IRI-90-57951, with industrial support from Siemens, and Texas Instruments

neglected. An example of the difficulties is determining the proper relative strengths for the various "forces" which is, at best, problematic because their natural units are not forces . What for example, is the proper conversion from image gradient magnitude to Netwons? What about the conversion for distance from a hand segmented image contours or SPAMM markers? The answer, unfortunately, is that the researcher must makes these transforms up so that the fitting behavior is acceptable – they invent a fantasy world where their algorithms appears to do the right thing. Because the physical analogy is weak, we have introduced the phrase, "physically-motivated techiques" to describe them.

We suggest a probabilistic paradigm for formulating and analyzing this recovery process. The important difference between this formulation and previous probabilistic formulations is that many of the newer "forces" and scheduling techniques used in the "physics-based" view can be interpreted as using robust statistical techniques as opposed to traditional Bayesian techniques.

It is computational intractable to model a shape's probability directly from all the pixels in an image and their correlations. It is also quite difficult to precisely define a meaningful "optimum" shape and even if one did, finding the globally optimum shape is only tractable for trivial cases. Thus, one must treat some aspects of the image as having unknown, possibly adversarial distributions producing image data which will lead to a false segmentations. The techniques of *robust statistics* are designed to minimize the damage caused by unmodeled data. This may involve reweighting our objective function so as to gradually trade robustness for accuracy as the optimization proceeds.

In the next section, we explore the basis of deformable modeling. We then discuss the shortcomings of existing practice. We then discuss a probabilistic formulation and show how to interpret known effects and existing practices in this framework, and suggest improved approaches.

2 What are Deformable Models?

This section provides an overview of deformable models. Those familiar with them can probably skip it.

Deformable (or active) models are curves or surfaces that iteratively deform from an initial state until they lie on or near structures in the 2D or 3D image data. The process that moves them combines "forces" that are determined from the image and from the current shape of the model. In the original formulation [KWT87], an active contour ("snake") is manually placed near a desired boundary in a 2D image. It changes shape and position, subject to internal stiffness forces and to attraction by nearby pixels of high gradient. It iteratively responds to these forces until it stabilizes at a position which is a compromise between sitting along the boundary (strongest edges) and maintaining smoothness.

A deformable model can be a discrete chain or mesh of node points connected by length constraints (spring, tension) and stiffness constraints (curvature). This is a finite difference model (FDM). It can also be a finite element model (FEM), which is made up of continuous parametric segments or patches, connectedf by continuity constraints. In an FDM, forces of image attraction act only on the nodes, whereas in an FEM, they can act on any point, or all points, of an element. The model may also be represented as a continuous parameterized global shape, as a sum of global shapes (modes), or as a global shape with patches representing local deviations [TM91].

A deformable model is usually formulated as a object with simple physical properties which define its potential energy. This energy is composed of various terms; those which do not depend on the image are divided into *internal energies*, *user energies*. and *external* or image energies. Typical internal energies are related to node distances, bending and curvature. These usually have their squared magnitude summed over the curve or surface. They often simulate the potential energies of mechanical entities such as springs, stiff rods or thin plates. User forces, *e.g. balloon forces* [Coh91], push the model outward or toward some point. Typical external energies assign strengths to image pixels or edgles, *e.g.* using intensity or gradient magnitude multiplied or divided by (often squared) distance. The form of the energy terms are often usually selected to make the solution of a differential equation quick or in partially closed form. The energy function's negative gradient, a vector in parameter space, is the "force" which moves the shape parameters to reduce the energy. In this way, a local energy minimum is sought. We must trust that the initial placement of the model was close enough that the local minimum we found was the "right" one.

Differential equations are set up to simulate the model's movement. These equations sometimes include such physical quantities as mass and viscosity, to stabilize the solution [LL93].

2.1 Analysis

In this section we motivate our approach to deformable models by analysing some of the problems that exist in current formulations. We believe the primary source of these problems is the ad-hoc analogy used for recovery. It manifests itself in many ways, of which we will discuss a few.

Inappropriate "Prior" Model Shapes: If there are no data forces, the process produces a shape which is its "prior" or rest state. Internal energy terms which penalize surface curvature may result in model prior states which in no way describe the object under recovery. Smoothing terms which seek to minimize the total size/bending of the finite elements composing the surface are extremely common in the deformable modeling paradigm [CC90, LL93, KWT87, TM91]. Unfortunately, one way to minimize these values is to shrink the element. As a result, the rest or prior shapes for many of the models are unlikely to resemble the object of interest. For many snake algorithms, *e.g.* [KWT87], the rest-shape is a single point. That is, given no data, the model disappears! And even in the presence of image data, there is always a force pulling inward, so the model will reach equilibrium somewhere short of the object boundary.

In a small subset of deformable model literature, *e.g.* [LL93, LC94], the internal stiffness force is modified to penalize each element based on its difference from an *a priori* specified "preferred" element size. By doing so the models will not disappear in the absence of data, but can be viewed as having an expected shape of either a straight line with fixed size links (for open snakes) or a circle formed from fixed-size links (for closed).

Large Perturbations of Linear FEMs: Almost all of the deformable object literature in vision and graphics has used FEMs, in particular linear FEMs. These are designed for "small" perturbations from an initial state.[3] However, in the case of weak priors, the model are required to undergo large deformations!

The linear FEM provides a first order approximation to the stiffness properties and is generally valid only in a local neighborhood [Bat82]. Inappropriately initializing a model and depending on, for example, balloon forces to push the model towards the data may result in an extreme distortion of even well calibrated model stiffness properties.

There has been more recent work on building global object models with FEM model on top of them. In particular, [TM91, PMY94] consider fitting an superquadric to an underlying model and then using an FEM surface to account for the remaining details. In [PS91], a global model using low order modes is recovered, and then a spline surface is fit on top to account for remaining details. This does reduce the magnititude of deformations, makine FEM models a better approximation. In both cases, however, the underlying model is a generalized blob that is a very rough approximation. For most objects, the resulting deformations are still moderatly large in size. In our recent work, [OFBG94], we build the FEM mesh ontop of a generalized cylinder to allow a more rest state with more detail. Still the deviations may be too large for a linear FEM.

Difficulties Using Residual Data Forces: We define residual forces as those which tend to zero as the distance between model and data tend to zero. An example of residual forces are segmented data forces (long range forces in [TM91]) which are weighted distance between a model point an a data point. Residual data forces also occur when the image energy is the magnitude of the Gaussian-blurred gradient through which the shape passes. The image force goes to zero as an image edge is approached.

Under existing deformable model paradigms if the model is endowed with material stiffness it becomes impossible for it to deform to interpolate the data in the presence of "residual" data forces, even if that data has no noise! In the case of noisy data, the answer is likely to be biased.

On careful reading, one might notice that in many of the deformable modeling papers the reported "stiffness" parameters are set to very small values,[4] which in turns allows the surface to nearly interpolate the data. It also means, however, that in regions of little data the surface is free to wobble and curve with little internal constraint.

Another way in which researchers have circumvented this problems is by "inventing" new forces such as a "balloon forces," [Coh91]. These forces are not predicted by the underlying physical analogy as they are neither material forces (stiffness) nor data (image) forces. Rather they are extra forces which, in an ad hoc fashion, force the equilibrium point to be "closer" to the data. (They are also used to provide an initial force to get the model near the data.)

How to determine weights For Data Fusion? The methods in the current literature for determining the scaling values for data forces and material properties are quite ad-hoc, when described at all. In the case of a stiff model being influenced by noisy data,

---

[3] Exceptions using non-linear FEMs include [TW93] and [HG93].

[4] For example in [TM91] the stiffness measures are of magnitude 10^{-6}.

how should the the data forces and material stiffnesses be scaled such that the "best" fit is arrived at? That is, how should the internal and external forces be related? These questions are by and large ignored by papers in the deformable model community. A related question is the fusion of information, how should different *types* of external forces, be related to one another in the "physically-based" paradigm? Determining the proper relative strengths for fusion of these external forces is, at best, problematic because their very units differ! Even the units of internal energy (stiffness) and external energy (data forces) don't match. What, for example, is the proper conversion from image gradient mangitude to Netwons?

In some of our prior work on surface modeling via regularization, the multi-sensor fusion problem was examined using stereo and texture information. The weighting of information directly determined the success of the integration, see [MB88]. Without a firm understanding of how the scales sould be determined, fusion is extremely difficult.

"Efficiency" Hacks: Various systems for deformable object models, including [TM91, OFBG94, OGB94] include code for "parameter scheduling." This code is used to insure that in the initial stages of the recovery, only global parameters are adjusted. When the global parameters have settled, adjustment of the local parameters begins. There are two reasons for this parameter scheduling. The first is to reduce computation time; the second is to provide robustness—if the local parameters are allowed to adjusted too soon they can "latch" onto a strong feature that is not part of the object of interest.

While it has advantages, is this method likely to give us the "best" solution? Without some solid foundation for judging the "goodness" of a fit it is impossible to tell what the efficiency costs in terms of fitting.

Poor Justification for Results: Given the ad-hoc nature of current approaches, justification of results to users such as radiologists becomes an issue. In most physically-motivated applications the stiffness and data forces are in reality selected based what works for some small test set of data. Thus it becomes impossible to state, for example, that a segmentation performed yield the "best" estimate of the object contour given the data and some prior knowledge of the domain. If a deformable modeling system is to be used in clinical medicine, having a justifiable interpretation of the process is not just an academic issue, it is a prerequisite to acceptance.

3 A Robust Statistical Approach to Deformable Modeling

In this section a novel view of deformable model recovery is presented. We claim that the weaknesses of existing methods outlined in the previous section may be overcome by recasting deformable object modeling as a robust statistics problem. In doing so it becomes possible to give meaning to both external and internal forces and thereby create a solid basis from which to interpret recovery results. In addition, results may be reported with confidence levels indicating the degree of success. Finally, the training necessary for any recovery system is integrated into the paradigm at a fundamental level.

In the context of robust statistics, more specifically M-estimators (see below), deformable modeling becomes a question of relating sensor models, data uncertainty, model uncertainty, and the uncertainty in our uncertainty distributions to arrive at a solution of maximum likelihood.

We believe this strategy will provide a solid foundation upon which to analyze existing assumptions and techniques and to propose new methods including new "force" formulations and novel "stiffness properties" rooted not on some preconceived notions about how the prior models *should* behave but rather on actual deviations from the expected model gained from training instances.

Giving Meaning to Forces and Stiffness: While there are lots of impressive equations supporting the "physically-based" paradigm, data points are *not* spring forces acting on a spring-like contour or surface. Furthermore, the material properties of the model may be completely unrelated to those of the object undergoing recovery – one can have a model deforming to describe a ceramic plate, despite the fact that the ceramic is rigid! Because the physical analogy does not hold, we prefer to have introduced the phrase, "physically-motivated techiques" to describe such approches.

As we shall see, the statistical interpretation is also very intuitive. A material that is "unlikely" to deviate from its base shape has an uncertainty measure that is more concentrated around the base shape, similar to a sharp energy peak. Data of which we are more certain will also have a more concentrated probability measure, hence larger (but shorter range) forces. If the data is known to be perfect the force would be arbitrarily strong resulting in the interpolation of the data. Similarly, the confidence in the correctness of that initial shape, (*i.e.*, its predilection for maintaining its initial shape) relates to stiffness of deformable models. If the object under recovery is unlikely to have a particular form the model stiffness will penalize a deformation towards that form.

Reporting Confidence Levels: In the probabilistic framework it is possible to report confidence levels for individual results thereby enhancing their value. In any experimental science a result lacking an error bound is a nearly useless result. Up to now errors in physically-motivated recovery have been demonstrated but not well analyzed. Such demonstrations give an *indication* of how well a similar case might do. By approaching recovery from a statistical point of view, however, it becomes possible to provide a confidence level appropriate for each individual experiment based on the confidences in the appropriateness of the prior model and the fidelity of the data for that experiment.

Such confidence levels provide several advantages. First, they create a means for determining the general predictability of a technique. Second, they lend a quantitative basis for disregarding certain recoveries as unlikely. Finally, they enable us to justify our results to end users. Doctors, for example, have knowledge of statistics and will be more likely to understand and embrace results based on statistics rather than on an imperfect analogy to the physical world.

Training can be integrated into both the internal and external force formulations at a fundamental level – the forces themselves are probability distributions to be calculated via training. The relative scaling of these forces becomes a problem of relating probabilities, a basic computation in statistics. And the units of these "forces" are no longer at issue since they are probabilities. Finally by invoking the powerful tools of robust statistics an appropriate prior model may be arrived at.

3.1 The Probability Framework

In this section we explore the probability formulation of deformable models. We shall discuss prior work in this area *after* discussing most of the issues, so that it can be examined in the light of these issues.

The probability we want to maximize with our deformable shape is $P(S \mid I)$, the probability of S being the "right" shape given that I is the image. Of course, since any single configuration has infinitesimal probability, P actually represents a probability density function. We are trying to find the following:

$$\arg\max_S P(S \mid I) = \arg\max_S \frac{P(S \wedge I)}{P(I)} = \arg\max_S \frac{P(I \mid S)P(S)}{P(I)}$$

When looking for the S that maximizes this, we can ignore the denominator, $P(I)$, because it is constant with respect to S. It is the probability of image I occurring, without knowing S. It is useful to decompose $P(S \wedge I)$ into $P(I \mid S)P(S)$ because each of the two factors can be efficiently approximated. Let us examine each factor separately.

$P(S)$ is an *a priori* shape model, a function which measures the inherent likelihood of the shape regardless of the image. It can penalize unlikely sizes, positions and curvatures. Note that this corresponds to the *internal energy*—all components of the energy which do not depend on the image.

$P(I \mid S)$ penalizes a shape if the image does not correspond to it. The *image energy* is derived from this. It is often simply and efficiently modeled as a function of only those image features that are close to shape S. This is a plausible move if nearby features (*e.g.* intensity and gradient at different resolutions) are found to have roughly the same distribution of strength *vs.* distance from S regardless of what S is. However, this assumption, along with what data is being thrown away, should be examined before this simplification is used.

3.2 Mapping Probability to Energy

In the framework of *maximum likelihood estimation* (MLE), the problem of maximizing a probability is often converted to the equivalent problem of minimizing its negative log. So, for instance, we may have a set of independent Gaussian observations y_i, each of known variance σ^2, generated from known x_i by a function $y(x; \mathbf{a})$ with unknown parameters $\mathbf{a}$. Finding $\mathbf{a}$ that maximizes the joint probability of the observations is equivalent to solving a least squares minimization:

$$\arg\max_{\mathbf{a}} P(Y_1 = y_1 \wedge Y_2 = y_2 \wedge \ldots) = \arg\max_{\mathbf{a}} \prod_i P(Y_i = y_i)$$

$$= \arg\max_{\mathbf{a}} \prod_i e^{-\frac{1}{2}(y_i - y(x_i;\mathbf{a}))^2/\sigma^2} = \arg\min_{\mathbf{a}} \sum_i \tfrac{1}{2}(y_i - y(x_i;\mathbf{a}))^2/\sigma^2$$

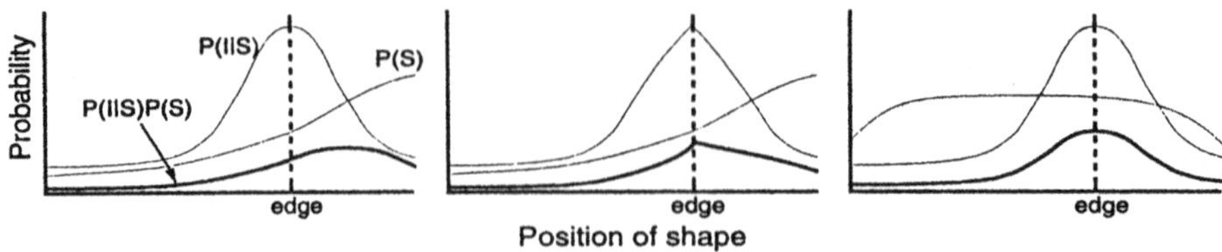

Fig. 1. *When a shape's prior probability is combined with the probability that its boundary could have produced the image edges, MLE may not lie on the edges. The cure is either an image probability that does not level off at its maximum, or a prior probability that is flat around the shape in question.*

Thus, minimizing a sum of energies can be equivalent to maximizing a product of probabilities if $E = -\ln \mathrm{P}$. It is useful to have a sum because it simplifies integrals and partial derivatives (which give us the force). This same probability-energy equivalence is well-known from the Gibbs distribution used for Markov random fields. It is useful to us as a way of formulating the energy of a deformable model, or of analyzing the probabilities implied by an energy formulation.

In all formulations we are aware of, several kinds of energy are added, and they need some kind of relative weighting. Since the units of the individual terms are unrelated, it is not at all clear what these weights should be! But with a probability-energy equivalence, there is no question of weighting. Once the separate probabilities are known, adjusting them with weights makes no sense. In other words, the "calibration" of relative energy strengths takes place during the estimation of probability distributions (perhaps from training data, perhaps from models of the domain and sensor). This framework makes it possible to intelligently integrate different kinds of information, for instance images from multiple sensors. (It may, however, require a goot bit of work.)

The energy minimum achieved is equivalent to a confidence level in the final answer. Thus, the absolute certainty of the locally minimum-energy shape is known, and poor segmentations can be rejected.

Addressing Stiffness Problems: Consider a shape S for which $\mathrm{P}(I \mid S)$ is at a local maximum but $\mathrm{P}(S)$ is not. If both are positive and differentiable then their product will not be at a local maximum. Thus, if S is the shape that best explains the data, but not the *a priori* most likely shape, then the segmentation process will not choose this shape. How far off the maximum will be depends on how sloped is $\mathrm{P}(S)$ and how sloped is $\mathrm{P}(I \mid S)$ in the neighborhood of the maximum.

What does this mean intuitively? That even for a high-likelihood, noiseless image, the segmentation will err. The problem is that a smooth energy maximum produces a force that goes to zero. When any other force is added, the total force at the (former) maximum now points elsewhere. There are two ways to fix this:

1) Make $\mathrm{P}(I \mid S)$ have a nondifferentiable peak when S approaches a suitably strong image edge. This way, imposing a bias from $\mathrm{P}(S)$ does not change the location of the

maximum; the image force does not approach zero at the edge.

2) Make sure P(*S*) is nearly flat for any reasonably likely shape *S*. This makes the internal force close to zero for such *S*'s.

Do we *want* P(*S*) to bias the maximum of P(*I* | *S*)? We probably do not want it to bias *S*'s position toward some *a priori* position—we want to trust the image edges. But we may want to bias the shape away from bumpiness. In this case, P(*S*) should be flat over a reasonable range of *positions* then fall of quiclky.[5] It should not necessarily be flat with respect to other parameters such as "excess" curvature.

Some formulations define the image energy as linear in the distance to the nearest edge. Such a linear potential satisfies the first of the two conditions above—it is pointy; its gradient, the force, does not approach zero near an edge. Inverse distance also has this desirable property. But squared distance lacks it.

We may get unsatisfactory segmentations if the prior shape model, P(*S*), is not good. For example, if (as is usual practice) the function penalizes any kind or amount of curvature, the model will be biased away from the object boundary in even the most exemplary image, unless that object happens to be a circle, sphere, line or plane. If there is an expected shape, the correct thing to penalize is deviation from that shape.

Finding a Gradient to Follow: If we were to accurately model the distribution of edge distance from underlying shape in a non-noisy image, we would usually find that the edges were all within a few pixels of the shape, unless they were caused by some other object. Thus, the probability of an edge, given a shape, would reach zero (or some constant nonzero noise level) a few pixels away from the shape. It is worth noting that a Gaussian would be a very inappropriate model. The optimal shape would need to lie very close to the image edges.

But we encounter a problem when we place an initial approximate shape in the image. Under such a probability distribution, the image force (probability or energy gradient) that is to deform the shape will be zero unless the initial approximation is correct to within a few pixels. Though we would like the shape that maximizes this probability, the image force will not get us there. A reasonable solution is to blur the image energy, at least initially, so the gradient from the strong energy minimum reaches the model. Later on, though, we will want to be using the correct energy function.

Thus, even if our probability model is correct, our minimization process may need to use knowledge about the properties of the domain, rather than simply following a gradient.

3.3 Robust Statistics

In many applications, image edges are not just caused by the object we seek or by sensor noise. There are usually other objects in the scene. Thus, an image energy which reacts to edges as if they were made by the object we seek becomes more or less accurate depending on the distribution of unrelated edges. To model image probability

[5] Obviously, if a probability is to have almost no slope over some range of a parameter but then fall of quickly, it cannot be Gaussian in that parameter. So the shape's position in the training images may not be well modelded as Gaussian.

correctly, image energy would have to discriminate between "bad" edges and "good" edges. While we could make some headway by incorporating expected color, gradient and other properties, we would still have little hope of reaching the desired energy minimum by simply following the energy gradient.

So instead of attempting to model extrinsic edges, we may treat them as largely unknown, and try to figure out how to limit the information that goes into image energy so that such "adversarial" edges can do minimal damage. This would make the image energy a *robust statistic.*

A robust statistic is an approximation of some other statistic which reduces sensitivity to "adversarial" (unmodeled) data. For instance, an average can be perturbed arbitrarily much by a single outlying datum; yet, in the absence of outliers, the average, or "central tendency," may be the statistic we seek. The median is a measure of central tendency that asymptotically approximates the average, but it is more robust—a fixed number of outliers can only change it by a fixed amount, no matter how bad those outliers are. We are interested in two qualities of a robust statistic: *Efficiency* which measures how well the robust statistic approximates the desired statistic. And the *breakdown point* measures what percentage of bad data it takes to degrade the result a certain amount. (The average has a breakdown point of zero.)

Robust techniques are generally used to deal with uncertainty in the underlying probability distribution or to handle "outliers", data that does not fit the uncertainty model. Recasting deformable object modeling as a robust M-estimation problem will help us to analyze the existing assumptions and techniques, and to propose new ones. Robust estimations are becoming used in vision, with M-estimates and "median" based techniques the most common, *e.g.* see [MMRK91, MB93].

Let us consider the definition of an M-estimate as defined in P. Huber's classic text ([Hub81, page 43]) on robust statistics:

Any estimate T_n, defined by an minimum problem of the form:

$$\sum \rho(x_i; T_n) = \min! \tag{1}$$

or by an implicit equation

$$\sum \psi(x_i; T_n) = \mathbf{0}, \tag{2}$$

where ρ is an arbitrary function, $\psi(x;\theta) = (\partial/\partial\theta)\rho(x;\theta)$, is called an M-estimate or maximum likelihood type estimate. [Note that for the choice $\rho(x;\theta) = -\log \mathrm{P}(x;\theta)$ gives the ordinary ML estimate].

In the above definition, T_n is the set of n dependent parameters of a model, each x_i is an independent parameter (data) and "min!" is a compact way of expressing a minimum over all dependent parameters, T_n.[6] The interpretation here is that ρ is a collection of uncertainties and one seeks to find the set of parameters with minimal cumulative uncertainty. The uncertainty captured in ρ could be data uncertainty, modeling uncertainty, or even uncertainty about uncertainty models used in the computation—the form is quite general. If one replaces $\rho(x;\theta)$ with $-\log \mathrm{P}(x;\theta)$ the formulation defines maximum like-

[6] The original definition denotes the data uncertainty distribution with the symbol f, but in this paper we will denote this probability distribution as P.

lihood estimation, hence the name maximum likelihood-type estimator or M-estimator for this more general formulation. Note that robust statistics books are quick to point out that the definition in terms of ψ may only find local minima. A classic example of an M-estimate is when ρ is the sum of squared distances between data points and associated model points. This is least squares modeling, which is the maximum-likelihood estimator for i.i.d. Gaussian noise. Robust M-estimators include weighted least-squares and trimmed least squares, which are sometimes called S-estimators).

We are not the first to note the relationship between deformable models (splines), energy formulations and uncertainties. The relation is over 25 years old—see [KW70]. The Gibbs energy formulation defines such a relationship and has been used in many areas including vision, especially in MAP/MRF studies; see [Sze89, ST89, GG91, VR93]. All past work (as known to the authors), however, relates energy directly to probabilities, and not to the robust statistical M-estimate formulation—the Gibbs energy formulations relates energy to probability by always assuming a maximum likelihood estimation (MLE), rather than robust M-estimation. The difference may be subtle, but we believe it is important. In particular, some of the ad-hoc but basically successful techniques introduced in deformable object models cannot be justified as model/data uncertainty, but they can be justified in terms of robust estimation.

The issues of robustness are related to accuracy of prior knowledge, and to outliers and their impact. For the current uses of deformable modeling, we do not think "uncertainty" models are sufficiently well understood. We believe that viewing them as a robust estimation will help. Outliers are data that cannot be accounted for with the uncertainty model—they lie outside the model. They are a real problem in non-rigid object modeling and can be caused by many things, such as edge features or texture inside the organ, other nearby organs, imager ghosting, etc. Thus the difference between formulation as robust M-estimation and pure Gibbs-based Bayesian estimation can be profound.

By far one of the most difficult issues in statistical estimation is "noise" that is not really noise but rather a signal from unmodeled data. For example, in ventricle modeling, edges from the papillary muscles or other heart chambers can be locally difficulty to distinguish from the inner heart wall, and have been shown to cause unwanted attractions and modeling errors. Object shadows are another example of difficult "noise." Such under-modeled problems wreak havoc with most regression techniques since the "noise" may be, from a statistical point of view, indistinguishable from a true signal. Only domain-specific knowledge, *e.g.* a partial model of the unmodeled data, can distinguish the two. By using *a priori* knowledge and training data, we believe that we can derive appropriate reweighting and trimming schemes to provide locally-tuned robust estimators.

Reinterpreting Data Forces: Let us now reinterpret common image forces by viewing them as robust M-estimates. The decompositions we give are not unique, but give the flavor of what can be learned by the reinterpretation.

A common long-range force, considered in many papers including [TM91, OFBG94], is the scaled distance between nodes and data-points, *assuming a given correspondence*. This force is commonly used for "pre-segmented" data, where the correspondences are computed (usually the nearest point) rather than being given by a user.

To convert from a force to an energy, we integrate. So if the force function for a singe point is $\beta_i d$, it yields a "quadratic energy" of the form $\sum \beta_i d^2 + \gamma$ for some constant of integration γ. Scaling, boundary conditions and choice of measurement units would determine gamma, but for simplicity let us assume $\gamma = 0$. This then defines the ρ function for the M-estimator. Thus this "force" model yields an MLE when the distribution between the data and the original node placements is given by $\log \mathrm{P} = \rho(u; T) = \sum \beta_i d^2$ which, after proper scaling, results in the conclusion that P is Gaussian! (If $\gamma \neq 0$, we have a Gaussian-like distribution with fatter or thiner tails depending on the sign of γ.) By taking the M-estimator point of view we see that a "linear" long distance force is a component of the MLE that assumes the data displacements are Gaussianly distributed (once again implying that we want to keep these displacements small.)

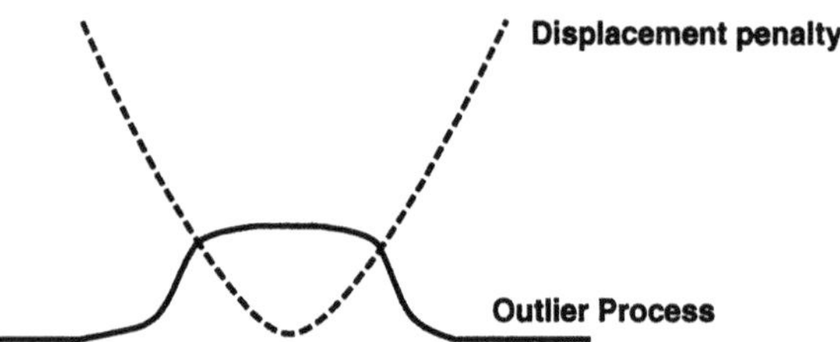

Fig. 2. *The penalty (energy) associated with edge displacement grows with distance from the boundary. However, an outlier process suggests penalties which go to zero because at greater distances it is likely that the edge is not associated with the model and is hence an outlier on the "edge-position" certainty function.*

The above derivation assumed that the point correspondences were given. The interpretation begins to get more complex as this assumption is relaxed, and this is often where the robust M-estimate, as opposed to ML estimates, comes in to play. If one looks at the "energy" for simple image forces, the most common is the sum (integral) of the gradient of a image evaluated on boundary of the shape. Recall that this measure is the probability of the data given the shape. Note that almost all deformable papers use the "edge" location, not the real data. The current techniques do not use edge orientation and use absolute strength (rather than expected edge strength). The likelihood of the edge given the surface, however, depends in a complex way on the edge shape (because of the blurring). The form of these forces, while moderately successful, has been completely ad hoc, being driven by what is convenient to compute. If viewed as a MLE (*i.e.* using Gibbs energy), it has some odd implications: That only the boundary of the model can cause edges—there are no edges inside of an object and no other (non-modeled) objects; that if two edges are nearby, it is more likely the model boundary generated the stronger edge; that the orientation of the edge does not affect its likelihood; and, because of blurring, that an image with a square outline is not a local maximum for a square shape. Note, however, that the forces in general resemble more traditional robust measures such as the Tukey bi-weight [Hub81], with a potential function growing locally but then decaying after the distance between data and model points becomes larger than some parameter; see figure 3. Their justification is not based on viewing the energy as probability, but a mixture of local probability measures and use of the current estimate to "robustly" eliminate outliers. The odd implications above occur because current models only consider the "edge" position, and because they mix data uncertainty (*e.g.* edge position) with uncertainty about the edge even being associated with the model (*i.e.* likelihood that the the edge an outlier). By properly recognizing the two components of the "force" as local estimate of certainty, and by robust reweighting, we can permit greater flexibility in determining the parameters and provide better interpretations to the results.

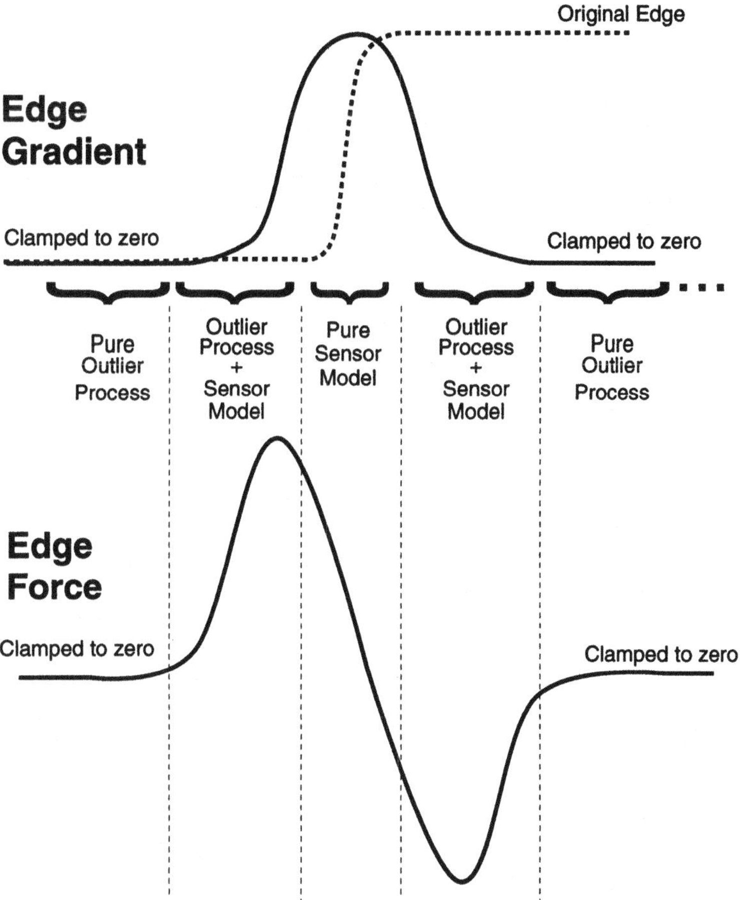

Fig. 3. *Current "image" edge forces use the gradient of a smoothed version of the image. These can be reinterpreted as "short-range" edge forces with robust bi-weight functions.*

Some suggestions this of this view is using expected values for edge strength, and incorporating edge orientation and sign or edge profile information. In terms of robustness, we should recognize that in regions with more clutter the local estimator should be required to have a higher statistical breakdown-point even if it must sacrifice some statistical efficiency [RL87].

It also suggests a simple idea, currently being explored, wherein one develops partial models of the regions surrounding the object of interest so that extraneous edges can be anticipated, even if they are not "modeled." These "partially modeled" edges will be used in conditioning matches and reweighting, but will not be "fit" by the modeling process. For example, if an edge is likely to occur near the edge of interest, *e.g.* the edge of a shadow is expected, we could either use its location to help determine the object edge (if we knew enough and were willing to compute it), or (more cheaply) we could simply insure that the "reweighting" function had sufficiently small support to ignore it.

We believe it is important to separate the sensor model and uncertainty modeling from the robust fitting aspects of the problem. Both are important but they are separate issues. Proper sensor and data uncertainty modeling is critical to allow meaningful statistical measures to be used—it directly affects the quality of the fitting results. The robustness is important to making the fitting process faster and more automated, and hence strongly effects the amount of effort required on the part of the user, but it only mildly effects the fit quality. We believe it is important to first understand the desired answer, and only then worry about approximations to increase the speed. For robust M-estimates of complex parameterizations, formal proofs of "optimality," or even proofs of convergence, are, unfortunately, quite difficult. Thus, while reformulation will provide a better foundation and better implementations, it will still be based on algorithms which require experimental verification.

New Data "Force" Models: A particular class of new force measures suggested by the statistical point of view is what we call conditional forces. Recall that the uncertainty measure we are estimating is $P(I \mid S)$, the probability of the data given the shape. It follows that the "force" between a data point and a node or element should be conditioned on the likelihood that these actually correspond. Simple conditioning of the match, *e.g.* requiring a particular sign of contrast, is just the beginning. We could/should use the shape and sensor models to generate "synthetic" data in a local region and compared this to the measured data. This would, however, require a very large amount of computation, so simpler approximations, such as using local contrast parameters, edge orientation, and curvature and intensity profile slope or curvature (edge sharpness), will after calibration, be used to define data forces.

We also point out that the uncertainty computation in which we are interested is global, and that simply summing local "forces" is tantamount to assuming the associated sources of uncertainty are independent (when conditioned on the current shape approximation). This assumption should be relaxed to include covariance computations of "neighboring" points; the cost/benefit of larger size neighborhoods will have to be studied. With advances in computing, larger and more accurate models will be feasible.

An area where it is clear that there is dependency in the edge information, not currently captured in the shape models, are areas of texture. In the area of medical imaging, in particular for internal organs, the image data is often highly textured.

An issue that will be of particular importance in these investigations is that of calibration. To make meaningful interpretations, we will need "models" of sensor error and prior model uncertainty. In the medical applications, careful experimentation will be necessary to derive such models. This may be the most important aspect of this reformulation—it provides an interpretation where the calibration/learning of model parameters is meaningful. This training will determine the strength of our foundation; if the training data is representative, we will have justifiable methods!

Reinterpreting Parameter Scheduling: As mentioned in section 2.1 various systems for deformable object models, including include code for "parameter scheduling." This code is used to insure that in the initial stages of the recovery, only global parameters are adjusted. When the global parameters have settled, adjustment of the local parameters begins.

From a robust statistical point of view this scheduling is tantamount to reducing the dimension of the approximate distribution until the "error" is sufficiently small to warrant a better approximation. The lower dimension also provides more efficient computation. The fixed terms/directions have effective weights of zero, and are thus ignored. We note that this is a simple form of iterative reweighting, and more effective schemes have been studied in the robust statistics literature [RL87, Hub81, MB93]. In particular, the reweighting should also depend on the inherent variability of the parameters. In addition, issues such as selection of the appropriate "order" (dimension) of the model have been addressed, see [Aka73, Boz87]. The iterated reweighting schemes should provide better quality and, hopefully, better speed. Since in the early stages we may also want to create an artificial gradient which does not really reflect likelihood but guides us toward the maximum likelihood, we may want to change the form of the distribution as well as its the weights.

3.4 Related Work

Cootes and Taylor have a simple Gaussian deformable model [CHTH93] which was put into a conceptually clearer probability framework by Baldwin [Bal94] in as-yet unpublished work. They label a fixed set of N feature points, defining a discrete "contour," in each 2D training image. Each 2D contour can be thought of as a single point p in $2N$-space, the parameter space of contours. They find the mean contour $\bar{p}$ and the covariance matrix C for the $2N$ variables. $\bar{p}$ and C represent the parameters of a Gaussian distribution of training contours in $2N$-space. C models the correlations between contour point coordinates. They then do a principal component decomposition on the distribution of contours, *i.e.* they find the eigenvectors of C. Each eigenvector represents a principal axis in the distribution of training contours in $2N$-space. The corresponding eigenvalue is the variance (spread) of the distribution along this axis. Any N-vertex contour is a point p in $2N$-space that has a *Mahalanobis distance* $(p - \bar{p})^{\mathrm{T}} C^{-1}(p - \bar{p})$, which can be directly related to the multi-dimensional Gaussian probability.

Cootes and Taylor look at the distributions of pixel values along one dimension near each of the N 2D points in a contour. They calculate covariance between pixels near each contour point, but not between those of different contour points. They then search for the contour that maximizes this grey-level Mahalanobis distance alone, $\mathrm{P}(I \mid p)$, while merely bounding the search by imposing an arbitrary maximum on the positional Mahalanobis distance, P(p). From a Bayesian point view, ignoring P(p) is inappropriate, but viewing from our perspective of robust statistics we note that have little confidence in the prior model and have replaced it with a "robust version" with a very broad support that is clamped outside some range.

Baldwin,[Bal94], however, uses P(p) as an internal energy which penalizes deviation from the correlations gathered during training. He models image energy, P(I | p), using the covariances of grey values at each contour vertex, at M different image resolutions. Thus it is another Gaussian, this one in MN-space. He maximizes the product P(p | I) of these two Gaussians—multiplying the probabilities is adding the Mahalanobis distances which are his internal and image energies. The assumption that the distributions are normal may cause the problems mentioned earlier, but otherwise this system is a simple and instructive example of the use of the probability formulation. Since the optimization simply relies on image gradients, it may not be robust to poor initial placement, but if started close it is robust with respect to extraneous edges. From a "robust" point of view, the algorithm might benefit by reweighting the various scales during maximization.

Kervrann and Heitz [KH94], extend Cootes' Gaussian modal representation of a discrete contour and track moving objects. They model the displacement from the mean contour as Gaussian, but they also model the difference between neighboring diplacements as a Gaussian Markov process. This is their prior shape energy. Their image energy is the number of pixels outside the contour which have changed by more than a threshold amount since the last frame (or since some reference frame), minus the same count for pixels inside the contour. Like Cootes, they estimate global model pose before allowing the model to deform. Unlike Cootes, they do this by sampling the parameters randomly, using relaxation *á la* simulated annealing. They alternately do the same for perturbations to the principal deformation components, using their Markov model. This is a pure probabilistic formulation with no special concession to robust statistics other than the inital pose computation. The hierarchy of rigid transformation followed by deformations in a few principal modes, stochastically sampled, provide robustness and avoid local minima.

The work of Vemuri [VR93, VR94] appears to use a probabilistic framework for continuous shape models. The models are based on the wavelet decomposition. The papers discuss the Gibbs formulation and provide rudimentary statistical interpretations of their energy formulations, including the idea of a "mean" shape. But they do not consider robust techniques, nor do they consider the statistical interpretation of their "image" forces, nor do they discuss calibration of data forces which were presumed to be Gaussian. They do, however, use a simple mean and variance computation to "train" the model's prior of the global parameters. Their prior was, unfortuantely, a very poor approximation to their object.

In [CSD94], a probabilistic framework is used to integrate the results of a region classification into a deformable model. Region classification is done using a Markov random field (MRF). Then contour is deformed based on a combination of two image energies—the line integral of a smoothed image gradient, and an area integral of the region inside the contour. This second term penalizes the contour for containing pixels classified as belonging outside, and vice versa. Despite equating these energies with the logs of probabilities $P(I_g \mid p)$ and $P(I_s \mid p)$, the authors still add them weighted by arbitrary constants. They say, "...K_1 and K_2 are the weighting constants which signifies the relative importance of the two terms in the above equation. Normally, one should choose them such that the contributions from each of the terms are comparable." When one treats the energies this way, they no longer represent probabilities in any quantitatively meaningful way. The feature of note here is that they gain robustness and tolerate poorer initial contour estimates by integrating two kinds of image information.

Although [NFSK94] does not use a probability formulation, they have a very clever method for overcoming the problem that image forces are only accurate very close to the desired object. They specify an initial position and orientation for their snake only at its endpoints, letting the rest hang where it may, unaffected by image force. Then they gradually turn on the image force starting at the ends and moving toward the middle. The image force does not attract the snake to the wrong object, since the force is only turned on at each snaxel once the previous one has settled, presumably close to the desired edge. This is an elegant example of a way to modify image forces gradually to make them robust against unrelated edges.

4 Conclusion

While various researchers have related model energy to probability, we believe that for many vision tasks, the use of "robust statistics," in particular M-estimators, is more appropriate that straightforward Bayesian analysis. This statistical framework provides a natural mechanism for "learning" the *a priori* model of objects to be segmented/recovered and provides standard measures of quality and significance of the recovered models.

Existing work using deformable models has shown them to be quite useful. However they suffer from various limitations, one of which is the "justification-by-analogy" of the formulation. By reinterpreting deformable object recovery from the view of robust statistics, we have found justification for various existing "ad hoc" but important aspects of deformable modeling techniques. In addition, this viewpoint suggests numerous enhancements to deformable modeling techniques.

References

[Aka73] H. Akaike. Information theory and an extension of the maximum likelihood principle. In *Proc. Second International Symposium on Information Theory*, pages 267–281, 1973.

[Bal94] Bernard Baldwin. personal conversations on thesis work, 1994. Mr. Baldwin is at the Courant Institute of New York University. Research in cooperation with the Memorial Sloan-Kettering Cancer Center.

[Bat82] K.J. Bathe. *Finite Element Procedures in Engineering Analysis*. Prentice-Hall, Englewood Cliffs, NJ, 1982.

[Boz87] H. Bozdogan. Model selection and akaike's information criterion (aic): General theory and its analytical extensions. *Pyschometrika*, 52(3):345–370, 1987.

[CC90] L.D. Cohen and I Cohen. A finite element method applied to new active contour models and 3d reconstruction from cross-sections. In *Proc. of the IEEE Int. Conf. on Computer Vision*, pages 587–591, Osaka, Japan, 1990. IEEE.

[CHTH93] T.F. Cootes, A. Hill, C.J. Taylor, and J. Haslam. The use of active shape models for locating structures in medical images. In *Proceedings of the 13th International Conference on Information Processing in Medical Imaging*, Flagstaff AZ, June 1993. Springer-Verlag.

[Coh91] L.D. Cohen. On active contour models and balloons. *Computer visoin graphics and image processing: Image Understanding (CVGIP:IU)*, pages 211–218, March 1991.

[CSD94] A. Chakraborty, L.H. Staib, and J.S. Duncan. Deformable boundary finding influenced by region homogeneity. In *Proc. IEEE Conf. on Computer Vision and Pattern Recognition*, pages 624–627, June 1994.

[GG91] D. Geiger and F. Girosi. Parallel and deterministic algorithms from mrf's: Surface reconstruction. *IEEE Trans. on Pattern Analysis and Machine Intelligence*, PAMI-13(5):674–693, May 1991.

[HG93] W.C. Huang and D. Goldgof. Nonridged motion analysis using non-linear fineite element modeling. In *Geometric Methods in Computer Vision*, volume 2031, pages 404–415. SPIE, 1993.

[Hub81] P. Huber. *Robust Statistics*. Wiley, New York, 1981.

[KH94] C. Kervrann and F. Heitz. A heirarchical stastical framework for the segmentation of deformable objects in image sequences. In *Proc. IEEE Conf. on Computer Vision and Pattern Recognition*, pages 724–727, June 1994.

[KW70] G.S. Kimeldorf and G. Wahba. A correspondence between bayesian estimation on stocastic processes and smoothing by splines. *Annals of Math. Stat.*, 41(2):495–502, 1970.

[KWT87] M. Kass, A. Witkin, and D. Terzopoulos. Snakes: Active contour models. In *Proc. of the IEEE Int. Conf. on Computer Vision*, pages 259–268, London UK, 1987. IEEE.

[LC94] K.F. Lai and R.T. Chin. Deformable contours: Modeling and extraction. In *Proc. IEEE Conf. on Computer Vision and Pattern Recognition*, pages 601–608, June 1994.

[LL93] F. Leymarie and M.D. Levine. Tracking deformable objects in the plane using an active contour model. *IEEE Trans. on Pattern Analysis and Machine Intelligence*, 15(6):617–633, June 1993.

[MB88] M.L. Moerdler and T.E. Boult. The integration of information from stereo and multiple shape-from texture algorithms. In *Proc. IEEE Conf. on Computer Vision and Pattern Recognition*, pages 514–529. IEEE, 1988.

[MB93] M.J. Mirza and K.L. Boyer. Performance evaluation of a class of m-estimators for surface parameter estimation in noisy range data. *IEEE Trans. on Robotics and Automation*, 9:75–85, 1993.

[MMRK91] P. Meer, D. Mintz, A. Rosenfeld, and D.Y. Kim. Robust regression methods for computer vision: A review. *Inter. J. Computer Vision*, 6(1):59–70, 1991.

[NFSK94] W. Neuenschwander, P. Fua, G. Székely, and O. Kübler. Initializing snakes. In *Proc. IEEE Conf. on Computer Vision and Pattern Recognition*, pages 658–663, June 1994.

[OFBG94] T. O'Donnell, X.S. Fang, T.E. Boult, and A. Gupta. The extruded generalized cylinder: A deformable model for object recovery. In *Proc. IEEE Conf. on Computer Vision and Pattern Recognition*, June 1994.

[OGB94] T. O'Donnell, A. Gupta, and T.E. Boult. A periodic generalized cylinder model with local deformations for tracking closed contours exhibiting repeating motion. In *The International Conference on Pattern Recognition*, Nov. 1994.

[PMY94] J. Park, D. Metaxas, and A. Young. Deformable modles with parameter functions: Application to heart-wall modeling. In *Proc. IEEE Conf. on Computer Vision and Pattern Recognition*, pages 437–442, June 1994.

[PS91] A. Pentland and S. Sclaroff. Closed-form solutions for physically based shape modeling and recognition. *IEEE Trans. on Pattern Analysis and Machine Intelligence*, PAMI-13(7):715–729, July 1991.

[RL87] P.J Rousseeuw and A.M. Leroy. *Robust Regression and Outlier Detection*. J. Wiley, New York, New York, 1987.

[ST89] R. Szeliski and D. Terzopoulos. Constrained fractals. *Computer Graphics*, 23(3):51–60, 1989. (SIGGRAPH).

[Sze89] R. Szeliski. *Baysian Modeling of Uncertanity in Low-level Vision*. Kluwer Academic Publishers, Boston, MA, 1989.

[TM91] D. Terzopoulos and D. Metaxas. Dynamic 3d models with local and global deformations: Deformable superquadrics. *IEEE Trans. on Pattern Analysis and Machine Intelligence*, PAMI-13(7):703–714, July 1991.

[TW93] D. Terzopoulos and K. Waters. Analysis and synthesis of facial images sequences using physical and anotomical models. *IEEE Trans. on Pattern Analysis and Machine Intelligence*, 15:569–579, 1993.

[VR93] B.C. Vemuri and A. Radisavljevic. From global to local, a continuum of shape models with fractal priors. In *Proc. IEEE Conf. on Computer Vision and Pattern Recognition*, pages 307–313, NYC NY, June 1993.

[VR94] B.C. Vemuri and A. Radisavljevic. Multiresolution stochastic hybrid spahe models with fractal priors. *ACM TOGS*, 1994. Special issue on Interactive Sculpting.

Towards A Robust Physics-Based Object Recognition System

Octavia I. Camps*

Dept. of Electrical Engineering
Dept. of Computer Science and Engineering
The Pennsylvania State University
University Park, PA 16802

Abstract. A successful 3D object recognition system must take into account imperfections in the input data, due for example to fragmentation or sensor noise. In this paper we propose a methodology for robust 3D object recognition using uncertain image data. In particular, we present a method capable of achieving acceptable performance in the presence of both segmentation problems and sensor uncertainty, thus eliminating the need for *ad hoc* heuristics. The proposed method is based upon the use of probabilistic models suggested by the underlying physics processes. These models are statistically validated and tested under controlled experimentation.

1 Introduction

Object recognition systems attempt to locate instances of objects in images. Most progress in this area has been made in industrial applications, such as robot manipulation and product inspection, where the visual environment can be controlled and the shape of the objects to be imaged is known in advance.

Many model-based systems find correspondences between model features and features detected in an image. Examples of features are points, edges, holes, junctions, or a combination of these. These correspondences are found using techniques such as interpretation trees [8, 6, 2], hashing [28, 4], alignment [14], and bipartite search [17]. The pairings are such that the features in the image can be obtained (approximately) by applying a geometric transformation to their corresponding model features. This transformation is usually referred as the *pose* of the object, that is the position of the object with respect to a coordinate system.

A successful 3D object recognition system must take into account imperfections in the input data, due for example to fragmentation or sensor noise. However, although there currently exists efficient model–based vision systems capable of recognizing and locating objects using nearly–perfect data, their performance degrades dramatically when confronted with real, non–perfect images. Recently some progress has been made in handling non-perfect data due to segmentation problems [1] and in analyzing the effect of sensor uncertainty [9].

* This work was supported in part by NSF grant IRI9309100 and in part by a Pennsylvania State University Research Initiation Grant.

In this paper we propose a methodology for robust 3D object recognition using uncertain image data. In particular, we present a method capable of achieving acceptable performance in the presence of both segmentation problems and sensor uncertainty, thus eliminating the need for *ad hoc* heuristics. The proposed method is based upon the use of probabilistic models suggested by the underlying physics processes. These models are statistically validated and tested under controlled experimentation.

2 Model Representation

The problem of describing the models is critical to the success of any object recognition system. *Characteristic views* [3, 15, 1] are commonly used to describe models for recognition purposes. A characteristic view is a representative view of a grouping of views or *view aspect* with similar properties.

The view aspect concept is very important in object recognition since it captures the topological characteristics of the views of an object. It allows a compact representation of the features of the models to be matched against the features in an image. Then, the object recognition/localization task can be divided into the following steps: (1) determine the correct view class; (2) find the correspondences between the features extracted from the image and those in the view class representation; and (3) use these correspondences and the links between the 3D features and the view class features to determine the pose of the object.

2.1 Characteristic Views

Characteristic views can be found by analytically partitioning a viewing sphere centered at the object into aspects [27, 5, 26]. The boundaries between these aspects are very accurate. However, the number of aspects tends to be large due to accidental viewpoints. An alternative approach, is to uniformly sample the viewing sphere around the object and to group together views that are "similar" [19]. This method results, in general, in a lesser number of aspects. However, the number of aspects will depend on the resolution of the sampling scheme and on the similarity measure used. In this paper, we use the later method, since it allows to limit the number of views and it is easy to implement.

The views obtained from the sampled viewing sphere are grouped into equivalence classes using a similarity metric. For this particular application we decided to cluster views depending on which model segments were observed in the views. Thus, each cluster had views in which roughly the same segments were observed. This is a simple but effective criteria for classification.

2.2 Probabilistic Prediction Models

In order to be robust to data uncertainty, physics-based knowledge of surface reflectance properties, light sources, sensor characteristics, and feature detector algorithms is incorporated to the view clusters.

For each view cluster C, we currently use the system PREMIO [1] to build a *probabilistic prediction model* combining hundreds of segmented views within the cluster. In the future, we will also use information obtained from real training images.

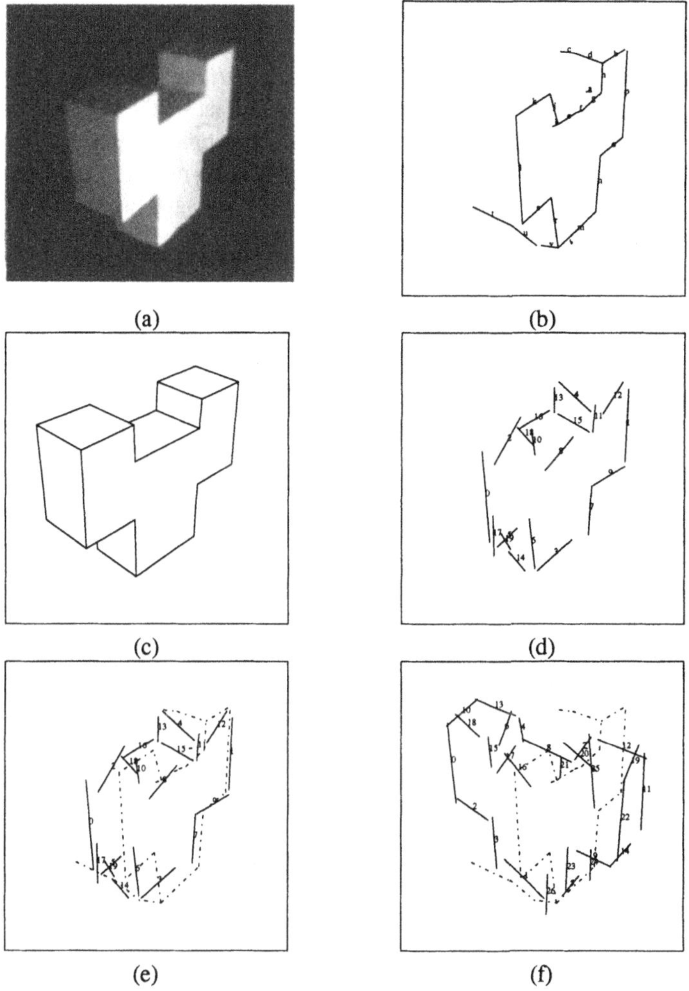

Fig. 1. (a) Fork image. (b) Segmented image. (c) Representative View. (d) Corresponding probabilistic model. (e) Alignment of (b) with (d). (f) Alignment of (b) with a different model.

A model M is represented by a quadruple $M = (L, R, f_L, g_R)$ where L is a set of model features or *labels*, R is a set of relational tuples of labels, f_L is the attribute value mapping that associates a value with each attribute of a label L, and g_R is the strength mapping that associates a strength with each relational tuple of R.

The set of labels L is formed by only those 2D features that have high enough probability of being detected for the given set of sensors and light sources. Furthermore, each feature in L has associated attributes which are given by the mean and the standard deviation of the attribute values of the feature for the n predictions. Similarly, the set of relational tuples R is formed for those relations among features in L such that they have high enough probability of holding. As with feature attributes, the relationship attribute values of the tuples in R are represented by the mean and standard deviation of the relational tuples for the n predictions.

The model M obtained in this way, is a *probabilistic model* of the object for the given set of configurations of sensors and lights. Note that neither all the features in L, nor all the relational tuples in R need to be present in a single prediction. Neither do all the features of a particular prediction need to be in L. The model M combines a group of predictions into a single model, which is a sort of "average" model. The differences between the model M and the individual predictions that were used to build the model are summarized in a set of statistics Θ [1].

Figures 1(a) and (b) show an image of an object and the corresponding segmentation. Figure 1 (c) shows a representative view for the object in (a). Figure 1 (d) shows a visualization of the corresponding probabilistic model, where the segments are drawn using their mean attribute values and the numbers by the segments indicate their relative detectability, with the lower the number, the higher the detectability.

3 View Classification

Given an image, the objective is to find which object and in particular which view class it was originated from. Let $C_1, C_2, \ldots, C_n$ be a set of potential view clusters. Given an image I, our aim then, is to select the cluster C_i to which the image will most likely belong to. To achieve this, we use a Bayesian approach.

Let $P(C)$ be the *a priori* probability that an image from cluster C will be observed, and let $P(I|C)$ be the probability that a given image I is captured when the object is observed from a viewpoint within cluster C.

Then, given an image I, we will classify it as coming from cluster C_m, if the *a posteriori* probability $P(C|I)$ is maximum for $C = C_m$.

The *a posteriori* probability $P(C|I)$ can be computed using the Bayes' Theorem:

$$P(C \mid I) = \frac{P(I \mid C)P(C)}{\sum_{i=1,\ldots,n} P(I \mid C_i)P(C_i)} \tag{1}$$

3.1 Probabilistic Model

In order to apply (1), we need to compute the involved probabilities.

The probabilities $P(C)$ can be estimated from the area that the corresponding cluster spans on the viewing sphere. The larger the area, the higher the probability. The probabilities $P(I|C)$ depend on the selected features comprising the model. We will model this probability as a multivariate Gaussian distribution with mean vector $\underline{\mu}$ and covariance matrix Σ of the form:

$$P(I|C) = N_{\underline{x}}\left(\underline{\mu}, \underline{\Sigma}\right) \tag{2}$$

where $\underline{x}$ is $(d{+}4) \times 1$ feature vector representing the image I. The feature vector $\underline{x}$ consists of: the number of segments in the image, the number of junctions of 2 segments, the number of junctions of 3 segments, the number of triples[2], and a feature metric error

[2] A triple is an ordered set of three lines, with the lines traced clockwise.

for the d most detectable segments. The mean and covariance matrix of the distribution can be estimated from a set of samples generated with the prediction module of the PREMIO system.

Figure 1(e) shows the segmented image (b) aligned with the probabilistic model with the highest *a posteriori* probability, while (f) shows it aligned with a second model with lower probability.

4 Feature Selection

Once the object and view cluster are identified, the system proceeds to match a set of image and model features. The use of a small set of features is recurrent in the literature [22, 12, 23, 13, 10]. Since these correspondences will be used to compute the pose of the object, the problem of selecting "good" features is of interest. We propose to select these features based on measurements of their detectability, reliability, and accuracy, using concepts similar to the ones introduced by Ikeuchi and Kanade in [16].

4.1 Feature Detectability and Reliability

Let M be a model, L be a set of model features, I be an image of M, U be a set of image features, and $h : H \longrightarrow U, H \subseteq L$ be the mapping that associates a set H of model features to the corresponding image features. Let l be a model feature, $l \in L$, and $u = h(l)$ be the corresponding image feature, $u \in U$. The feature *detectability* of l, denoted $D_M(l)$ is given by:

$$D_M(l) = P\left(l \in H \mid M\right) \quad . \tag{3}$$

Because a model can have several features with similar attributes, all of which could potentially be matched with the same image feature, feature detectability alone is not sufficient to determine which correspondence is correct. In order to determine the correctness of a match, the matching routine needs to know not only how detectable a model feature is, but also how "reliable" it is.

The feature *reliability* of l, denoted as $R_M(l)$ is defined as the probability that a hypothesized correspondence between a model and image feature will be correct given that the model feature is detected in the image:

$$R_M(l) = P\left(l \longrightarrow h(l) \mid l \in H, M\right) \quad . \tag{4}$$

Feature reliability is an extension of feature detectability in the sense that a model feature must be detected in the image in order for a correspondence to be hypothesized and only then the correctness of the match can be assessed.

Computing feature detectability is rather straightforward. PREMIO approximates the detectability of a feature by the frequency at which it appears in a set of images. However, computing feature reliability is a more complex problem since it involves the process by which a correspondence is hypothesized. Next, we will discuss the theory supporting a method of computing feature reliability and illustrate its use in an object recognition system.

Let $\mathcal{I}$ be the set of all possible images of M, then $R_M(l)$ can be expressed as the feature reliability contributions for l integrated (summed) over all images of $\mathcal{I}$. Formally,

$$R_M(l) = \int_{\mathcal{I}} P\left(l \rightarrow h(l) \mid I, l \in H, M\right) . P\left(I \mid l \in H, M\right) dI \tag{5}$$

The reliability depends on how correspondence hypotheses are made. We assume that the matching strategy will hypothesize a correspondence between a label and a unit only if they are "sufficiently similar".

Let $\rho(l,u)$ be a metric that measures the similarity between the feature attributes of l and $h(l)$, where $h(l)$ is the observation of l in an image. Based on our experiments and the Central Limit Theorem, the similarity between the label l and its corresponding unit can be modeled as a Gaussian distribution, denoted $P_{\rho_l}(\rho_l) = N_{\rho_l}(\mu_{\rho_l}, \sigma_{\rho_l})$.

Given $P_{\rho_l}(\rho_l)$, it is natural to hypothesize a match between l and u only if

$$\|\rho(l,u) - \mu_{\rho_l}\|^2_{\sigma_{\rho_l}} \leq Th \tag{6}$$

where $\|\rho(l,u) - \mu_{\rho_l}\|^2_{\sigma_{\rho_l}}$ is the squared Mahalonabis distance between $\rho(l,u)$ and the expected value of $\rho(l,h(l))$ and Th is a threshold measuring the maximum allowed difference between the observed metric value and its expected value.

Consider the set of image features, U, and a label $l \in L$. Then, the subset of units that could potentially be matched with l is the subset of units, C, such that they satisfy (6):

$$C = \left\{ u \;\middle|\; u \in U, \|\rho(l,u) - \mu_{\rho_l}\|^2_{\sigma_{\rho_l}} \leq Th \right\} .$$

Thus, the first factor in the right side of (5) can be expressed as

$$P\left(l \rightarrow h(l) \mid I, l \in H, M\right) = \begin{cases} 0 & \text{if } h(l) \notin C \\ \frac{1}{\#C} & \text{otherwise} \end{cases} \tag{7}$$

where # denotes cardinality.

The second factor in the right side of (5) can be expressed using Bayes rule and (3) as:

$$P\left(I \mid l \in H, M\right) = \frac{P\left(l \in H \mid I, M\right) P\left(I \mid M\right)}{D_M(l)} .$$

The probability $P\left(l \in H \mid I, M\right)$ is the probability that label l is detected given that the observed image is I and the model is M, and it can be modeled as

$$P\left(l \in H \mid I, M\right) = \begin{cases} 0 \text{ if } \#C = 0 \\ 1 \text{ otherwise} \end{cases}$$

Finally, $P\left(I \mid M\right)$ is provided by PREMIO's statistics θ [2].

RD Heuristic

Feature reliability and detectability can be combined to form a joint ("RD") heuristic in order to produce a matching routine which is more efficient than a routine using detectability alone. The RD joint heuristic can be stated as the joint probability of feature reliability $R_M(l)$ and detectability $D_M(l)$ denoted as $P\left(l \rightarrow h(l), l \in H \mid M\right)$. By definition of conditional probability and (3) and (4):

$$\begin{aligned} &P(l \rightarrow h(l), l \in H|M) = \\ &= P\left(l \rightarrow h(l) \mid l \in H, M\right) P\left(l \in H \mid M\right) = \\ &= R_M(l) D_M(l) . \end{aligned}$$

The probability $P\left(l \rightarrow h(l), l \in H \mid M\right)$ can then be used to rank the labels to be matched. Thus, a matching routine can use these rankings to make more reliable hypotheses reducing potential mistakes and expensive backtracking.

Figure 2 (a) shows the new rankings for the model given in Figure 1 (d) when the RD heuristic is used. Figures 2 (b) to (d) compare the performance in terms of their CPU time, probability of false alarm (finding the wrong correspondences) and probability of misdetection (not finding a set of correspondences) of an iterative deepening search algorithm [2] when only detectability or both detectability and reliability heuristics are used. It is observed that both CPU time and false alarms are always less for the RD heuristic. The probability of misdetection is also better for the RD heuristic up to 13 correspondences.

4.2 Feature Accuracy

Most methods to compute the pose use a *few* point-to-point [11, 8] or line-to-line [20, 21] correspondences. If the data is perfect with no sensor uncertainty and with no incorrect correspondences, then the pose is exact, and the transformed model features exactly coincide with the image features. However, in most real cases the noise in the data will propagate into the pose. Moreover, the extent of the effect of the uncertainty depends on the correspondences used to compute it.

Currently, we are investigating the following problem:

> Let N be the number of model features and $n \leq N$ be the (small) number of features that will be assigned a correspondence and will be actually used to compute the pose of the object. Then, find the subset of n model features such that the effect of the data uncertainty in the estimation of the pose is minimized.

In [7] we reported a suboptimal solution for this problem for the special case when the features are points, the pose estimation algorithm is an iterative least square procedure, and the translation of the camera is constrained. In this case, the sensitivity of the pose estimation algorithm to the noise in the data is given by the amount of perturbation on the rotation matrix due to a small perturbation in the data. Using a sensitivity analysis similar to the one presented in [11] we showed that the trace of the matrix $(J'J)^{-1}$, where J is the Jacobian matrix of the image points with respect to the incremental correction

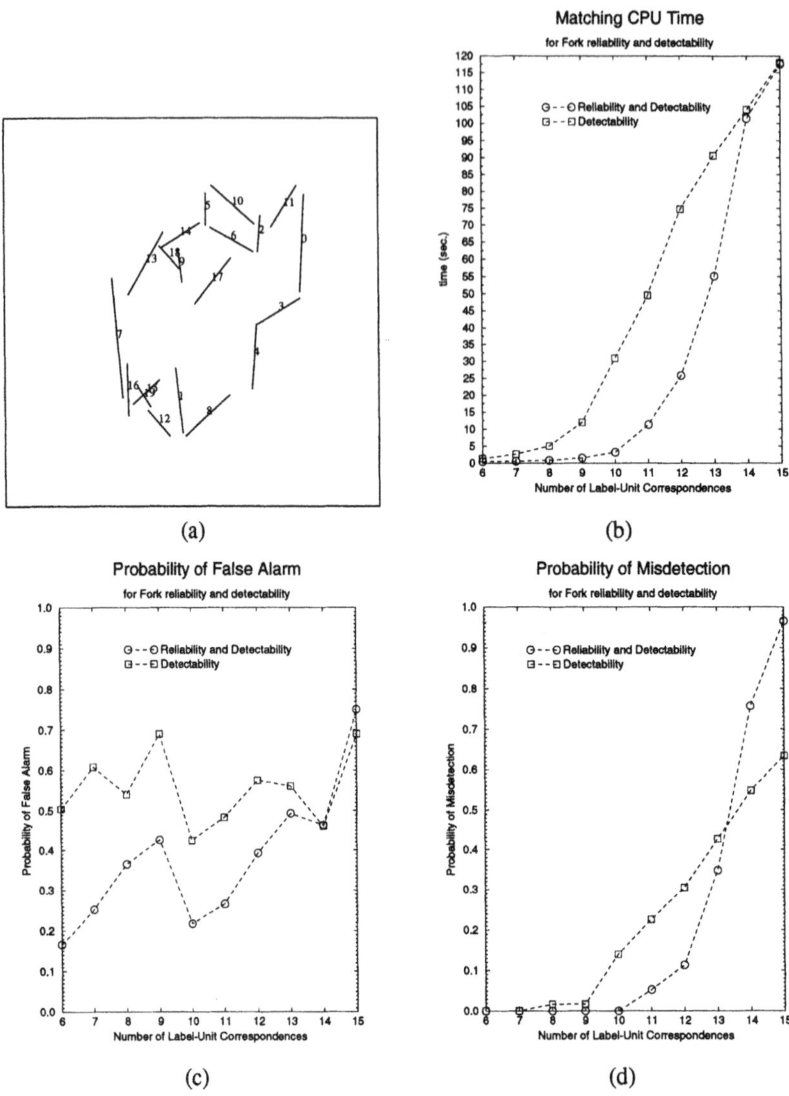

Fig. 2. (a) RD segment ranking. (b) Matching CPU time. (c) Probability of False Alarm. (d) Probability of Misdetection.

on the pose, can be used as a measure of the sensitivity of the pose to the noise in the data. Thus, the subset of n points that minimizes the trace of the above matrix:

$$\min_{\text{Subsets of } n \text{ points}} \left\{ \text{trace } (J'J)^{-1} \right\} \tag{8}$$

is a good choice from these considerations. Unfortunately, the number of possible subsets of n points is, in general, too large to attempt to solve this minimization problem directly. However, a suboptimal solution to this problem can be obtained by using an incremental approach:

The Greedy Algorithm

> Given a subset $\mathcal{P}_k$ of k model points, select a model point m such that the trace of $(J'J)^{-1}$ is minimized for the extended subset $\mathcal{P}_{k+1} = \mathcal{P}_k \cup \{m\}$.

The proposed algorithm is a *greedy* algorithm that finds a model point such that when it is added to the points used so far, the computed pose is robust to the noise present in the data. The algorithm is suboptimal since it has a limited horizon of one point at a time.

Selection of initial points. The pose estimation algorithm starts with an initial rotation $R^{(0)}$ and then iterates to refine this pose. In order to compute the initial rotation $R^{(0)}$, a minimum of two points correspondences are required. Let m_1 and m_2 be two model points. It can be easily shown that if m_1, m_2, and the origin of the world reference frame are aligned, i.e. $m_2 = \alpha.m_1$, the estimated pose is not unique. Furthermore, if the points m_1 and m_2 are close to the origin, a small perturbation in the coordinates of the corresponding image points leads to a large change in the estimated pose.This suggests the heuristic rule that the initial points should be selected such that the area of the triangle formed by the two model points and the origin is maximum.

Selection of subsequent points. Once k correspondences have been found the problem of selecting the next correspondence such that the estimated pose is robust reduces to selecting the model point that minimizes the trace of the 3×3 matrix $(J'J)^{-1}$ that can be computed incrementally [7].

Handling outliers. The initial rotation $R^{(0)}$ is found by solving a system with four equations and three unknowns such that a least square error criterion is minimized for the two initial model points. If the error of this fit is too high, at least one of the points is likely to be an outlier and a newpair of points is selected.

When a subsequent point is added, one can use the current rotation $R^{(k)}$ to project the model points currently used and compare their location with their corresponding image points. If the distance between these is higher than a multiple of the standard deviation of the noise, then the point is rejected as an outlier.

If at a given point, too many points are classified as outliers, the initial points are suspected as outliers and the process starts again for a new pair of initial points.

Fig. 3 shows the results obtained with an image of a bookend. Fig. 3(a) shows a grayscale image of the bookend with the model points highlighted. Fig. 3(b) shows the back projection of the model when all the model points are used to compute the pose. Fig. 3(c) shows the back projection of the model onto the image when four random points (circled on the figure) are used to compute the pose. Finally, Fig. 3(d) shows the back projection of the model when four points are selected using the greedy algorithm (circled on the figure). Clearly, the greedy solution is better than the random one, and comparable to the one obtained using all the points.

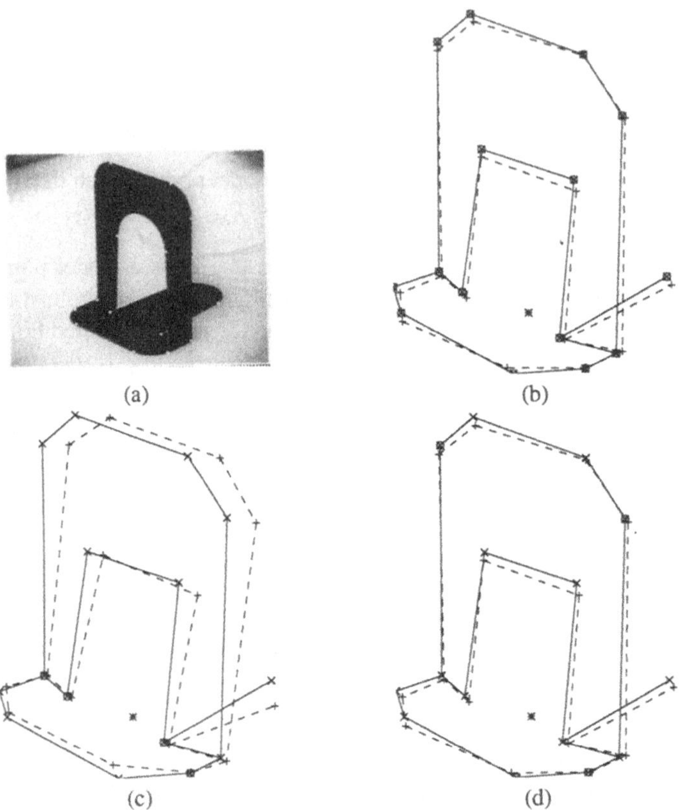

Fig. 3. Bookend image. (a) Grayscale image with model points highlighted. (b) Back projection using all the model points. (c) Back projection using four random model points (circled on the figure). (d) Back projection using four points selected using the greedy algorithm (circled on the figure).

We are currently working on the generalization of this algorithm to the case when the features are line segments. We plan to combine the new results with our current detectability and reliability heuristic to form a joint heuristic to guide the matching routine in the selection of the next correspondence to be sought.

5 Performance Evaluation

The importance of controlled experiments has only recently been stressed in computer vision. Controlled experiments are essential in order to illustrate the validity of a solution presented. We tested all the modules described here using artificially generated data as well as real images. The experimental protocol used has six steps: 1) modeling of the ideal inputs and the random perturbation processes; 2) annotating ground truth data; 3) estimating the free parameters of the random models; 4) statistically validating these models; 5) testing the algorithms; and 6) analyzing the results.

Appendix A gives the experimental protocol used to evaluate the performance of the RD heuristic. A detailed description of the protocol for the view classification module is given in [25], and for the greedy algorithm to select features based on pose accuracy is given in [7].

6 Discussion

A successful 3D object recognition system must take into account imperfections in the input data, due for example to fragmentation or sensor noise. However, although there currently exists efficient model–based vision systems capable of recognizing and locating objects using nearly–perfect data, their performance degrades dramatically when confronted with non–perfect images. We believe that to overcome these problems we must develop a *robust* 3D object recognition paradigm. Specifically, we need to:

1. Develop mathematical models for robust 3D object recognition from uncertain 2D image data.
2. Develop matching schemes that use these models to robustly recognize and compute the pose of an object. These schemes should also provide levels of confidence for the hypothesis made.
3. Develop thorough experimental protocols to characterize the performance and robustness of the systems.

We believe that techniques from robust statistics coupled with physics-based knowledge in a Bayesian framework are promising tools to achieve these goals. Our preliminary results show that they naturally lead to rigorous models capturing the underlying physical processes and that they are subject to experimental validation.

References

1. O. I. Camps, L. G. Shapiro, and R. M. Haralick. Image prediction for computer vision. In Jain A.K. and P.J. Flynn, editors, *Three-dimensional Object Recogntion Systems*. Elsevier Science Publishers BV, 1993.
2. O. I. Camps, L. G. Shapiro, and R. M. Haralick. A probabilistic matching algorithm for computer vision. *Annals of Mathematics and Artificial Intelligence*, 10, 1994.
3. I. Chakravarty and H. Freeman. Characteristic views as a basis for three-dimensional object recognition. In *SPIE 336 (Robot Vision)*, pages 37–45, 1982.
4. M. Costa, R.M. Haralick, and L.G. Shapiro. Optimal affine-invariant point matching. In *Proc. of the International Conference on Pattern Recognition*, pages 233–236, Atlantic City, New Jersey, June 1990.
5. D. Eggert. *Aspect Graphs of Solids of Revolution*. PhD thesis, Department of Computer Science and Engineering, University of South Florida, Tampa, Florida, 1991.
6. P.J. Flynn. *CAD-Based Computer Vision: Modeling and Recognition Strategies*. PhD thesis, Michigan State University, 1990.
7. T. L. Gandhi and O. I. Camps. Robust feature selection for object recognition using uncertain 2D image data. In *IEEE Computer Society Conference on Computer Vision and Pattern Recognition*, pages 281–287, June 1994.

8. W.E.L. Grimson. *Object Recognition by Computer: The role of geometric constraints*. MIT Press, Cambridge, MA, 1990.
9. W.E.L. Grimson. Recognizing 3D Objects from 2D Images: An Error Analysis. Technical Report A.I. Memo No. 1362, MIT, Cambridge,MA, November 1992.
10. Charles D. Hansen. *CAGD-Based Computer Vision: The Automatic Generation of Recognition Strategies*. PhD thesis, The University of Utah, 1988.
11. R.M. Haralick and L.G. Shapiro. *Computer and Robot Vision*. Addison-Wesley, 1992.
12. J. Henikoff and L. Shapiro. Interesting patterns for model–based matching. In *ICCV*, 1990.
13. P. Horaud and R.C. Bolles. 3DPO: A system for matching 3-D objects in range data. In A.P. Pentland, editor, *From Pixels to Predicates*, pages 359–370. Ablex Publishing Corporation, Norwood, New Jersey, 1986.
14. D.P. Huttenlocher and S. Ullman. Object recognition using alignment. In *Proceedings of the First International Conference on Computer Vision*, pages 102–111, 1987.
15. K. Ikeuchi. Generating an interpretation tree from a CAD model for 3D-Object recognition in bin-picking tasks. *Int. J. Comp. Vision*, 1(2):145–165, 1987.
16. K. Ikeuchi and T. Kanade. Modelling sensor detectability and reliability in the configuration space for model-based vision. Technical Report CMU-CS-87-144, Carnegie-Mellon University, Computer Science Department, July 1987.
17. W.Y. Kim and A.C. Kak. 3D Object Recogntion Using Bipartite Matching Embedded in Discrete Relaxation. *IEEE Trans. on Pattern Analysis and Machine Intelligence*, 13(3):224–251, March 1991.
18. D.E. Knuth. *The Art of Computer Programming, vol. 2*. Addison-Wesley, 1969.
19. M. Korn and C. Dyer. 3D-Multiview Object Representations for Model–Based Object Recognition. *Pattern Recognition*, 20(1):91–103, 1987.
20. R. Kumar and R. Hanson. Analysis of different robust methods for pose estimation. In *Proc. of IEEE Workshop on Robust Computer Vision*, October 1990.
21. C.N. Lee and R.M. Haralick. Exterior orientation from line-to-line correspondences – a Bayesian approach. In *Proc. of the IEEE Computer Vision and Pattern Recogntion*, June 1993.
22. D. G. Lowe. Three–dimensional object recognition from single two–dimensional images. *Artificial Intelligence*, 31:355–395, 1987.
23. R. Mohan and R. Nevatia. Using perceptual organization to extract 3d structures. *IEEE Trans. on Pattern Analysis and Machine Intelligence*, 11(11):1121 – 1139, November 1989.
24. A. Papoulis. *Probability, Random Variables, and Stochastic Processes*. McGraw-Hill, third edition, 1991.
25. A. Pathak and O. I. Camps. Bayesian view class determination. In *IEEE Computer Society Conference on Computer Vision and Pattern Recognition*, June 1993.
26. J. Ponce and D. Kriegman. Computing exact aspect graphs or curved objects: parametric surfaces. In *8th National Conference on AI*, pages 340–350, 1987.
27. J.H. Stewman. *Viewer Centered Representations for Polyhedral Objects*. PhD thesis, Department of Computer Science and Engineering, University of South Florida, Tampa, Florida, 1991.
28. H.J. Wolfson. Model based object recognition by geometric hashing. In *First Europ. Conf. on Comp. Vis.*, pages 526–536, 1990.

A RD Heuristic Experimental Protocol

A.1 Probabilistic Prediction Model Generation

Model Generation. PREMIO generates the model M using the following steps:

1. Select a region $\mathcal{V}$ of the viewing space. The region $\mathcal{V}$ is a spherical sector between two spheres. It is specified by the range of the longitude $(\Phi_{v\min}, \Phi_{v\max})$ and latitude $(\theta_{v\min}, \theta_{v\max})$ angles and the radius of the viewing sphere $(R_{v\min}, R_{v\max})$.
2. Select a region $\mathcal{I}$ of the illumination space. This region is specified by the range of the longitude $(\Phi_{i\min}, \Phi_{i\max})$ and latitude $(\theta_{i\min}, \theta_{i\max})$ angles and the radius of the illumination sphere $(R_{i\min}, R_{i\max})$.
3. Given the desired number of samples N_v and N_i, sample the viewing and illumination space regions. Let $\mathcal{V}_s$ be the set of the sampled viewing positions and $\mathcal{I}_s$ the set of the sampled lighting positions.
4. Generate the predictions. For each pair $(v, i) \in \mathcal{V}_s \times \mathcal{I}_s$, predict the subset of detectable labels L_{vi}, its associated attribute mapping $f_{L_{vi}}$, the subset of detectable relational tuples R_{vi}, and its associated relationship strength mapping $g_{R_{vi}}$. Also generate the corresponding set of units U_{vi}, the associated attribute mapping $f_{U_{vi}}$, the set of relational tuples S_{vi}, and the associated strength mapping $g_{S_{vi}}$.
5. Obtain the detectability values for each label l. The previous step produced $N_v \times N_i$ different predictions. Approximate the probability of a label/relationship being detected, given that the view and light are in the specified regions $\mathcal{V}$ and $\mathcal{I}$, by the observed frequency rate of their detectability in the generated predictions. These approximations are based on the fact that the predictions were made from uniformly sampled camera and light positions as well as on the central limit theorem assuming that N_v and N_i are large enough.
6. Select the desired minimum label detectability t_f and the minimum relational detectability t_r.
7. Combine the $N_v \times N_i$ predictions into a single model $M = (L, R, f_L, g_R)$ such that the labels in L have a detectability greater than t_f and the relational tuples in R have a detectability greater than t_R.
8. Compute the *similarity probability distribution* $P_\rho(l)$ for each label in L.
9. Obtain the reliability values for each label in L.
10. Compute the joint reliability and detectability values for each label in L.

Statistics Generation. PREMIO generates the statistics Θ as follows:

1. For each generated prediction, PREMIO finds the true observation mapping between the predicted image and the obtained model M, $h_{vi} : H_{vi} \longrightarrow U_{vi}$, with $H_{vi} \subseteq L$, and $(v, i) \in \mathcal{V}_s \times \mathcal{I}_s$. These observation mappings only include correspondences with units that were originated from labels with detectability greater than or equal to t_f.
2. Obtain matching errors. The previous step produced $N_v \times N_i$ true observation mappings, h_{vi}. For each prediction PREMIO computes the quantities: $\#L + \#U_{vi} - 2\#H_{vi}$, $\#(R - S_{vi} \circ h_{vi}^{-1}) + \#(S_{vi} - R \circ h_{vi})$, $\rho(f_{U_{vi}} \circ h_{vi}, f_{L|H_{vi}})$, and $\rho(g_{S_{vi}} \circ h_{vi}, g_{R|H_{vi}})$.
3. PREMIO uses the matching errors generated in the previous step to estimate the parameters of the four Gaussian distributions P_U, P_S, P_{f_U}, and P_{g_S}: μ_f, σ_f, μ_R, σ_R, μ_{f_U}, σ_{f_U}, μ_{g_S}, and σ_{g_S} by using the sample means and variances.

Model Validation. The PPM model obtained must be validated using statistical tests such as the Kolgomorov Smirnov test and the Chi-square test [24, 18]. If the model does not pass the tests, it can be rectified by one or more of the following methods:

1. Increase the number of predictions by increasing N_v and/or N_i. By increasing the number of predictions, the confidence interval of the statistics Θ, is narrowed, and hence a better estimation of the model is obtained.
2. Reduce the extension of the viewing region $\mathcal{V}$ and/or the illumination region $\mathcal{I}$. The failure to pass the test may be due to large dissimilarities among the views used to generate the model. By reducing the extensions of the regions $\mathcal{V}$ and $\mathcal{I}$, we can increase the similarity between these views.
3. Try different probability distributions, such as truncated Gaussian or double exponential distributions. The error distributions are defined only for *positive* errors. Hence, the approximation of an error distribution to a Gaussian distribution is only valid if its mean is more than two standard deviations away from the origin. If that is not the case, an asymmetric distribution, such as a truncated Gaussian or a truncated double exponential distribution, should be used.

A.2 Image Generation

PREMIO's matching routine was tested on simulated images using the joint feature reliability and feature detectability of the PPM labels as a heuristic to determine which labels should be matched first. PREMIO generates simulated images using the following steps:

1. Given the desired number of samples N'_v and N'_i, PREMIO uniformly samples the viewing and illumination space regions $\mathcal{V}$ and $\mathcal{I}$. Let $\mathcal{V}'_s$ be the new set of the sampled viewing positions and $\mathcal{I}'_s$ the new set of the sampled lighting positions.
2. Generate the images. For each pair $(v, i) \in \mathcal{V}'_s \times \mathcal{I}'_s$, PREMIO predicts the subset of detectable labels L_{vi}, its associated attribute mapping $f_{L_{vi}}$, the subset of detectable relational tuples R_{vi}, and its associated relationship strength mapping $g_{R_{vi}}$. PREMIO also generates the corresponding set of units U_{vi}, the associated attribute mapping $f_{U_{vi}}$, the set of relational tuples S_{vi}, and the associated strength mapping $g_{S_{vi}}$.

A.3 Matching Routine

The effect of the joint feature reliability and feature detectability values assigned to the PPM labels was tested using PREMIO's matching routine. PREMIO attempts to match the randomly-generated images against the PPM using the joint feature reliability and feature detectability values as a heuristic to determine which labels the matching routine should attempt to match first. The matching algorithm was run several times with each successive run attempting to match more correspondences. The performance of the matching routine was evaluated by generating a *receiver operating curve* plot in which the probability of *misdetection error* was plotted against the probability of *false alarm error* over the range of correspondences sought. The following steps summarize PREMIO's matching routine:

1. Choose the number n of correspondences to match.
2. For each of the generated images:
   (a) Run the matching routine to search for an observation mapping having n correspondences between the PPM and the image.
   (b) If the observation mapping is found in a reasonable amount of time, then calculate the number of incorrect correspondences present in the mapping.
   (c) If the number of incorrect correspondences was calculated, then calculate the ratio of the number of incorrect correspondences to n for the given observation mapping. This ratio, denoted f, is the *false alarm ratio* and is defined as:

$$f = \frac{\#\text{Incorrect Correspondences}}{n} . \tag{9}$$

   The ratio f is inversely proportional to the strength of the mapping found.

A.4 Evaluation of Matching Results

If PREMIO does not find an observation mapping for the given number of n correspondences, then the experiment is termed a misdetection error (ME). The probability of a misdetection error, given n, is defined as:

$$P(\text{ME} \mid n) = \frac{\#\text{Mappings not Found}}{\#\mathit{Images}} \tag{10}$$

If PREMIO finds an observation mapping, then the correctness of the observation mapping is determined by comparing the false alarm ratio f to the false alarm threshold ratio F. If f is larger than F, then the observation mapping is incorrect and the experiment is termed a false alarm error (FAE). F was varied in order to test the performance of the system. The probability of a false alarm error, given that a mapping m_n with n correspondences was found and given F, is defined as:

$$P(\text{FAE} \mid m_n, F) = \frac{\#\text{Incorrect Mappings}}{\#\text{Mappings Found}} \tag{11}$$

Equations (10) and (11) measure the performance of the matching routine. The performance of the matching routine for various values of n and F can be shown on receiver operating curves (ROC) where probability of misdetection is plotted against probability of false alarm error.

Toward Non-Parametric Digital Shape Representation and Recovery

Ari Gross[1] and Longin Latecki[2]

[1] University Graduate Center and Queens College, City University of New York, Flushing NY 11367, USA
[2] University of Hamburg, Vogt-Kölln-Str. 30, 22527 Hamburg, Germany

Abstract. In this paper, the authors argue that shape representation and recovery can be accomplished using digital non-parametric models and methods. First, we describe some recent results in digital topology and digital geometry. It is shown, for example, that for compact, twice-differentiable objects, which we refer to as parallel regular, a digitization resolution always exists that preserves topology and that a topology-preserving digitization resolution r always preserves the qualitative differential geometry of the object surface. Moreover, it is shown that if an object is parallel regular, very few digital boundary patterns are realizable and that each such digital neighborhood has a well-defined geometric interpretation with respect to tangent direction. We then define the set of digital boundary curves as either an adjacency graph or a grammar, and show how to recursively enumerate the set of such curves. Finally, we describe how a generalized torus can be recovered from a single intensity image using digital non-parametric models and methods.

1 Introduction

The real world consists of many objects, both natural and man-made. It is very different than a robotic car-assembly plant or a light bulb factory. If computer vision representation and recovery methods are to work outside of such-tightly controlled environments, there is a need for representations that are not as well-defined as parametric surfaces. Consider the question: "What is the most general set of shape models that a vision system could use such that shape recovery from a digital monocular intensity image is still possible?". The answer is probably not a set of parametric models since the human vision system works in a much wider object domain.

Another problem with parametric shape models as the basis for developing recovery methods is that the models themselves are continuous while the image is discrete. Trying to recover continuous features in a digital domain is difficult for several reasons. One reason is that invariants are not necessarily digitally invariant, e.g., the centroid of a planar figure is affine invariant but not digitally affine invariant. Another problem is one of localization: a digital inflection point is hard to detect in an image.

Aside from the lack of robustness, the authors also believe that current parametric recovery methods are slow. This is a result of trying to make the parametric model as general as possible by increasing the number of parameters to recover. We are often left trying to recover a model in some very large parameter space that minimizes some error-of-fit function, which may or may not minimize an actual geometric property of the object surface.

This paper starts from a very different point of view. The authors first consider the effects of digitization on object topology and geometry. In particular, if an object is parallel regular then it is shown that a digitization resolution r always exists that is guaranteed to preserve its topology. Also, this digitization resolution preserves qualitative differential geometry, i.e., a boundary point which is locally convex cannot be digitized to a locally concave pixel and a boundary point which is locally concave cannot be digitized to a locally convex pixel. Proofs of the theorems in this paper are given in [3].

The most important result shown in this paper, however, is that the number of topologically realizable digital boundary neighborhoods is very small and that each neighborhood has a well-defined geometric interpretation with respect to the tangent direction. This allows us to define the set of digital boundary curves as a graph or a grammar and to recursively enumerate the set of such curves. It also demonstrates that shape representation may have a lot more to do with formal language theory than once believed.

2 Previous Work and Definitions

Serra [11] considered different kinds of digitizations. He showed that, for a certain class of planar sets, their digitizations preserve homotopy, which implies that there is a complete correspondence between connected components of the planar set and its digitization and their complements. However, he proved this only for digitizations in hexagonal grids, where a digitization of a set in R^2 is the set of points in Z^2 which are contained in the set.

Pavlidis [8] tried to generalize to two-dimensions Shannon's Sampling Theorem, which is well-known in one-dimensional signal processing. However, this theorem is not true, see [3].

We now define a class of subsets of the plane or 3D space representing "real objects", which we will call par(r)-regular sets. We assume that A is a closed subset of the plane or space such that its boundary bdA is compact and, for every point $a \in bdA$, the tangent line (or plane) at a is well-defined. We denote the tangent line (or plane) at a by $t(a)$ and the normal line at a by $nl(a)$. Note that A, as well as bdA, does not have to be connected. All the results in this paper, unless explicitly stated otherwise, hold in 2D as well as in 3D.

Definition: For every $a \in bdA$, let $n(a,r)$ denote the **normal vector** at a of length r pointing towards the outside of A. We also view $n(a,r)$ as the set of points located on it. We denote the normal vector at a of length r pointing towards the inside of A by $-n(a,r)$.

Definition: A set A will be called **par(r,+)-regular** if, for every two distinct points $x, y \in bdA$, $n(x,r)$ and $n(y,r)$ do not intersect. For example, in Figure 1, set X is not par(r,+)-regular while set Y is par(r,+)-regular, where r is the length of the depicted vectors. A set A will be called **par(r,-)-regular** if, for every two distinct points $x, y \in bdA$, $-n(x,r)$ and $-n(y,r)$ do not intersect. A set A will be called **par(r)-regular** if it is par(r,+)-regular and par(r,-)-regular. A set A will be called **parallel regular** if there exists a constant r such that A is par(r)-regular.

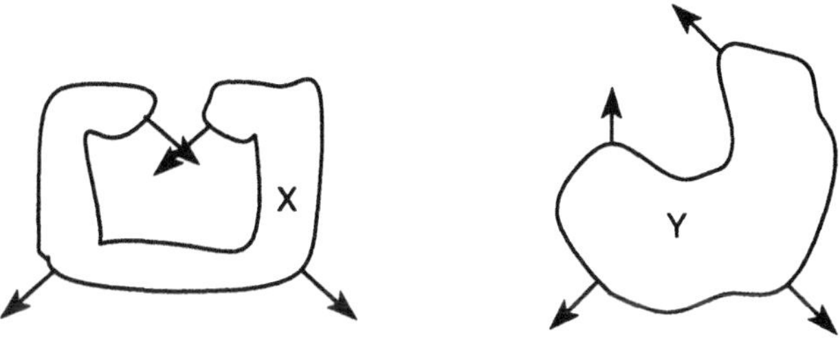

Fig. 1. X is not par(r,+)-regular, but Y is par(r,+)-regular.

We will sometimes call parallel regular sets **objects**. Assuming an object is parallel regular, we now want to define the notion of a parallel set by adding the normal vectors of fixed length to the original set.

Definition: The **parallel set** of A with distance r is given by $Par(A,r) = A \cup \bigcup\{n(a,r) : a \in bdA\}$. For $-r$, we have $Par(A,-r) = cl(A - \bigcup\{-n(a,r) : a \in bdA\})$, where cl denotes the usual set closure in R^2 and R^3.

Definition: $B(x,r)$ denotes the closed ball of radius r centered at a point x. $Dil(A,r)$ denotes the dilation of A with balls of radius r, i.e. $Dil(A,r) = \bigcup\{B(x,r) : x \in A\}$.

The following definition generalizes the concept of the osculating circle.

Definition: Let $a \in bdA$ and let $n(a,e)$ be the normal vector at a of length e pointing towards the outside of A. For every $a \in bdA$, we define an **inside osculating ball** $iob(a,e)$ at a with radius e to be the open ball centered at point $a - n(a,e)$ with radius e. For every $a \in bdA$, we define an **outside osculating ball** $oob(a,e)$ at a with radius e to be the open ball centered at point $a + n(a,e)$ with radius e.

Note that the following holds for every $z \in bdA$:
$iob(z,e) \cap oob(z,e) = \emptyset$, $t(z) \cap iob(z,e) = \emptyset$ and, $t(z) \cap oob(z,e) = \emptyset$.

The following theorem [3] justifies the definition of osculating balls for parallel regular sets.

Theorem 1 *Let A be par(r)-regular. Then $iob(x,r) \subseteq A$ and $oob(x,r) \subseteq A^c$ for every $x \in bdA$.*

3 Digitization Preserving Topology

Our definition of a digitization models a real digitization process. Consistent with real sensor output, a digitization is defined with respect to a grid of squares (or cubes in 3D), where each square (or cube) has diameter r. A square (or cube) is a black pixel (or voxel) iff the ratio of the area (volume) of the object to the area (volume) of the field "seen" by the corresponding sensor is greater than some constant threshold value. For any threshold value, we show that the digitization with diameter r of a par(r)-regular set A will be homotopy equivalent to A.

Definition: Let X be any set in the plane (or space). Let Q be a cover of the plane (space) by squares (cubes) with diameter r such that the intersection of two squares is either empty, a corner point or an edge (or a face). Such a cover is called a square grid (cubical grid) with diameter r. Each square (cube) in Q is either white or black. If we treat the squares of Q as points in Z^2 (Z^3) with the corresponding colors, we obtain a digital picture, which will be called a **digitization** of X with diameter r. We will also identify the digitization of X with the union of closed black squares (cubes), i.e. the digitization of X is a closed subset of the plane (or space). Thus, the digitization of X refers to either the digital picture or the union of closed black squares (cubes).

In the following, we define some important digitization classes.

Definition: Let X be any set in the plane (or space). A square (cube) of $p \in Q$ is black iff $p \cap X \neq \emptyset$, and white otherwise. We will call such a digitization an **intersection digitization** with diameter r of set X. We will denote this digitization by $Dig_{\cap}(X, r)$, namely $Dig_{\cap}(X, r) = \bigcup\{p : p \cap X \neq \emptyset\}$. $Dig_{\cap}(X, r)$ either denotes the digital picture or the union of closed black squares (cubes). See Figure 2, for example, where the union of all depicted squares represents an intersection digitization of an ellipse.

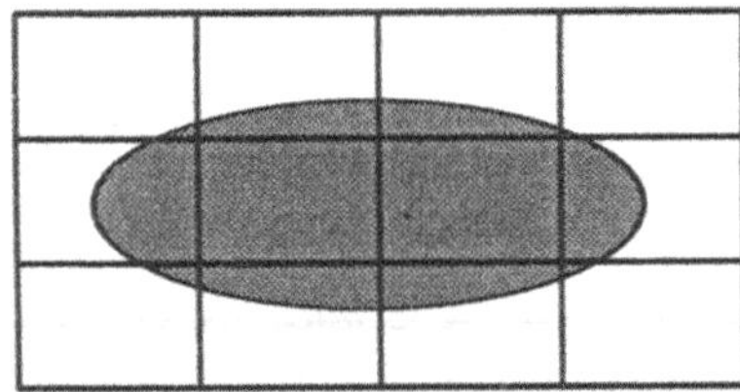

Fig. 2. The union of all squares represents an intersection digitization of an ellipse.

With respect to real camera digitization, an intersection digitization corresponds to the procedure of coloring a pixel (or voxel) black iff there is part of the object A in the field "seen" by the corresponding sensor.

Now we consider digitizations corresponding to the procedure of coloring a pixel (or voxel) black iff the object X fills the whole field "seen" by the corresponding sensor. For such digitizations, a square p is black iff $p \subseteq X$ and white otherwise. Note that this statement is equivalent to the following rule: a square p is white iff $p \cap X^c \neq \emptyset$

and black otherwise, where X^c denotes the complement of set X. We will refer to such digitization of a set X as **subset digitization** and denote by $Dig_{\subset}(X, r)$, where $Dig_{\subset}(X, r) = \bigcup\{p : p \subseteq X\}$. In Figure 3, the two black squares represent $Dig_{\subset}(X, r)$, where X is an ellipse.

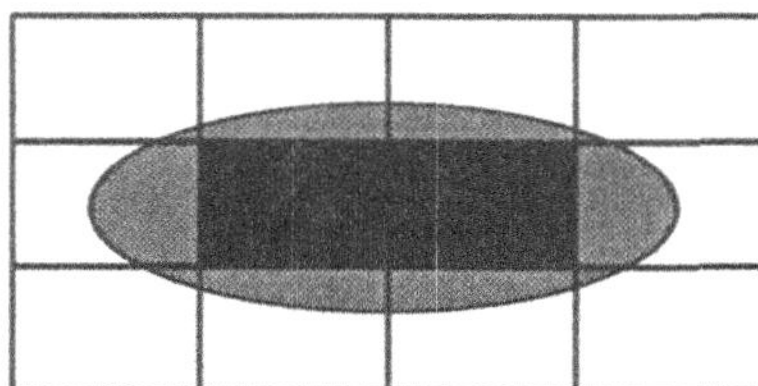

Fig. 3. The two black squares represent a subset digitization of an ellipse.

Next, let us consider digitizations corresponding to a procedure in which a pixel is colored black iff the ratio of the area of the object "seen" by the sensor to the area of the entire field "seen" by the same sensor is greater than some constant threshold value v. An example is given in Figure 4, where the gray pixels represent a digitization of the ellipse with the ratio equal to 1/5.

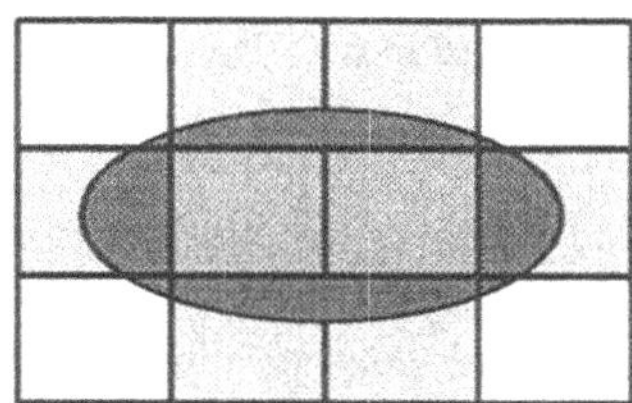

Fig. 4. The gray pixels represent a digitization of an ellipse with the area ratio equal to 1/5.

Let X be any set in the plane (or space). In the 2D case, square $p \in \mathcal{Q}$ is black iff $area(p \cap X)/area(p) > v$ and white otherwise, and in the 3D case, cube $p \in \mathcal{Q}$ is black iff $volume(p \cap X)/volume(p) > v$ and white otherwise, where $0 \leq v < 1$ is a constant. If we treat the squares of $\mathcal{Q}$ as points in Z^2 (Z^3) with the corresponding black or white values, we obtain a digital picture, which will be called the **v-digitization** of X with diameter r. We will denote such digitizations by $Dig_v(X, r)$. We recall that we also identify the digitization of X with the union of black closed squares (cubes). Thus $Dig_v(X, r)$ either denotes the digital picture or the union of black closed squares (cubes). We will also denote $Dig_1(X, r)$ as the digitization in which the ratio of the area (volume) is equal to 1.

We have the following inclusions $Dig_{\subset}(X, r) \subseteq Dig_v(X, r) \subseteq Dig_{\cap}(X, r)$ for every $v \in [0, 1]$ and $Dig_v(X, r) \subseteq Dig_w(X, r)$ if $w \leq v$ for every $v, w \in [0, 1]$. Since our results apply to any of these digitizations, we will hereafter use $Dig(X, r)$ without subscript to denote $Dig_\cap(X, r)$, $Dig_\subset(X, r)$, and $Dig_v(X, r)$ for every $v \in [0, 1]$.

In the following, we briefly review the concept of homotopy equivalence.

Definition: Let X and Y be two topological spaces. Two functions $f, g : X \to Y$ are said to be **homotopic** if there exists a continuous function $H : X \times [0, 1] \to Y$, where $[0, 1]$ is the unit interval, with $H(x, 0) = f(x)$ and $H(x, 1) = g(x)$ for all $x \in X$. The function H is called a **homotopy** from f to g. X and Y are called **homotopy equivalent** or of the same **homotopy type** if there exist two functions $f : X \to Y$ and $g : Y \to X$ such that $g \circ f$ is homotopic with the identity over X (id_X) and $f \circ g$ is homotopic with the identity over Y (id_Y).

Intuitively, a homotopy H represents a continuous deformation of the map f to g. As a consequence of the properties of homotopy equivalence, there is a complete correspondence between connected components of X and Y and their complements if X and Y are pathwise connected. The Euler characteristic, as well as the fundamental groups of X and Y, are the same (see [7]). Therefore, we will use homotopy equivalence as a definition for topology preserving.

Definition: We will say that a digitization $Dig(X, r)$ of some set X is **topology preserving** if X and $Dig(X, r)$ are homotopy equivalent.

Now we are ready to state our main theorems about topology preservation. For proofs see [3]. We recall that $Dig(A, r)$ denotes $Dig_\cap(X, r)$, $Dig_\subset(X, r)$, and $Dig_v(X, r)$ for every $v \in [0, 1]$, and these digital images model the output of many real digitization processes.

Theorem 2 *Let A be a par(2r)-regular set. Then A and Dig(A, r) are homotopy equivalent.*

Theorem 3 *Let A be a closed subset of the plane or space with a compact boundary such that the second derivative exists at every point $x \in bdA$. Then there always exists a digitization resolution $r > 0$ such that every digitization Dig(A, r) of A is topology preserving.*

If a set is not par(2r)-regular, then its digitization $Dig_\cap(A, r)$ can have different topological structure, as illustrated in Figure 5(a), where set A is simple connected, but $Dig_\cap(A, r)$ represented by gray squares is not simple connected, since there is a white "hole" in it. Of course, one can always find a set X having some special shape which is not par(2r)-regular, yet X and $Dig_\cap(X, r)$ are homotopy equivalent, like the set presented in Figure 5(b). Although topology was preserved in digitizing the set shown in Figure 5(b), it is clear that important shape properties were lost. However, as we will show in the next section, if a set A is par(2r)-regular, then $Dig_\cap(A, r)$ will never significantly change its geometric shape properties.

There are many important object classes used in computer vision and medical imaging that are par(2r)-regular for some r and for which the calculation of r is straightforward. One such class is that of zero-torsion generalized tubes, defined in [2]. In [3] , for example, it is shown how to calculate r for a given torus T such that each $Dig(T, r)$ is topology preserving.

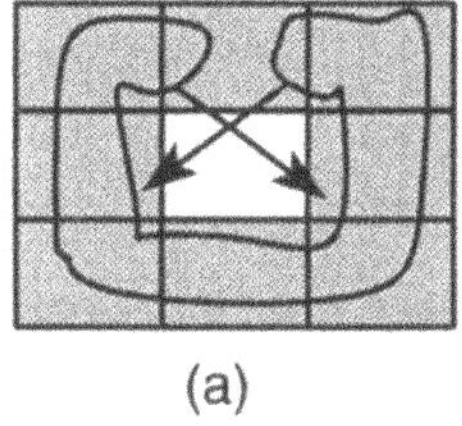

(a)

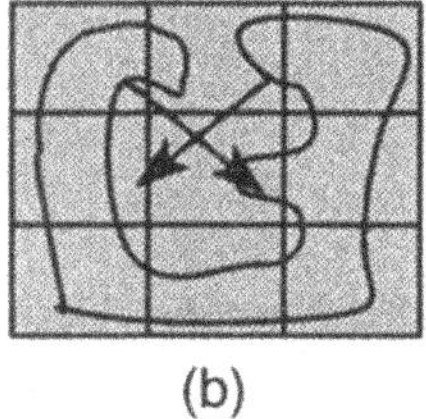

(b)

Fig. 5. (a) A and $Dig_{\cap}(A, r)$ are not homotopy equivalent. (b) X and $Dig_{\cap}(X, r)$ are homotopy equivalent.

4 Digital Differential Geometry

In this section, we show that if A is a par(2r)-regular set, then only a few digital patterns can occur as neighborhoods of boundary points in its digitization $Dig_{\cap}(A, r)$. This is very useful for noise detection, since if the neighborhood of a boundary point does not match one of these patters, it must be due to noise. So, if in a practical application the resolution r of the digitization is such that the parts of the object which have to be preserved under the digitization form a par(2r)-regular set, then our results allow for efficient noise detection.

We also show that the digitization $Dig_{\cap}(A, r)$ of a par(2r)-regular set A will not change the qualitative differential geometric properties of the boundary of A, i.e. a boundary point which is locally convex cannot be digitized to a locally concave pixel and a boundary point which is locally concave cannot be digitized to a locally convex pixel.

First, we state a basic fact about the local connectedness of par(r)-regular sets. For the proof see [3].

Theorem 4 *Let A be par(r)-regular. Then $A \cap B(x, t)$ is connected for every $t \leq r$ and $x \in R^2$ (or $x \in R^3$).*

Now we review some definitions from digital topology which are based on [9] and [5]. As usual in digital topology, we assume that all sets are subsets of $\mathbb{Z}^2$ or $\mathbb{Z}^3$. The points in a set S will be termed *black* or *foreground* points, while those of the complement S^c will be termed *white* or *background* points.

The *4-neighbors* (or *direct neighbors*) of a point (x, y) in $\mathbb{Z}^2$ are its four horizontal and vertical neighbors $(x+1, y)$, $(x-1, y)$ and $(x, y+1)$, $(x, y-1)$. The *8-neighbors* of a point (x, y) in $\mathbb{Z}^2$ are its four horizontal and vertical neighbors together with its four diagonal neighbors $(x+1, y+1)$, $(x+1, y-1)$ and $(x-1, y+1)$, $(x-1, y-1)$. The *n-neighborhood* of a point P is the set $N_n(P)$ consisting of P and its n-neighbors. $N_n^*(P)$ is the set of all neighbors of P without P itself, i.e., $N_n^*(P) = N_n(P) - \{P\}$. In the 2D case, $N_8(p)$ is also referred to as $N(p)$ and called the *neighborhood* of p. The points in $N_8^*(P)$ are numbered 0 to 7 according to the following scheme:

$N_3(P)$	$N_2(P)$	$N_1(P)$
$N_4(P)$	P	$N_0(P)$
$N_5(P)$	$N_6(P)$	$N_7(P)$

Definition: In the 2D case, a black point p is a **boundary point** if one of its 4-neighbors is white.

Theorem 5 *(2D case) Let A be par(2r, +)-regular. Then the following pattern and its 90° rotation cannot occur in $Dig_\cap(A, r)$.*

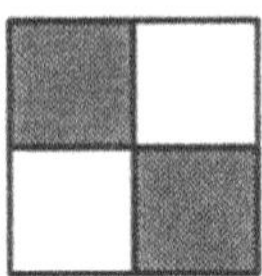

By this theorem $Dig_\cap(A, r)$ is well-composed, i.e. every 8-component of black points, as well as of white points, is also a 4-component. Well-composed sets have very nice digital topological properties [6], in particular, the Jordan Curve Theorem holds and the Euler characteristic is locally computable. These results imply that many algorithms for digital image processing can be simpler and faster.

Theorem 6 *(2D case) Let A be par(2r)-regular. Then the following patterns and their 90° rotations cannot occur in $Dig_\cap(A, r)$, where the light gray points can be either black or white:*

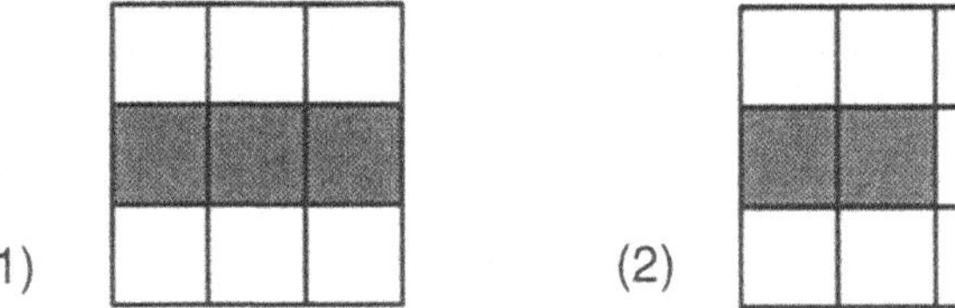

The following theorem is a very important result for recovering differential geometric properties of an object and for noise detection. If a template occurs that is different from those enumerated in the theorem, we can be sure that it must be due to noise.

Theorem 7 *(2D case) Let A be par(2r)-regular. Then the neighborhood N(p) of a (4-) boundary black point $p \in Dig_\cap(A, r)$ can have only one of the configurations presented in Figure 6 (modulo reflection and 90° rotation).*

Observe that the set of realizable patterns given in Theorem 7 is very small. Moreover, this set is minimal, i.e. the number of patterns which can occur as the neighborhood of a (4-) boundary black point in $Dig_\cap(A, r)$ cannot be further reduced, since it is easy to construct a par(2r)-regular set A such that $Dig_\cap(A, r)$ contains each of the patterns in Figure 6. Knowing that these seven configurations constitute all possible configurations (modulo reflection and rotation) which can occur on the boundary of the digitization $Dig_\cap(A, r)$ of a par(2r)-regular set A, it is now a simple task to classify each of these configurations with respect to their differential geometric properties. First, we identify all possible boundary configurations that are digitizations of a half plane.

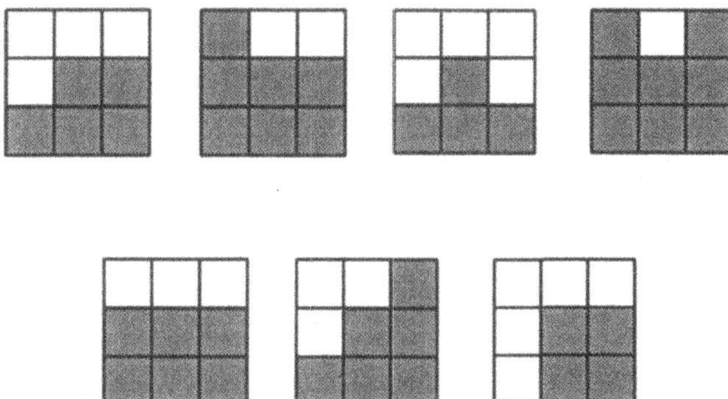

Fig. 6. All possible configurations modulo reflection and rotation which can occur on the boundary of the digitization $Dig_{\cap}(A, r)$ of a par(2r)-regular set A.

Theorem 8 *(2D case) Let H be a closed half plane. Then p is a boundary point of* $Dig_{\cap}(H, r)$ *iff the neighborhood of p has one of the patterns shown in Figure 7 (modulo reflection and* 90° *rotation).*

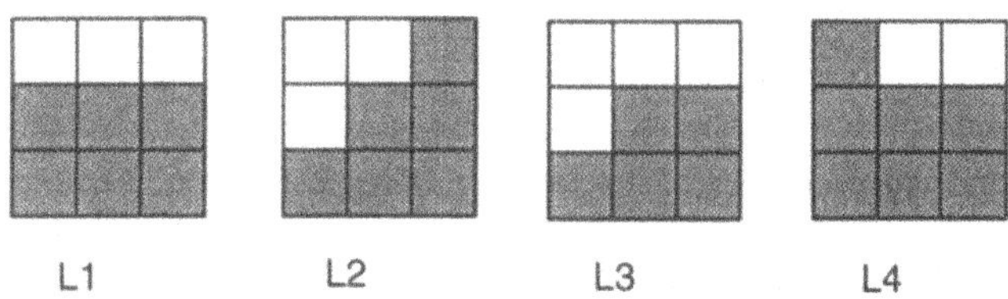

Fig. 7. Linear digital neighborhoods.

Definition: Based on Theorem 8, we define a **linear digital neighborhood** of p as any neighborhood configuration $N(p)$ corresponding to L1, L2, L3, L4 in Figure 7 modulo reflection and 90° rotation. Next we define a **convex digital neighborhood** of p as any neighborhood configuration $N(p)$ corresponding to Convex1 or Convex2 in Figure 8 modulo reflection and 90° rotation. We define a **concave digital neighborhood** of p as any neighborhood configuration $N(p)$ corresponding to Concave1 in Figure 8 modulo reflection and 90° rotation.

Theorem 9 *Let x be a boundary point of a planar set A such that* $A \cap B(x, 2r)$ *is convex, and let p be the square of* $Dig_{\cap}(A, r)$ *containing x. If p is a boundary point of* $Dig_{\cap}(A, r)$*, then* $N(p)$ *cannot be digitally concave.*

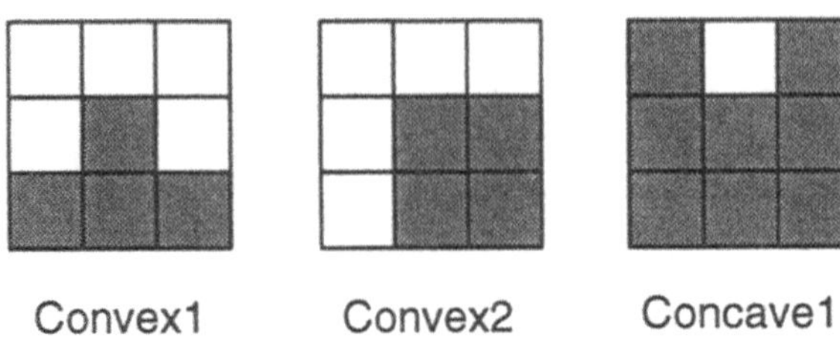

Fig. 8. Convex and concave digital neighborhoods.

Theorem 10 *Let x be a boundary point of a planar set A such that $A \cap B(x, 2r)$ is concave, and let p be the square of $Dig_{\cap}(A, r)$ containing point x. If p is a boundary point of $Dig_{\cap}(A, r)$, then N(p) cannot be digitally convex.*

Our results, although stated for parallel regular objects, are also helpful to analyze the shape properties of non-parallel regular objects, such as piecewise par(2r)-regular sets, as shown in [3].

5 Non-Parametric Representation and Recovery

The results obtained in the previous Section can be used to both represent and recover objects. All of the possible digital boundaries were shown in Figure 6 modulo reflection and 90° rotation. When the set of digital boundaries is fully enumerated, there are 32 patterns. Each pattern is the digitization of a local section of curve, with a range of tangent directions. We can now define a **right adjacency graph** for the digital boundary curve of a parallel regular set as a graph where the nodes are the 32 digital boundary patterns and there is a directed arc from node *A* to node *B* whenever pattern *B* is a possible right-hand boundary neighbor of pattern *A*. A **left adjacency graph** is analogously defined. Part of the right adjacency graph is shown in Figure 9. It can be seen that the *outdegree* = 4 for the portion of the graph shown. In fact, the entire graph has *outdegree* = 4. This graph is easily converted to a grammar and the resulting grammar can then be used to generate sets of digital boundary curves.

The adjacency graph can be used directly in 3D shape recovery. Let a generalized torus be a generalization of the classic torus, where the cross-section curve (i.e., inner circle) is now allowed to be any simple, closed, twice-differentiable curve, while the outer curve is still restricted to being a circle. An example of a generalized torus is shown in Figure 10.

Given that the object shown in Figure 10 is a generalized torus, how can we recover its 3D shape from its intensity image? Since we are not given a parametrization for the cross-section curve, recovering the shape of the curve is not accomplished simply by solving a system of equations or applying non-linear minimization methods. Unlike recovering the shape of a straight homogeneous generalized cylinder, the projected cross-section curve of a generalized torus is not known *apriori.*

Since the object to be recovered is a generalized torus, there is only one curve to recover, the cross-section curve. In [2] it was shown that the contour of a generalized tube is insufficient to constrain the underlying 3D shape and that there is at least one degree of freedom. An exhaustive algorithm to recover a digital approximation to the underlying 3D generalized torus requires recursively enumerating the set of digital cross-section curves and selecting the one that yields a digital generalized torùs that best fits the one in the image. For any digital cross-section curve, we can construct a digital generalized torus. For each digital generalized torus, we can test if its contour is compatible with the image contour. Assuming that the contours are compatible, we then want to find the digital generalized torus that minimizes some intensity-based error-of-fit measure.

Fortunately, there are additional constraints on this problem so that finding the correct digital cross-section curve need not be done in an exhaustive manner. It is reasonable to assume that, given the generalized torus image, the maximal length of a digital cross-section curve can be bounded by some value n. This means that an exhaustive solution requires trying all paths in the adjacency graph of $length \leq n$, which has an exponential time bound of $O(4^n)$. Clearly, even though the search space is large, there are few paths in the adjacency graph that actually need to be explored. For example, a valid path in the adjacency graph must be digitally non-intersecting and closed. Also, any path that takes the resulting object outside of the image contour cannot be part of a valid solution. In addition, digital contour points must satisfy the constraint that their surface normals are approximately orthogonal to the viewing direction. Initial results at recovering the digital cross-section curve of a generalized torus from its intensity image are encouraging and lead the authors to believe that the actual time bound for recovering the digital cross-section curve may be linear.

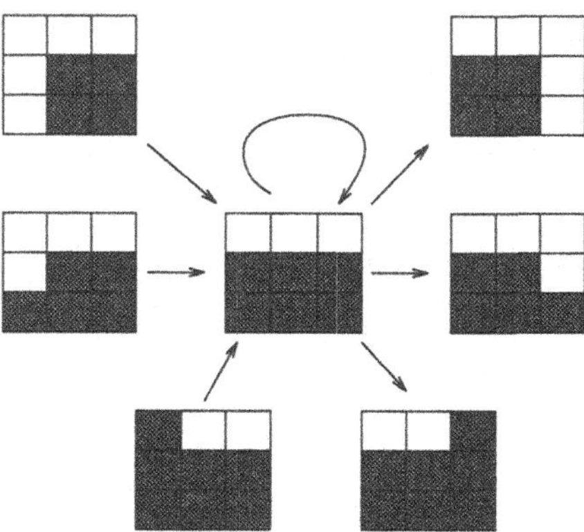

Fig. 9. Part of the adjacency graph for digital boundary configurations.

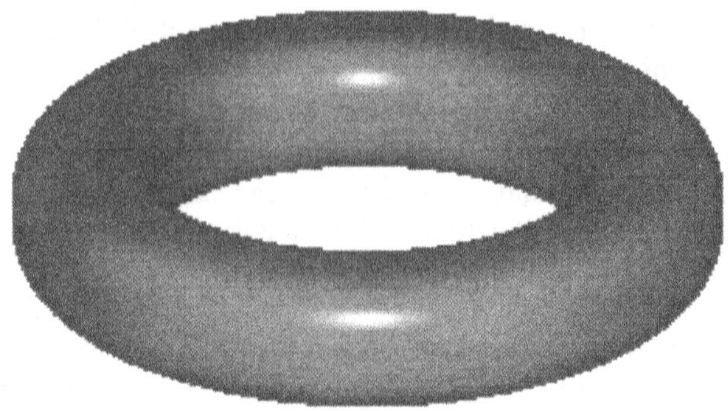

Fig. 10. Intensity image of a generalized torus.

6 Conclusions and Future Work

In this paper, we showed that topological and differential geometric properties are preserved under digitization. For a par(2r)-regular set A in the plane and 3D space, digitizations $Dig_{\cap}(A,r)$, $Dig_{\subset}(A,r)$, and $Dig_{v}(A,r)$ are topology preserving for every $0 \leq v \leq 1$. This result is important for practical applications, since these digitizations model the output of many real digitization processes, and for a large class of real objects, including medical objects, a constant r can always be computed such that the object is par(2r)-regular. Also, the authors showed that only a few digital patterns can occur as neighborhoods of a boundary point in $Dig_{\cap}(A,r)$ of a par(2r)-regular 2D set A, and that the digitization $Dig_{\cap}(A,r)$ of a par(2r)-regular 2D set A does not change the qualitative differential geometric properties of the boundary of A. Finally, we described a method to recover the 3D shape of a generalized torus using the digital boundary adjacency graph.

7 Acknowledgments

The authors would like to acknowledge the support of the National Science Foundation Grant IRI-93-02041 and NASA Grant 449547.

References

1. M. P. do Carmo, *Differential Geometry of Curves and Surfaces*, Prentice-Hall, Englewood Cliffs, New Jersey, 1976.
2. A. Gross, Analyzing Generalized Tubes, *Proceedings of SPIE's Conference on Intelligent Robots and Computer Vision*, Boston, 1994.
3. A. Gross and L. Latecki, Digitizations Preserving Topological and Differential Geometric Properties, to appear in *Computer Vision, Graphics, and Image Processing: Image Understanding*.
4. A. Gray, *Modern Differential Geometry of Curves and Surfaces*, CRC Press, Boca Raton, 1993.

5. T. Y. Kong and A. Rosenfeld, Digital topology: Introduction and survey. *Computer Vision, Graphics, and Image Processing*, 48:357–393, 1989.
6. L. Latecki, U. Eckhardt, and A. Rosenfeld, Well-Composed Sets. *Computer Vision, Graphics, and Image Processing: Image Understanding*, Vol. 60, No. 3, 1994.
7. G. L. Naber, *Topological methods in Euclidean spaces*, Cambridge University Press, Cambridge, 1980.
8. T. Pavlidis, *Algorithms for Graphics and Image Processing*, Springer–Verlag, Berlin, 1982.
9. A. Rosenfeld, Digital topology. *American Mathematical Monthly*, 86:621–630, 1979.
10. A. Rosenfeld and A.C. Kak, *Digital Picture Processing*, Academic Press, New York, 1982.
11. J. Serra, *Image Analysis and Mathematical Morphology*, Academic Press, New York, 1982.
12. M. Spivak, *A Comprehensive Introduction to Differential Geometry*, Publish or Perish, Vol. 3, 2nd Edition, Berkeley, 1979.

Spherical Representations: from EGI to SAI*

Katsushi Ikeuchi and Martial Hebert

School of Computer Science
Carnegie Mellon University
Pittsburgh PA 15213

Abstract. One of the fundamental problems in representing a curved surface is how to define an intrinsic, i.e., viewer independent, coordinate system over the surface. More precisely, in order to establish point matching between model and observed feature distributions over curved surfaces, we need to set up a coordinate system that maps a point on a curved surface to a point on a standard coordinate system. This mapping should be independent of the viewing direction. Since the boundary of a 3-D object forms a closed surface, a coordinate system defined on the sphere is preferred. We have been exploring several intrinsic mappings from an object surface to a spherical surface. We have investigated several representations including: the EGI (Extended Gaussian Image), the DEGI (Distributed Extended Gaussian Image), the CEGI (Complex Extended Gaussian Image), and the SAI (Spherical Attribute Image). This paper describes each representation and the lessons that we have learned by using those representations in recognition systems.

1 Introduction

One of the fundamental problems in object recognition the representation of object models. The representation governs the performance of recognition systems. We restrict ourselves here to surface-based representations since those are the most relevant in computer vision. A surface-based representation describes an object as a collection of visible "faces" of the object. Representative surface-based representations include edge-and/or face-based invariants, aspect graphs, and spherical representations.

The simplest type of object representation is based on planar faces. A planar face has a clear boundary of surface orientation discontinuity and its internal pixels provide less information. A polyhedron, consisting of planar faces, effectively represents the relations between faces for recognition. Early works by Oshima-Shirai [1] and Bolles [2] effectively use such graphs of visible face relations.

One of the basic problems in using such visible graphs was determining the number of different graphs required to represent one object. Koendering's aspect answers this question [3]. The aspect representation specifies an object as a

* The research is sponsored in part by National Science Foundation under Grant IRI-9224521, and in part by the Advanced Research Projects Agency under the Army Research Office under Grant DAAH04-94-G-0006

collection of all possible topologically different relational graphs of visible faces. Our earlier work on the vision algorithm compiler used this aspect representation for object recognition [7]. Some recent works on aspect graph generation are reviewed in [8].

For curved object recognition, the boundary of a curved surface patch is often ill-defined; the relative relationships between faces are unreliable. On the other hand, points on the surface do carry useful information, such as curvature.

To specify each internal point requires a coordinate system over the curved surface. We can describe invariant feature distributions, including curvature distributions, along a particular coordinate system.

This coordinate system should be

- *intrinsic:* The coordinate system is defined using intrinsic characteristics of a surface so that we can recover such coordinate system from the surface itself, independent on the viewing direction.
- *covering:* It is desirable to define a coordinate system over the entire object surface in a uniform manner because it is difficult to reliably segment a curved surface into regions.

For example, let us consider a simple coordinate system on a spherical surface using two parameters, longitude and latitude. The coordinate system can be defined over the spherical surface. Thus, it is covering. The coordinate system depends on the direction of an imaginary line between the North and South poles. Since the North and South poles are not defined by the surface characteristics, it is not intrinsic.

Several attempts have been made at recovering such coordinate systems. Brady and Ponce proposed a coordinate system based on the lines of curvatures [5]. The lines of curvatures are defined as the lines connecting the directions of extreme at each point over a curved surface. Since usually at each point the extrema directions are defined uniquely from the surface itself, the coordinate system is intrinsic. We can define the lines of curvatures over the entire surface of an object (*covering*). Thus, this representation satisfies the previous two criteria. However, the lines of curvatures cannot be defined uniquely at umbilical points. Unfortunately, such common surfaces as spherical surfaces and planar surfaces consists of only umbilical points, and thus cannot be represented using the proposed coordinate system. Besides this theoretical drawback, the coordinate system is unreliable because of the noise introduced by the second derivatives for obtaining the lines of curvatures from observed data. Stein and Medioni proposed a coordinate system referred to as the Splash [6]. The Splash is a geodesic polar coordinate system defined around a high curvature point as its origin. As its characteristics of the geodesic polar coordinate system, the representation is suitable for a small patch but not for an entire surface.

We have been working on a spherical representation to map the surface of an object to a standard coordinate system (a unit sphere). Objects usually handled by vision systems have closed surfaces: topologically equivalent to a spherical surface. Thus, we began our effort by developing a mapping method from an arbitrary object surface to a spherical surface and store invariants over the spherical surface. In this paper, we briefly describe our earlier efforts on such spherical representations. We begin with a discussion of the Extended Gaussian Image developed around 1980, and, continue on to describe our recent work on the Spherical Attribute Image.

2 Gauss Mapping and Related Representations

Gauss derived a mapping method that uses surface orientations to map points on an arbitrary curved surface onto points on the sphere [9]. Let us assume a surface patch p on a curved surface. At that surface patch, we can define the surface normal $N(p)$ uniquely (Figure 1). Regardless of the position of p, we can translate the normal vector so that its origin coincides with the origin of the coordinate system. The end point of the unit normal lies on a unit sphere. The mapping that associates this point on the unit sphere with the patch is referred to as a Gauss map and the sphere is referred to as a Gaussian sphere.

Formally, let $S \subset R^3$ be a surface with an orientation N. The map $N : S \mapsto R^3$ takes its values in the unit sphere S^2. The map N, thus, defined, is called the Gauss map of S. See Fig. 1.

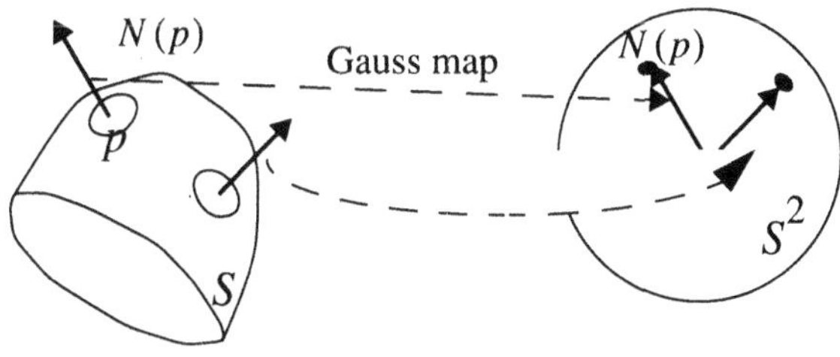

Fig. 1. The Gauss Map

This Gauss map is intrinsic and covering. Although surface orientations depend on the viewing direction, the relative relationships of normals among patches are intrinsic. Thus, the distribution of normals on the Gaussian sphere is independent on the viewing direction. We can represent the entire surface of an object on the Gaussian sphere. Thus, it is covering.

Extended Gaussian Image (EGI). Let us assume that an object surface is evenly sampled into patches. At each surface patch, we can define a surface normal with a single unit of mass. Each surface normal is assumed to be able to vote the mass to the corresponding point on the Gaussian sphere. From the voting by the all surface patches over the object surface, we can observe a distribution of mass over the Gaussian sphere. This mass distribution depends on the shape of the object and is referred to as the Extended Gaussian Image (EGI) of the object.

This mapping does not incorporate the original spatial relationship between surface patches. Thus, it seems that the original shape information is somehow reduced. However, Minkowski proved that, if two convex objects have the same EGIs, those objects are congruent [10]. Other important characteristics of EGI [11] are:

- The EGI mass on the sphere is the inverse of Gaussian curvature on the object surface,
- The mass center of the EGI is at the origin of the sphere,
- As an object rotates, its EGI also rotates in the exact same way.

Using the Minkowski theorem, we can develop an object recognition system. From an object model, we sample its surface evenly, calculate surface normals, and obtain the model EGI. After obtaining the surface normal distribution of an observed object, we can repeat the same process to build an observed EGI. By examining which model EGI has a distribution similar to the observed one, we can recognize the observed object; by examining which part of the EGI distribution corresponds to the observed partial EGI, we can determine from which direction we are observing it. Some of the earlier work on object recognition using EGI are found in [12, 13, 14]

2.1 Distributed EGI

The applicability of the original EGI is strictly limited to convex objects. A non-convex object has more than two separated regions with the same surface orientations. Such areas, though physically separated (for example, regions A and B in Fig. 2), will be mapped to the same point on the Gaussian sphere. However, when an object is observed from a particular viewing direction, some of the area (such as region A) may be occluded when viewed from this direction and may not contribute EGI mass on the Gaussian sphere. Thus, the observed EGI is not same as the corresponding part of the model EGI on the Gaussian sphere.

In order to avoid this effect, we have derived a distributed EGI [15]. This method recalculates a partial EGI for each viewing direction as shown in Fig. 3.

We can represent all possible viewing directions using a viewing sphere. Thus, we sample a viewing sphere into sampling viewing directions. At each sampling viewing direction, we recalculate EGI by taking into account the effect of self occlusion. Since this effect is accommodated in the model, we can determine the attitude of a non-convex object using this distributed EGI.

2.2 Complex EGI

Definition. One of the problems with the EGI is that we can determine the orientation of an object but cannot determine the position of the object. In

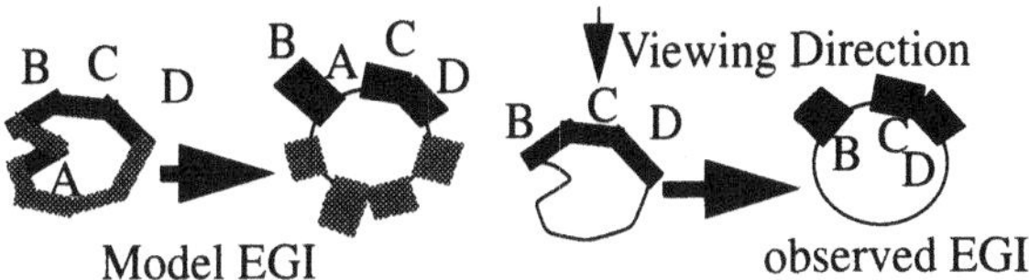

Fig. 2. EGI of an non-convex object

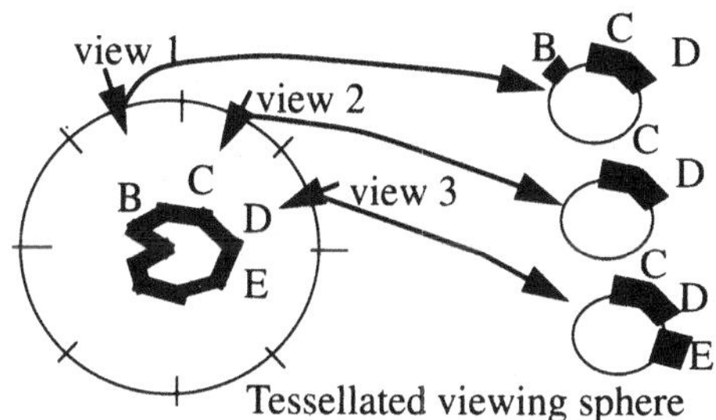

Fig. 3. Definition of Distributed EGI (DEGI)

order to recover translation, we have introduced the complex EGI to encode positional information [16].

The basic approach is, for each surface patch, to store at the corresponding point of the Gaussian sphere the distance between the path and a fixed origin. The CEGI weight at each point on the Gaussian sphere is a complex number whose magnitude is the surface area and whose phase is the distance information. When an object translates, the magnitude of the complex mass remains the same while its phase changes accordingly.

Object recognition is accomplished by EGI matching using the magnitude only. The translation component is computed by using the phase difference.

Formally, the complex weight associated with a surface patch is $A\exp(jd)$, where A is the area of a patch with surface normal $\mathbf{n}$, the normal distance d to a fixed origin (Fig. 4). The distance d is positive if the perpendicular vector from the origin to the patches is in the same direction as the outward facing normal of the patch.

For any given point in the CEGI corresponding to normal $\mathbf{n}$, the magnitude of the point's weight is $A\exp(jd)$. A is independent of the normal distance, and if the object is convex, the distribution of A corresponds to the conventional EGI representation. If the object is not convex, the magnitude of each weight will not necessarily be equal to the weight in the corresponding EGI.

The translation invariant property of the weight magnitude applies even if

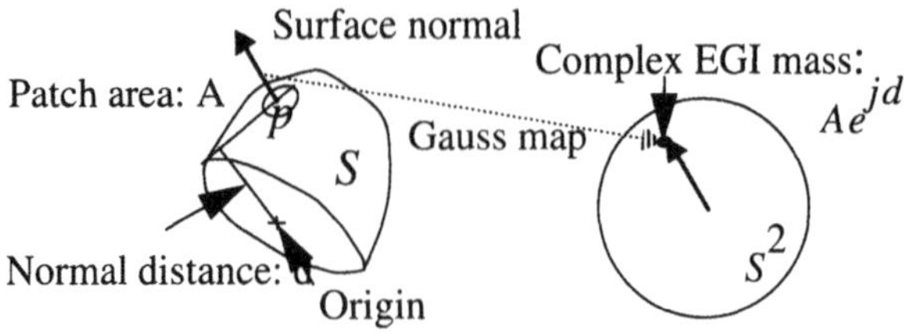

Fig. 4. Complex EGI (CEGI)

there is more than one surface patches with the same normal. Consider the surface patches whose normals are $\mathbf{n}$ shown in Fig. 5 (the distribution $d_1, \ldots, d_k$ is henceforth referred to as the surface normal distance distribution).

Before translation, the corresponding complex weight is

$$p = \sum_{i=1}^{k} A_i \exp(jd_i) \ . \tag{1}$$

After a translation along a vector $\mathbf{T}$, the complex weight becomes:

$$P' = \sum_{i=1}^{k} A_i \exp(j(d_i + \mathbf{T} \cdot \mathbf{n})) = \exp(j(\mathbf{T} \cdot \mathbf{n}))P \ . \tag{2}$$

Hence, for each point of the CEGI, the magnitude of the weight is independent of the translation. The complex number wraps around for every translation distance of 2π. Therefore, the computed translation is known only up to 2π. To eliminate this ambiguity, all distances are normalized such that the greatest expected change in translation distance is π.

Pose Determination Strategy. Given a prototype CEGI and a partial CEGI of an unknown object, we can recognize the object and determine its orientation as follows: First, we calculate the magnitude distributions of both CEGI's, and second, we match the resulting distribution with that of the prototype. Once both the object and its orientation with respect to the stored model are identified, the object translation can be calculated by using the suitably oriented CEGI's.

The translation parameters can be determined by applying a least-squares techniques as follows: Suppose that the object has been translated by δx, δy, and δz in the x, y, and z, respectively. Then, the weight $A\exp(jd)$ of a surface patch becomes $A\exp(j(d+\delta\mathbf{d}\cdot\mathbf{n}))$ after translation, where $\delta\mathbf{d} = \delta x\mathbf{i}+\delta y\mathbf{j}+\delta z\mathbf{k}$ and $\mathbf{n} = n_x\mathbf{i}+n_y\mathbf{j}+n_z\mathbf{k}$. Then, for each matched weight $P'_{\mathbf{n}_j}$ in the object CEGI corresponding to the weight $P_{\mathbf{n}_j}$ in the model CEGI, let

$$\omega_i = arg\frac{P'_{\mathbf{n}_i}}{P_{\mathbf{n}_i}} \tag{3}$$

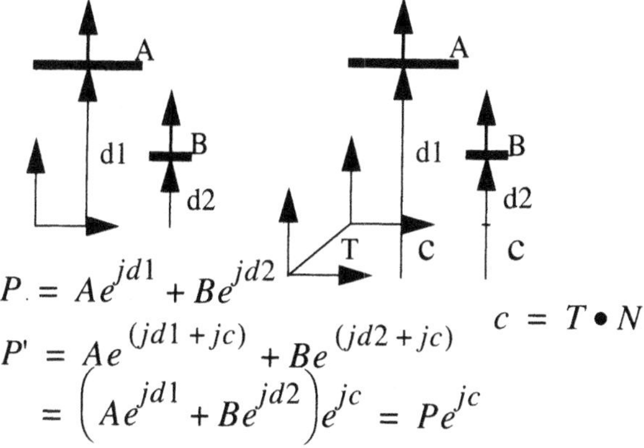

Fig. 5. Translation effect on CEGI

$$= arg(\frac{A_{\mathbf{n}_i} \exp j(d_i + \delta\mathbf{d} \cdot \mathbf{n}_i)}{A_{\mathbf{n}_i} \exp jd_i}) \tag{4}$$

$$= \delta x n_{ix} + \delta y n_{jy} + \delta z n_{kz} \quad . \tag{5}$$

The translation is computed by minimizing:

$$\epsilon = \sum_{i=1}^{N_{\text{visible}}} (\omega_i - n_{ix}\delta x - n_{jy}\delta y - n_{kz}\delta z)^2 \quad , \tag{6}$$

where N_{visible} is the total number of visible surface patches on the object.

Figure 6 shows the localization results of a non-convex curved object using CEGI.

Similar representations have been proposed by Nalwa [17]. He proposed that a surfaces be represented by their Gaussian images augmented by the support function. This support function is the signed distance of the oriented tangent plane from a predefined origin. He proposes to ascribe to each different surface a separate support function value. This means that, in general, the proposed variant of the Gauss map of a surface is not globally one to one. Although it is less compact it can uniquely determine a surface. A method to determine object pose based on this representation was not presented in Nalwa's paper.

Roach et al. [18] encode positional information by expressing the equation of a surface patch in dual space. The resulting encoded representation is called the spherical dual image. A point in the dual space represents both the orientation and position of a patch; edges are explicitly described as connections between dual points. A drawback of this approach is that planes passing near or through the designated origin cannot be dualized properly since they map to infinity.

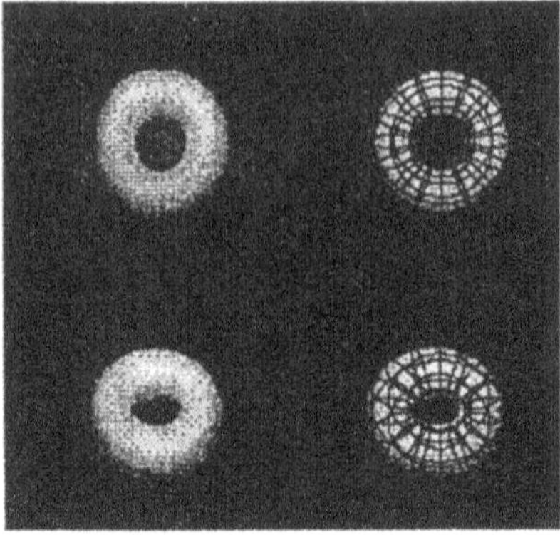

Fig. 6. Localization results using CEGI

3 Spherical Attribute Image

3.1 A Novel Mapping Based On Deformable Surface

The fundamental problem of the EGI family is that it depends on the Gauss mapping. For that reason, more than two parts on an object surface may be mapped on the same point of the sphere. More than two objects may have the same EGI. Further, a partial EGI from a part of an object is not same as a part of EGI from the whole object. Thus, under occlusion, we cannot perform the EGI matching. This problem is due to the fact that the Gauss mapping is not unique for non-convex objects.

We have derived a novel method to make a one-to-one mapping between a non-convex object surface and a spherical surface [19, 21]. The method uses a deformable surface. We first prepare a semi-regularly tessellated geodesic dome (a tessellated unit sphere). Then, we deform the geodesic dome onto an object surface as close as possible (data force) while maintaining the local regularity constraint (regularization force): to ensure that tessellations have a similar area and the same topology as one another. The final representation is given as the equilibrium between the data force and the regularization force. By doing so, the object surface is semi-uniformly tessellated, and each tessellation on the object surface has a counterpart on the undeformed geodesic dome (unit sphere); thus, we can establish a one-to-one mapping between the object surface and the unit sphere. The mapping is referred to as deformable surface mapping (DSP) (see Fig. 7).

This DSP is an intrinsic and covering map. The local topology of patches is preserved on the sphere under the DSP independently on the viewing directions, and the DSP can cover the entire surface of an object.

At each tessellation on the object surface, we calculate invariants such as Gaussian curvature or surface albedo, and map them to the corresponding original tessellation of the geodesic dome. We can observe a distribution of invariants on the unit sphere. Among the possible invariants, Gaussian curvature is the

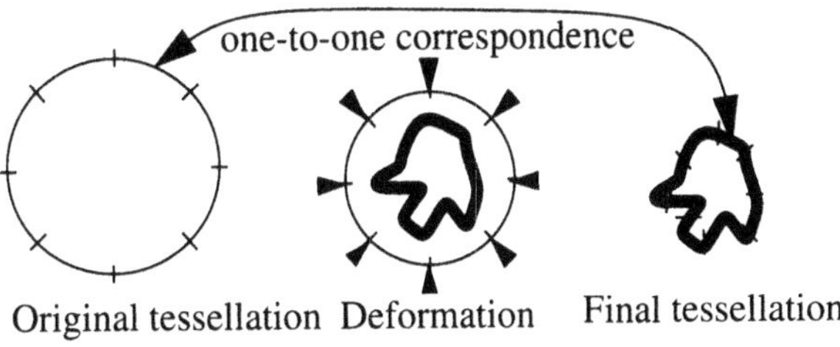

Fig. 7. Deformable surface mapping (DSP)

most important. The distribution based on the Gaussian curvature is referred to as the Spherical Attribute Image (SAI).

In the following section, we will briefly describe the SAI. First, we explain how to tessellate an arbitrary surface into a semi-regular mesh, and how to calculate the simplex angle (discretized Gaussian curvature), a variation of curvature, at the nodes of the mesh, and how to map the mesh to a spherical image. Finally, we discuss how to handle partial views of 3-D objects.

3.2 Semi-Regular Tessellation

A natural discrete representation of a surface is a graph of points, or tessellation, such that each node is connected to each of its closest neighbor by an arc of the graph. It is desirable for many algorithms to have a constant number of neighbors at each node. We use a class of tessellations such that each node has exactly three neighbors. Such a tessellation can be constructed as the dual of a triangulation of the surface.

Let us first consider tessellations of the unit sphere. A regular tessellation would be a tessellation covering a complete spherical surface such that the distance between vertices is constant and each node has exactly three neighbors. It is well known that only approximate global regularity can be achieved. Specifically, the approach that we use is to first build a triangulation by subdividing each triangular face of a 20-face icosahedron into N smaller triangles. The final tessellation is built by taking the dual of the N^2 faces triangulation, yielding a tessellation with the same number of nodes. This tessellation of a sphere, a geodesic dome, is the starting point of our technique. Figure 8 shows an example of a 1620-face semi-regular geodesic dome.

3.3 Regularity Constraint

In order to obtain a mesh of an arbitrary surface, we deform a tessellated surface until it is a good approximation of the object surface. We need to add another constraint in order to build meshes suitable for matching. Specifically, we need

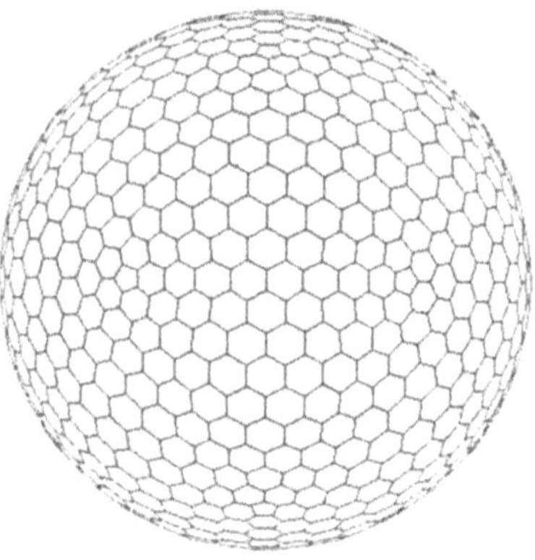

Fig. 8. Semi-regular geodesic dome

to make sure that the distribution of mesh nodes on the surface is invariant with respect to rotation, translation and scale.

Let us consider the 2-D case first. If all the edges of the mesh have the same length, the tessellation is regular. Namely, the length $\mathbf{PP}_1$ should be same as the length $\mathbf{PP}_2$. This condition is same as that $\mathbf{P}$'s projection to the $\mathbf{P}_1\mathbf{P}_2$, Q is at the center $\mathbf{G}$.

We can extend this definition of the regularity to the 3D case. Let $\mathbf{P}$ be a node of the tessellation, $\mathbf{P}_1$, , $\mathbf{P}_3$ be its three neighbors, $\mathbf{G}$ be the centroid of the three points, and $\mathbf{Q}$ be the projection of $\mathbf{P}$ on the plane defined by $\mathbf{P}_1$, $\mathbf{P}_2$, and $\mathbf{P}_3$ (See Figure 9). The local regularity condition simply states that $\mathbf{Q}$ coincides with $\mathbf{G}$.

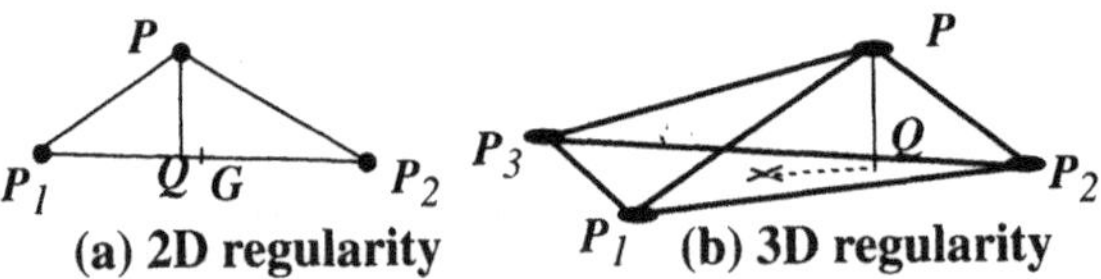

Fig. 9. Regularity constraint

3.4 Discrete Curvature Measure

The next step in building a discrete surface representation is to define a measure of curvature that can be computed from a tessellation. Instead of estimating surface curvature by locally fitting a surface or by estimating first and second derivatives, we proposed a measure of curvature computed at every node from the relative positions of its three neighbors [19]. We call this measure of curvature the simplex angle and we denote its value at node $\mathbf{P}$ by $g(\mathbf{P})$. Although $g(\mathbf{P})$ is neither the Gaussian nor the mean curvature at $\mathbf{P}$, it behaves as a qualitative measure of curvature which is sufficient for matching purposes. Finally, $g(\mathbf{P})$ is invariant with respect to rotation, translation, and scaling.

3.5 Deformable Surface Mapping

A regular mesh drawn on a closed surface can be mapped to a spherical mesh in a natural way. For a given number of nodes K, we can associate with each node a unique index which depends only on the topology of the mesh and which is independent of the shape of the underlying surface. This numbering of the nodes defines a natural mapping h between any mesh M and a reference mesh S on the unit sphere with the same number of nodes: $h(\mathbf{P})$ is the node of S with the same index as $\mathbf{P}$.

Given h, we can store at each node $\mathbf{P}$ of S the simplex angle of the corresponding node on the surface $g(h(\mathbf{P}))$. The resulting structure is a spherical image, that is, a tessellation on the unit sphere, each node being associated with the simplex angle of a point on the original surface. We call this representation the Spherical Attribute Image (SAI).

If the original mesh M satisfies the local regularity constraint, then the corresponding SAI has several invariance properties:

1. For a given number of nodes, the SAI is invariant by translation and scaling of the original object.
2. The SAI represents an object unambiguously up to a rotation. More precisely, if M and M' are two tessellations of the same object with the same number of nodes, then the corresponding SAIs, S and S', are identical up to a rotation of the unit sphere. One consequence of this property is that two SAIs represent the same object if one is the rotated version of the other. It is this property which will allow us to match surfaces that differ by arbitrary rigid transformations.
3. A connected patch of the surface maps to a connected patch of the spherical image. It is this property that allows us to work with non-convex objects and to manipulate models of partial surface, neither of which are possible with conventional spherical representations.

Figure 10(a) and (b) show an intensity image and the corresponding set of points from the range image. In this example, we use the dual of the 9th subdivision of a 20-face icosahedron, (1620 faces). This initial mesh is deformed and reaches the stable state shown in Fig. 11(a). The corresponding SAI data is shown in Fig. 11(b). In the SAI display, the distance from each vertex to the origin is proportional to the simplex angle.

In general, parts of the surface may be occluded by other parts of the object in the range image. The surface fitting algorithm interpolates smoothly across regions of occluded data. In addition, nodes of the mesh are flagged as interpolated or non-interpolated depending on their distances from the closest data

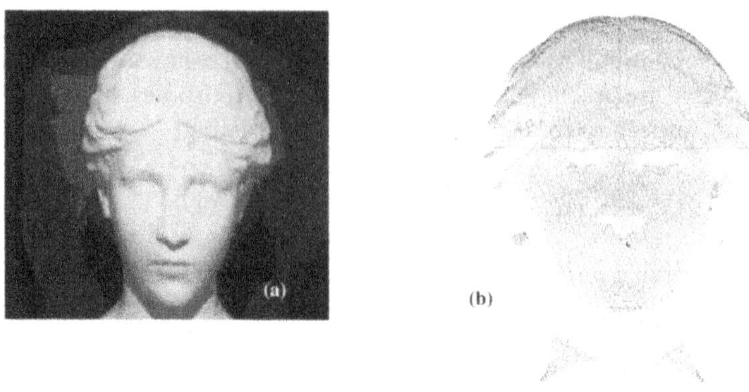

Fig. 10. Input data: (a) Intensity image, (b) Range image

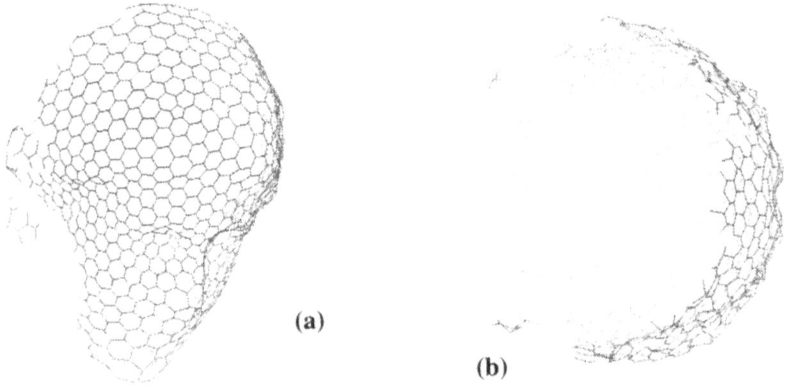

Fig. 11. (a) Deformable mesh, (b) SAI representation on the unit sphere

point. Specifically, a node is marked as "interpolated" if the closest data point is at a distance greater than a threshold. The matching procedure then uses the interpolation flags to determine which nodes should be included in the matching function. The same mechanism is used in order to deal with backfacing regions of the surface. Additional issues on matching partial surfaces are discussed below.

3.6 Matching Surface Models

We now address the matching problem: Given two SAIs, determine the rotation between them, and then find the rigid transformation between the two original sets of points. The representations of a single object with respect to two different viewing directions are related by a rotation of the underlying sphere. Therefore, the most straightforward approach is to compute a distance measure between two SAIs. Once the rotation yielding minimum distance is determined, the full 3-D transformation can be computed.

In the following discussion, we will consider only the vertices of the SAIs that correspond to visible parts of the surface. Let S and S' be the SAIs of two views. The SAI S and S' are representations of the same area of the object if there exists a rotation $\mathbf{R}$ such that $g(\mathbf{P}) = g(\mathbf{RP})$ for every point $\mathbf{P}$ of S.

The problem now is to find this rotation using the discrete representation of S and S'. This is done by defining a distance $D(S, S', \mathbf{R})$ between SAIs as the sum of squared differences between the simplex angles at the nodes of one of the spheres and at the nodes of the rotated sphere. Formally, the distance is defined as:

$$D(S, S', \mathbf{R}) = \sum (g(\mathbf{P}) - g(\mathbf{RP}))^2 \ . \tag{7}$$

The minimum of D corresponds to the best rotation that brings S and S' in correspondence.

Figure 12 shows the result of matching two views of a head. Figure 12(a) shows the intensity images of the two views of the object. Figure 12(b) shows the corresponding SAIs. Figure 12(c) shows the distribution of D as a function of two of the rotation angles, ϕ and θ. The graph exhibits a sharp minimum corresponding to the best rotation between the two spherical maps.

Figure 12 shows the graph of D as function of ϕ and θ obtained by sampling the space of all possible rotations, represented by three angles (θ, ϕ, ψ). Although convenient, this approach is too expensive to be practical.

An alternative matching algorithm is based on the observation that the only rotations for which $D(S, S', \mathbf{R})$should be evaluated are the ones that correspond to a valid list of correspondences $(\mathbf{P}_i, \mathbf{P'}_j)$ between the noes $\mathbf{P}_i$ of S and the nodes $\mathbf{P'}_j$ of S'. Figure 13(a) illustrates the correspondences between nodes: Node $\mathbf{P}_1$ of the first SAI is put in correspondence with node $\mathbf{P'}_{i1}$ of S' and its two neighbors, $\mathbf{P}_2$ and $\mathbf{P}_3$, are put in correspondence with two neighbors of $\mathbf{P'}_{i1}$, $\mathbf{P'}_{i2}$ and $\mathbf{P'}_{i3}$, respectively. This set of three correspondences defines a unique rotation of the spherical image. It also defines a unique assignment for the other nodes, that is, there is a unique node $\mathbf{P'}_{ij}$ corresponding to a node $\mathbf{P}_i$ of S, given the initial correspondences. Moreover, there is only a small number of such initial correspondences.

Based on this observation, we developed an SAI matching algorithm decomposed into two stages: a pre-processing phase and a run-time phase. During pre-processing, we generate the data structure shown in Fig. 13(b). The data structure is a two dimensional array in which each row corresponds to a possible rotation of the SAI and in which column j of row i is the index of the node $\mathbf{P}_{ij}$ corresponding to node $\mathbf{P}_j$ and correspondence number i. At run-time, the distance is evaluated for each row of the array:

$$D_i(S, S', \mathbf{R}) = \sum (g(\mathbf{P}_j) - g(\mathbf{P}_{ij}))^2 \ . \tag{8}$$

The row that produces the minimum D_i gives the best correspondence between nodes of the mesh, $(\mathbf{P}_j, \mathbf{P'}_{ij})$, which is used for computing the full transformation between the object meshes as described in the next section. This algorithm is guaranteed to find the global optimum of D and it does not require an initial estimate of the transformation. It is efficient because all that is

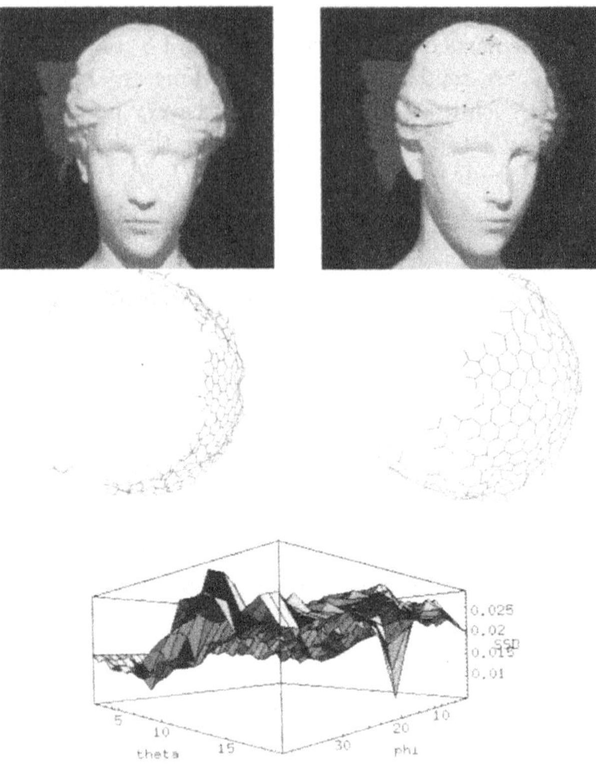

Fig. 12. Matching two SAIs

required at run time is to look up the correspondence table, to compute sum of square differences of corresponding nodes and to add them up.

3.7 Computing the Full Transformation

The rotation of the SAIs is not the same as the rotation of the original objects; it is the rotation of the spherical representations. An additional step is needed to compute the actual transformation between objects as described below.

The last step in matching objects is to derive the transformation between the actual objects, given the rotation between their SAIs (See Fig. 14). The rotational part of the transformation is denoted by $\mathbf{R}_o$, the translational part by $\mathbf{T}_o$. Given a SAI rotation R, we know the corresponding node $\mathbf{P}'$ of each node $\mathbf{P}$ of S. Let $\mathbf{M}$, and $\mathbf{M}'$, be the point on the view corresponding to the node $\mathbf{P}$ of S, and $\mathbf{P}'$ of S', respectively. A first estimate of the transformation is computed by minimizing the sum of the squared distances between the points $\mathbf{M}$ of the first view and the corresponding points $\mathbf{R}_o\mathbf{M}' + \mathbf{T}_o$ of the second view. The optimum transformation for E can be computed by using standard quaternion-based techniques [22].

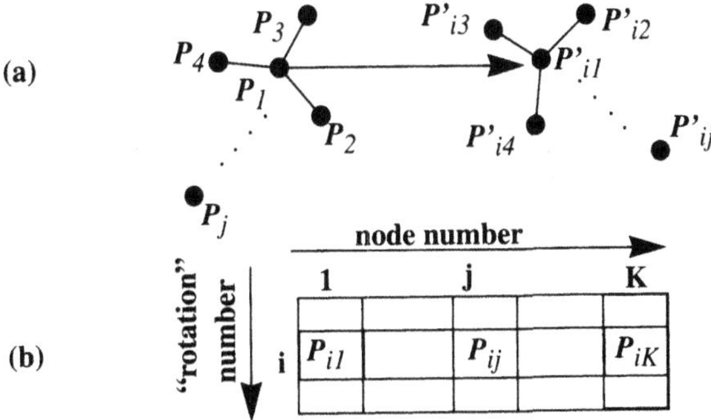

Fig. 13. Efficient matching algorithm; (a) Valid correspondence between nodes; (b) Table of correspondences

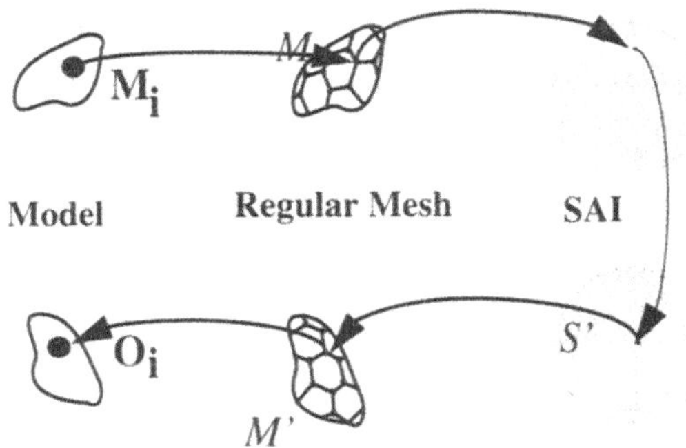

Fig. 14. Computing the full transformation

3.8 Matching Surface Models

In order to compare SAIs computed from different views, we need to adjust the number of nodes because the relative sizes of the visible and hidden areas vary depending on the viewing direction. As mentioned before, the nodes which are in regions of the object where no data points were presented are explicitly marked as "interpolated". As a result, the size of the visible and interpolated parts of the mesh can be easily computed.

Let us consider the problem of merging two views, V_1 and V_2. Let S_1 and S_2 be the number of nodes that would be visible from V_1 and V_2 if we had a complete model of the object. Let the visible areas of the object surface be A_1 and A_2 for V_1 and V_2, respectively. The ratio of the number of visible SAI nodes to the total number of SAI nodes, S_o is equal to the ratio of the visible area to the entire object area, A_o:

$$\frac{S_1}{S_o} = \frac{A_1}{A_o} \tag{9}$$

$$\frac{S_2}{S_o} = \frac{A_2}{A_o} \ . \tag{10}$$

However, we do not know A_o since we have only partial views of the object, but we can estimate A_1 and A_2 from each of the views. Eliminating So from these equations, we obtain

$$S_2 = S_1 \frac{A_2}{A_1} \ . \tag{11}$$

This equation enables us to modify the SAIs from different views so that the distribution of nodes in the visible area is consistent between views. More precisely, we compute the scale factor $\frac{A_2}{A_1}$ from the estimated visible areas from each of the images, and move the nodes of the SAI from V_2 so that the equation is satisfied. See Fig. 15.

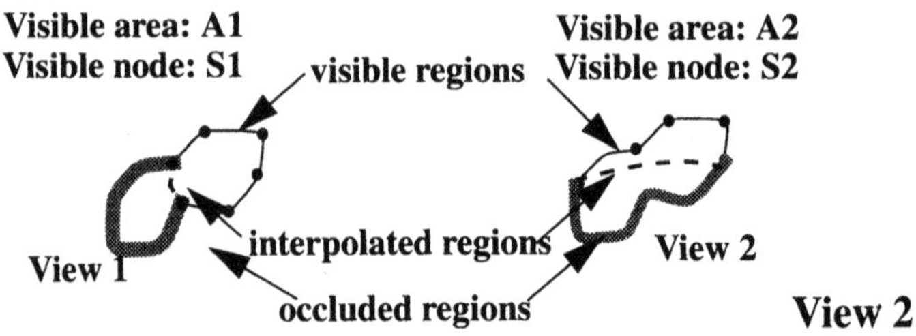

Fig. 15. Matching partial views

The key in this procedure is the connectivity conservation property of the SAI. Specifically, if a connected patch of the surface is visible, then its corresponding image on the SAI is also a connected patch on the sphere. This property allows us to bring the two connected patches into correspondence using a simple spherical scaling. This property is the fundamental difference between the SAI and the spherical representations which cannot deal easily with partial views.

Figure 16 shows the final result of computing the transformation between the two views. Figure 16(a) shows the superimposition of the data points from the two range images before computing the transformation. Figure 16(b) shows the same combined data set using the transformation computed using the algorithm above. This display shows that the two views are registered correctly.

4 Conclusion

In this paper, we argued that a critical issue in representing general objects is to be able to define an intrinsic coordinate system on a surface, onto which properties such as curvature may be mapped. A convenient way of addressing

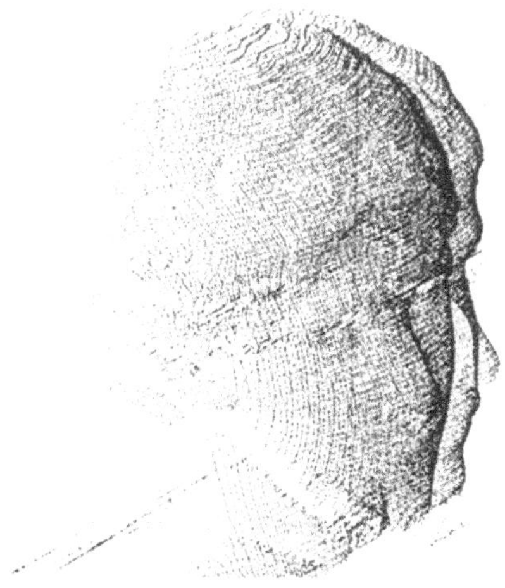

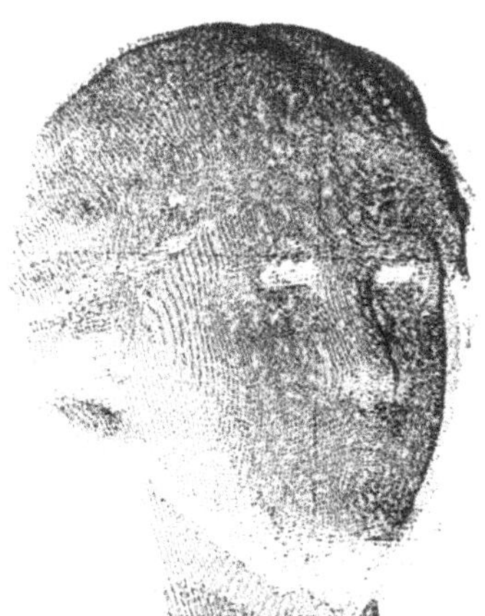

Fig. 16. Merging two views; (a) Overlaid views before registration; (b) Overlaid views after registration.

this problem is to define an intrinsic mapping between a closed surface and the unit sphere.

Although we are still far from a completely satisfactory solution, we have made significant progress. Starting with the EGI, which can only handle convex objects under rotation, we have introduced the DEGI and the CEGI which can deal with translations and, to some extent, with non-convexity.

Finally, the SAI relaxes many of the constraints of the EGI-like representations by preserving the connectivity of the surface, that is, a connected path on the surface maps to a connected path on the sphere. This property allows us to deal with non-convex objects and with general transformations.

We are still a long way from a general solution, however. First of all, the SAI is limited to objects with a genus 0 topology. Second, the algorithm used for extracting the underlying deformable surface does have limitations with respect to the variation in the object shape.

Nevertheless, we believe that intrinsic coordinate maps are a fundamental tool for general object matching and we working toward improving the SAI to handle more general cases.

5 Acknowledgments

The research described in this paper was done with our former and current students, Dr. Sing Bing Kang of DEC Cambridge Research Lab, Dr. Herve Delingette of INRIA Sophia-Antipolis, Dr. Kazunori Higuchi of Toyota Central Research Lab, and Harry Shum of CMU as well as Katsushi Ikeuchi's former advisor, Prof. B.K.P. Horn of MIT. Their contributions are greatly appreciated.

References

1. Oshima, M. and Shirai, Y., "Object recognition using three-dimensional information," *IEEE Trans. Pattern Analysis and Machine Intelligence*, PAMI-5,4, July 1983.
2. Bolles, R.C. and Horaud, P., "3DPO: A three-dimensional part orientation system," in *Three-Dimensional Machine Vision*, Kanade (ed.) Kluwer, Boston, MA, 1987.
3. Koenderink, J.J. and Van Doorn, A.J., "The Singularities of the Visual Mapping" ,Biological Cybernetics, 24(1), 1976
4. Chin, R.T. and Dyer, C.R. "Model-based recognition in robot vision," *ACM Computing Surveys*, 18(1), 1986.
5. Brady, M. Ponce, J., Yuille, A. and Asada, M., "Describing surfaces," *Robotics Research: 2*, Hanafusa, H. and Inoue, H. (eds.), MIT Press, 1985.
6. Stein, F. and Medioni, G., "Structural indexing: Efficient 3-D object recognition," *IEEE Trans Pattern Analysis and Machine Intelligence*, 14, 2, pp.125-145, February 1992.
7. Ikeuchi, K., "Generating an Interpretation Tree from a CAD Model for 3-D Object Recognition in Bin-Picking Tasks," *Intern. Journal of Computer Vision*, Vol. 1, Nol. 2, pp.145-165, 1987.
8. Bowyer, K. W. and Dyer, C. R., "Aspect graphs: an introduction and survey of recent results," *Inter. Journal of Imaging Systems and Technology*, 2, pp.315-328, 1990.
9. Gauss, K. F., *General Investigations of Curved Surfaces*, Raven Press, New York, 1965.
10. Lysternik, L.A., *Convex Figures and Polyhedra*, Dover Publications, New York, 1963.
11. Horn, B.K.P., "Extended Gaussian Image," *Proc. of IEEE*, Vol. 72, No. 12, pp.1671-1686, December 1984.
12. Ikeuchi, K., "Recognition of 3-D Objects using the Extended Gaussian Image," *Proc. of Intern. Joint Conf on Artificial Intelligence*, Vancouver, B.C., pp.595-600, August 1981.
13. Brou, P., "Using the Gaussian Image to Find the Orientation of Objects," *Intern. Journal of Robotics Research*, Vol. 3, No. 4, pp. 89-125, Winter 1984.
14. Little, J. J., "Determining Object Attitude from Extended Gaussian Image," *Proc. of Intern. Joint Conf on Artificial Intelligence*, Los Angeles, California, pp. 960-963, August 1985.
15. Ikeuchi, K. and Horn, B.K.P., "Picking up an Object from a Pile of Objects," in *Robotics Research: The First International Symposium*, J. M. Brady and R. Paul (eds.), MIT Press, Cambridge, Massachusetts, pp. 139-162, 1984.
16. Kang, S.B. and Ikeuchi, K., "The Complex EGI: New Representation for 3-D Pose Determination," *IEEE Trans Pattern Analysis and Machine Intelligence*, Vol. 15, No. 7, pp. 707-721, July 1993.
17. Nalwa, V. S., "Representing Oriented Piecewise C2 Surfaces," *Proc. 2nd Intern. Conf. on Computer Vision*, pp. 40-51, December 1988.
18. Roach, J. W. Wright, J.S., and Ramesh, V., "Spherical Dual Images: A 3-D Representation Method for Solid Objects that Combines Dual Space and Gaussian Sphere," *Proc. IEEE Conf on Computer Vision and Pattern Recognition*, pp. 236-241, June 1986.

19. Dellingette, H., Hebert, M., and Ikeuchi, K., "A Spherical Representation for the Recognition of Curved Objects," *Proc. Intern. Conf on Computer Vision*, pp.103-112, May 1993.
20. Besl, P. and Kay, N.D., "A Method for Registration of 3-D Shapes," *IEEE Trans Patern Analysis Machine Intelligence*, Vol. 14, No.2, Feb 1992.
21. Higuchi, K. Delingette, H. Hebert, M. and Ikeuchi, K., "Merging Multiple Views using a Spherical Representation," *Proc. IEEE Workshop on CAD-based Vision*, pp.124-131, Feb 1993.
22. Faugeras, O. and Hebert, M., "The representation, recognition, and locating of 3-D objects," *Intern. Journal of Robotics Research*, 5, 3, pp.27-52, 1986.

From Physics-Based Representation to Functional Modeling of Highly Complex Objects

Demetri Terzopoulos

Department of Computer Science, University of Toronto
10 King's College Road, Toronto, ON, Canada, M5S 1A4

Abstract. The physics-based modeling paradigm augments standard geometric representations with the principles of physical dynamics. It yields powerful models that unify the representation of object shape (geometry) and motion (dynamics) within a single computational framework. Thus, physics-based object representation transforms abstract geometry into real world, object-oriented Geometry++ with potentially enormous benefits for computer vision. In this paper, I will first review some of the physics-based models for vision that we have developed in recent years, including deformable models, physics-based recursive estimators, and dynamic splines, plus some applications to medical image analysis and CAGD. I will then preview a promising future direction for the physics-based modeling approach—the functional simulation of complex, living things and the use of sophisticated models of animals as virtual robots for the synthesis of active vision systems.

1 Introduction

Geometry is fine for modeling static objects; however, it is woefully inadequate, for modeling objects that can undergo rigid, articulated, or nonrigid motion—that is to say, most of the objects of interest to computer vision researchers. The physics-based modeling paradigm, born circa mid 1980s in vision (and almost simultaneously in computer graphics), augments standard geometric representations with the principles of physical dynamics. This results in dramatically more powerful models which unify the representation of object shape (geometry) and motion (dynamics) within a single computational framework.

Physics-based modeling incorporates physical principles into conventional geometric models. These principles control the creation and behavior of the models through forces, torques, internal strain energies, heat, and other physical quantities. Mathematically, the approach entails the use of systems of dynamic (ordinary and partial) differential equations of motion that govern model behavior. These equations, which may be simulated using standard numerical techniques, make the modeling primitives responsive to simulated physical environments, physical constraints, user interaction, and a variety of input data sets.

Employing Lagrangian mechanics we can systematically transform conventional parameterized and free-form geometries into tractable dynamic primitives for use in vision (and also in graphics and CAGD). Primitives can be connected with physical constraints to form composite, multipart dynamic models. Shape and motion estimation algorithms stem from the coupling of dynamic models to visual data through simulated forces. Furthermore, the physics-based paradigm forges a link between geometry and stochastic estimation theory through the notion of continuous, extended Kalman filtering. In this context, dynamic models serve as Kalman system models, leading to a new class of recursive on-line shape and motion estimation algorithms that account formally for uncertainty and incomplete visual information. These algorithms define rigorous analysis-by-synthesis processes.

Thus the physics-based object representation paradigm transforms abstract geometry into real-world, "object-oriented" geometry, with potentially enormous benefits for computer vision. In this paper, I will first review some of the physics-based models for vision that we have developed in recent years and some of their areas of application outside of mainstream vision. I will then preview a promising future direction of the physics-based modeling approach—the simulation of complex, living things and the use of sophisticated, functional models of animals as virtual robots for the synthesis of active vision systems.[1]

2 Physics-based modeling (Geometry++ :-)

2.1 Deformable models for vision and graphics

Deformable models are a powerful class of physics-based modeling primitives inspired by the behavior of natural materials. These models move rigidly or deform in response to applied forces in accordance with the principles of continuum mechanics. For the purposes of computer graphics, realistic images and animations of elastic inelastic, and thermoelastic objects may be synthesized when the applied forces arise from animation control systems and the interplay of deformable models and simulated physical environments [1]. Our results demonstrate that physics-based modeling primitives offer a much greater degree of realism in representing most real-world objects for considerably less effort than is possible with conventional, geometric models.

In applications to computer vision, deformable models may be used to infer image disparity and flow fields or to infer the shapes and motions of objects from their images [2, 3]. In this context, we apply to the models external forces that impose constraints derived from image data. The forces actively shape and move models to achieve maximal consistency with imaged objects of interest and to maintain the consistency over time.

2.2 Nonrigid shape and motion analysis

A crucial feature of the physics-based approach is that we can make use of Lagrangian mechanics to formulate various dynamic models and couple them to acquired image streams through time-varying force fields. We have demonstrated the use of dynamic deformable surfaces for 3D object reconstruction, stereo-motion estimation, and nonrigid

[1] Since space limitations preclude a serious survey of physics-based modeling work, I will cite only my own papers in the ensuing discussion. The interested reader may peruse these papers for more complete motivational and background discussions, including many references to related work by other researchers.

object tracking using deformable generalized cylinders with weak symmetry constraints [4, 2] and dynamic models with local and global deformations [5]. We have also developed discrete dynamic models that may be used to reconstruct and track time-varying surfaces in a spatially adaptive manner [6, 7]. Recently, we forged a link between these dynamic models and recursive estimation theory to arrive at Kalman-filter based shape and motion estimation algorithms for use in real-time vision systems [8, 9]. These algorithms can recover the nonrigid motion of flexible single or multipart 3D objects from image sequences and dynamic 3D data [10]. An important special case of dynamic deformable models is the active contour or *snakes* model for extracting extended image features, such as bright and dark curves and edges, and the tracking of moving features in image sequences [11, 8].

2.3 Applications to medical image analysis

The rapid development of computer and noninvasive sensor technologies is revolutionizing medicine, offering scientists and physicians new investigative and diagnostic tools. Computer-based medical image analysis is an especially challenging task which has not kept pace with the ability to acquire digital images in any of several modalities. We have developed a system which applies snakes to interactive segmentation and reconstruction of 3D models of anatomical structures from serial sections [12]. Other groups have applied snakes to CT, MRI, PET, confocal microscopy, and ultrasound images. Recently we have developed finite-element surface models for reconstructing and tracking the left ventricular cavity in dynamic tomographic images during the cardiac cycle [13]. In other work, we have developed dynamic oriented particle systems for reconstructing biological objects of complex, unknown topologies from medical images [14]. Snakes may also be generalized to make them topologically flexible in the reconstruction of anatomical structures with unpredictable topologies [15].

2.4 Physics-based splines for CAD modeling

The use of physics as a metaphor for shape design offers a significantly richer variety of modeling scenarios than do the purely geometric primitives employed in conventional shape design systems. The natural behavior of physics-based models makes them functionally intuitive from a user's point of view; however, an important advantage of a physical simulation over the real world is that users may readily adjust the behavior of models at will, and even suspend or amend physical laws according to the requirements of the modeling task. Such flexibility gives impetus to a new, physics-based view of computer aided geometric design. The "computational modeling clay" metaphor is suggestive [1]. We have developed a dynamic generalization of industry-standard, non-uniform rational B-splines (NURBS) and associated constraint methods for physics-based interactive sculpting [16, 17]. Dynamic NURBS (D-NURBS) promise to provide the ability to capture 3D CAD models of manufactured objects automatically from image sensor input, to manipulate models interactively in simulated physical environments, and to visualize the results of model-based designs in real time on graphics workstations.

3 Q: Where do we go from here? A: Functional modeling

The physics-based models mentioned in the above review tend to be significantly more complex than the raw geometric models upon which they are built. Although their dynamic behavior makes them generally useful for shape and motion estimation in vision, medical image analysis, and computer-aided geometric design, they are merely models of abstract physical materials, such as elastic, or viscoelastic materials, or inanimate unibody or multibody physical objects. It is important to realize, however, that the potential of the physics-based modeling paradigm extends far beyond this plateau.

It is easy to imagine physics-based models of incredibly greater sophistication. A promising direction of the physics-based paradigm, which we are actively exploring, is the simulation of living systems. In the sequel we will describe two of the models that we are currently developing which are capable of realistically simulating important aspects of humans and other animals. We will give an indication of the impact that such *functional models* can have in areas such as visual communication and active vision.

4 Human facial modeling for communication

Facial image analysis and synthesis is desirable for numerous applications. Among them is low bandwidth teleconferencing which may involve the real-time extraction of facial control parameters from live video at the transmission site and the reconstruction of a dynamic facsimile of the subject's face at a remote receiver. Teleconferencing and other applications require facial models that are computationally efficient and also realistic enough to accurately synthesize the various nuances of facial structure and motion. We have argued that the anatomy and physics of the human face, especially the arrangement and actions of the primary facial muscles, provide a good basis for facial image analysis and synthesis [18].

We have developed a highly automated approach to constructing realistic, functional models of human heads [19]. These physics-based models are anatomically accurate and may be made to conform closely to specific individuals. Currently, we begin by scanning a subject with a laser sensor which circles around the head to acquire detailed range and reflectance information. Next, an automatic conformation algorithm adapts a triangulated face mesh of predetermined topological structure to these data. The generic mesh, which is reusable with different individuals, reduces the range data to an efficient, polygonal approximation of the facial geometry and supports a high-resolution texture mapping of the skin reflectivity.

The conformed polygonal mesh forms the epidermal layer of a physics-based model of facial tissue. An automatic algorithm constructs the multilayer synthetic skin and estimates an underlying skull substructure with a jointed jaw. Finally, the algorithm inserts synthetic muscles into the deepest layer of the facial tissue. These contractile actuators, which emulate the primary muscles of facial expression, generate forces that deform the synthetic tissue into meaningful expressions. To increase realism, we include constraints to emulate tissue incompressibility and to enable the tissue to slide over the skull without penetrating into it.

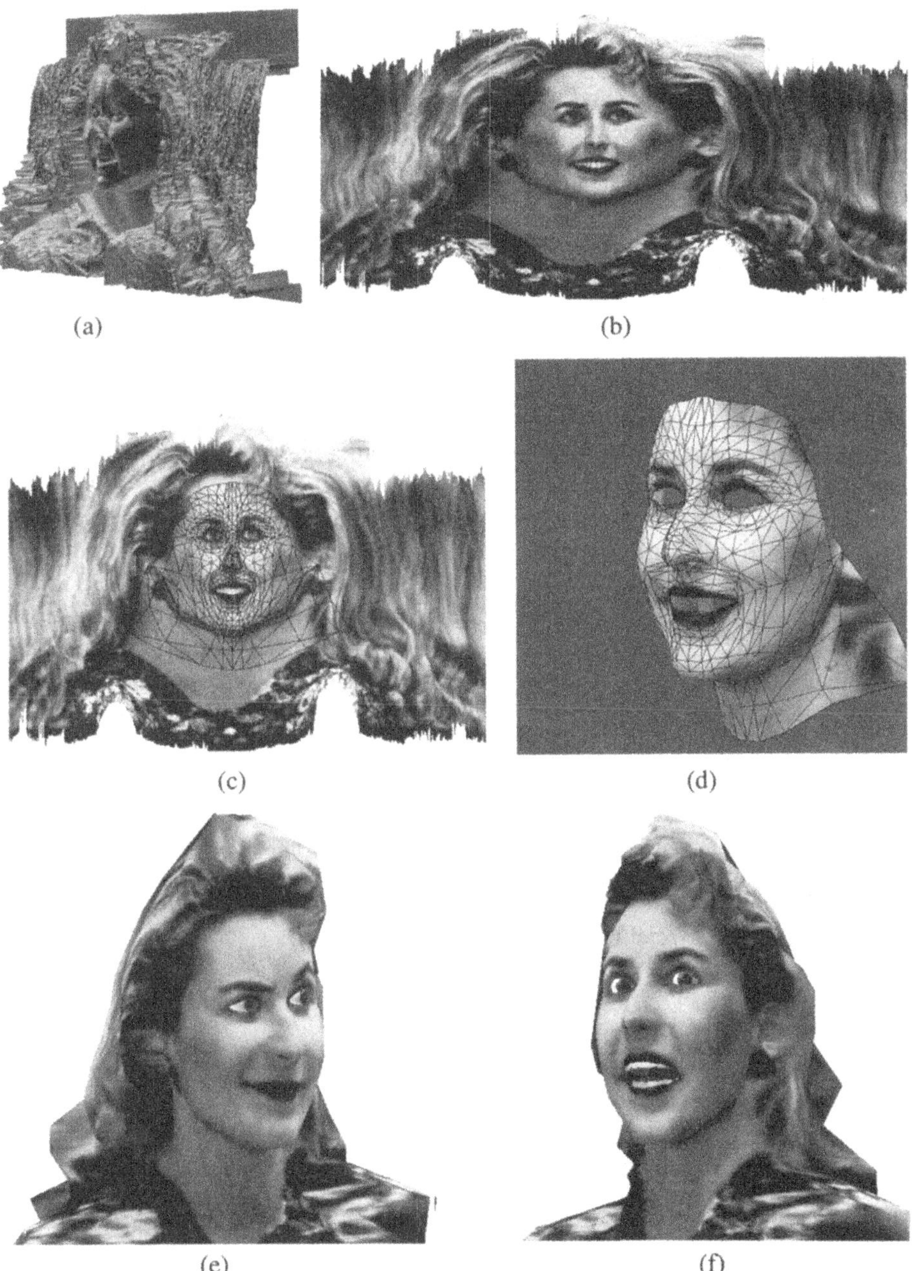

Fig. 1. Facial modeling using scanned data. (a) Radial range map. (b) RGB photometric map. (c) RGB map with conformed epidermal mesh overlayed. (d) 3D mesh and texture mapped triangles. (e–f) Animate face model.

Fig. 1 illustrates the aforementioned steps. The figure shows a 360° head-to-shoulder scan of a woman, "Heidi," acquired by Cyberware, Inc., using their Color 3D Digitizer. The data set consists of a radial range map (Fig. 1(a)) and a registered RGB photometric map (Fig. 1(b)). The range and RGB maps are high-resolution 512 × 256 arrays in cylindrical coordinates, where the x axis is the latitudinal angle around the head and the y axis is vertical distance. Fig. 1(c) shows the generic mesh projected into the 2D cylindrical domain and overlayed on the RGB map. The triangle edges in the mesh are elastic springs, and the mesh has been conformed semi-interactively to the woman's face using both the range and RGB maps. The nodes of the conformed mesh serve as sample points in the range map. Their cylindrical coordinates and the sampled range values are employed to compute 3D Euclidean space coordinates for the polygon vertices. In addition, the nodal coordinates serve as polygon vertex texture map coordinates into the RGB map. Fig. 1(d) shows the 3D facial mesh with the texture mapped photometric data. Once we have reduced the scanned data to the 3D epidermal mesh of Fig. 1(d), we can assemble a physics-based face model of Heidi. Fig. 1(e,f) demonstrates that we can animate the resulting face model by activating muscles.

This physics-based anatomically motivated facial model has allowed us to develop a new approach to the analysis of dynamic facial images for the purposes of estimating and resynthesizing dynamic facial expressions [18]. Part of the difficulty of facial image analysis is that the face is highly deformable, particularly around the forehead, eyes, and mouth, and these deformations convey a great deal of meaningful information. Techniques for tracking the deformation of facial features include snakes [11]. Motivated by the anatomically consistent musculature in our model, we have considered the estimation of dynamic facial muscle contractions from video sequences of expressive faces. We have developed an analysis technique that uses snakes to track the nonrigid motions of facial features in video. Features of interest include the eyebrows, nasal furrows, mouth, and jaw in the image plane. We are able to estimate dynamic facial muscle contractions directly from the snake state variables. These estimates make appropriate control parameters for resynthesizing facial expressions through our face model. The model resynthesizes facial images at real time rates.

5 Artificial animals and active vision

A provocative area of artificial life research is the computer implementation of animats.[2] Animats are computer models of animals situated in their natural habitats. They can help researchers unravel some of the mysteries of the most complex living systems known. By extending the physics-based modeling paradigm, we have been able to develop sophisticated animats patterned after animals as evolved as those in the superclass Pisces [20].

Imagine a virtual world inhabited by a variety of realistic fishes (Fig. 2). In the presence of underwater currents, the fishes employ their muscles and fins to swim gracefully around immobile obstacles and among moving aquatic plants and other fishes. They autonomously explore their dynamic world in search of food. Large, hungry

[2] See S. Levy, *Artificial Life* (Pantheon, 1992) for a popular exposition of the artificial life field.

Fig. 2. Artificial fishes in their physics-based virtual world. Various artificial fishes foraging and performing courtship displays (top). Predator shark stalking school of prey fish (bottom).

predator fishes stalk smaller prey fishes in the deceptively peaceful habitat. Prey fishes swim around contentedly until the sight of predators compels them to take evasive action. When a dangerous predator appears in the distance, similar species of prey form schools to improve their chances of survival. As the predator nears a school, the fishes scatter in terror. A chase ensues in which the predator selects victims and consumes them until satiated. Some species of fishes seem untroubled by predators. They find comfortable niches and forage on floating plankton when they get hungry. Driven by healthy libidos, they perform elaborate courtship rituals to secure mates.

To simulate living systems with the level of complexity of those depicted above, we have taken the bottom-up, compositional modeling approach. In our work, we model not just 3D form and appearance, but also the basic physics of the animal and its environment.

On this simulated physics substrate, we can effectively model the animal's means of locomotion. This, in turn, places us in a position to model the animal's perceptual awareness of its world, its behavior, and its ability to learn. A nontrivial behavioral repertoire was achieved in stages. First, primitive reflexive behaviors, such as obstacle avoidance, that directly couple perception to action were put in place. Then the primitive behaviors were combined into motivational behaviors whose activation depends also on the artificial animal's mental state, including hunger, libido, and fear. This bottom-up, holistic approach to modeling the animal situated in its world is crucial to achieving realism.

Thus, our artificial fishes are autonomous agents situated in a simulated physical world. The agent has (i) a three-dimensional body with internal muscle actuators and functional fins, which deforms and locomotes in accordance with biomechanic and hydrodynamic principles, (ii) sensors, including eyes that can image the environment, and (iii) a brain with motor, perception, behavior, and learning centers. Fig. 3 illustrates the artificial fish model.

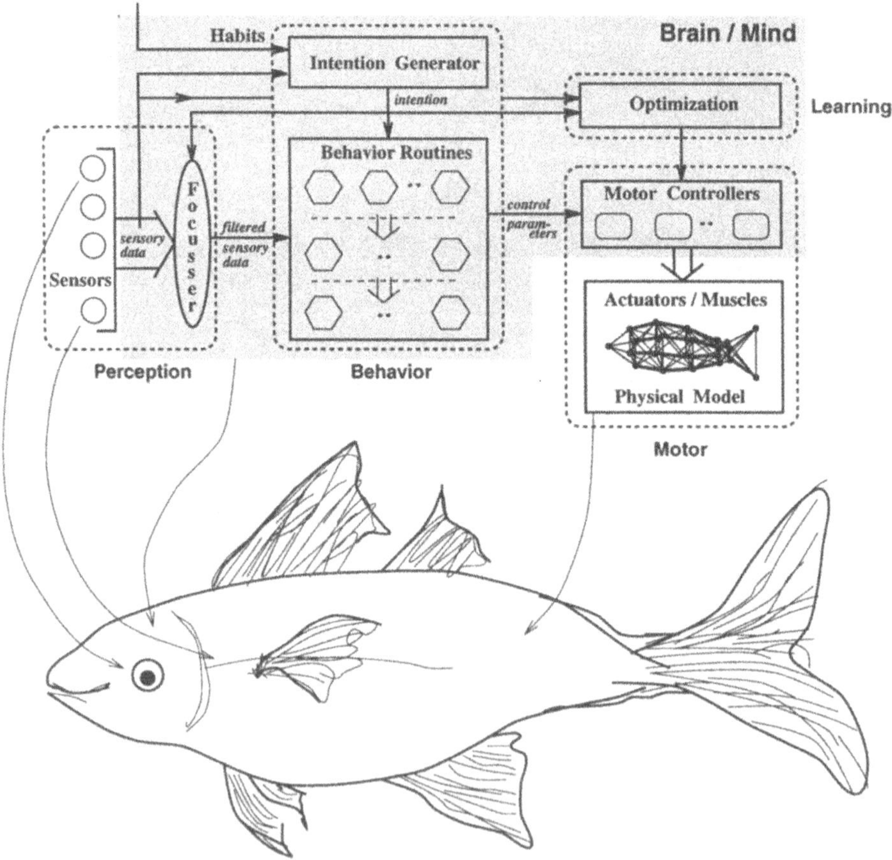

Fig. 3. Control and information flow in artificial fish.

The motor system, comprising the actuators and a set of motor controllers, controls the dynamic model of the animal. The mechanical body model represents a good compromise between anatomical consistency, hence realism, and computational efficiency. The model is rich enough to permit motor controllers to be designed by gleaning information from the biomechanics literature. The motor controllers are parameterized procedures. Each is dedicated to carrying out a specific motor function. They translate natural control parameters such as the forward speed of locomotion or angle of a turn into detailed muscle actions that execute the motor function.

The perception system relies on a set of on-board virtual sensors to provide sensory information about the dynamic environment. The brain's perception center includes a perceptual attention mechanism which allows the artificial animal to sense the world in a task-specific way, hence filtering out sensory information superfluous to its current behavioral needs. For example, the artificial fish attends to sensory information about nearby food sources when foraging.

The behavior center of the brain mediates between the perception system and the motor system. An intention generator, the animat's cognitive faculty, harnesses the dynamics of the perception-action cycle. The innate character of the animal is established by a set of habits that determine whether or not it is male or female, predator or prey, etc. The intention generator combines the habits with the incoming stream of sensory information to generate dynamic goals for the animal, such as to hunt and feed on prey. It ensures that goals have some persistence by exploiting a single-item memory. The intention generator also controls the perceptual attention mechanism and invokes the appropriate motor controllers to enable the animal to carry out the current intention. Primitive behavior routines, such as obstacle avoidance, and more sophisticated motivational behavior routines, such as mating, implement the behavioral repertoire of the artificial animal.

The learning center of the brain enables the artificial animal to learn how to locomote through practice and sensory reinforcement. The motor learning algorithms use reinforcement based optimization to discover muscle controllers that produce efficient locomotion. The learning algorithms also enable the artificial animals to train themselves to accomplish higher level, perceptually guided motor tasks, such as maneuvering to reach a visible target or leaping out of the water to perform the sort of stunts that marine mammals perform at SeaWorld.

5.1 Animat vision

The advance in modeling sophistication that we have made with the artificial fish animat prompts us to propose a new approach to computer vision research that we call *animat vision*. Animat vision prescribes the use of artificial animals as autonomous virtual vehicles for the synthesis of active vision systems. The approach has been made possible by the confluence of

1. advanced physics-based artificial life modeling of natural animals
2. photorealistic computer graphics rendering, especially as implemented in modern 3D graphics workstations, and
3. active vision algorithms.

The late psychologist J.J. Gibson, in a sense the "grandfather" of active vision, stressed in pre-computational terms the importance of modeling the active observer situated in the dynamic environment. In recent years the active vision approach has developed into the dominant paradigm for computer vision research that it is today.

The prevailing mindset about active vision systems is that they should be grounded in robot hardware. Efforts to do this have certainly led to a better understanding of sensorimotor processing. Undeniably, however, efforts to equip real-time mobile robots with vision have been hampered by the hardware. In most cases, the hardware/processor constraints would permit only the simplest active vision algorithms. Furthermore, although active vision systems ostensibly reflect biological perceptual mechanisms, few researchers believe that the robot systems available to them today are reasonable models of any biological animal—animals do not have CCD chip eyes, electric motor muscles, and wheel legs. The animat vision methodology that we propose has the potential to liberate a considerable portion of the vision research community from the tyranny of robot hardware. It caters to researchers who are driven to reverse engineer and ultimately understanding the powerful vision systems that nature has evolved.[3]

The basic animat vision idea in a nutshell is to implement realistic artificial animals and to imbue them with the ability to perceive and visually understand the realistic virtual worlds in which they are situated so they can function individually and socially within these worlds. Each animat is an autonomous agent. As exemplified by the artificial fish described above, the agent has a functional body that can locomote realistically and a mind with perception, motor, and behavior centers. The animat is endowed with functional eyes that can image the dynamic 3D virtual world onto 2D virtual retinas. Retinal imaging is accomplished by photorealistic graphics rendering of the world from the animat's point of view as it moves through the world. The perceptual center of the animat's brain exploits active vision algorithms to process the incoming stream of dynamic retinal images on-line and understand what is being seen.

The advantages of animat vision go beyond the aforementioned fact that researchers need not mess with finicky hardware that is a poor excuse of an artifical animal. In particular, *unlike the physical world*:

- The quantitative photometric, geometric, and dynamic information that is needed to render the virtual world is available explicitly. Generally, the animats are privy to none of this environmental ground truth data, but must extract all of their visual information the "hard way"—from the retinal image streams. However, the ground truth data can be extremely useful in assaying the effective accuracy of the vision algorithms or modules under development.
- One can slow down the "cosmic clock" of the virtual world relative to the cycle time of the computer CPU on which it is being simulated. This increases the amount of computation that each agent can consume between clock ticks without retarding the agent's responses in cosmic time. This in turn permits the development and evaluation of new and/or computationally complex vision algorithms in an active vision context that are not presently implementable in real-time hardware.

[3] To be sure, many applications-driven researchers are motivated to build robot vision systems for inspection, surveillance, navigation, etc., and they must bite the bullet vis-a-vis hardware.

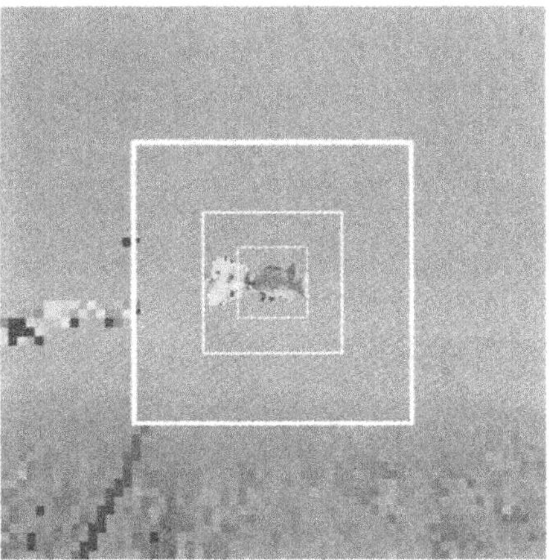

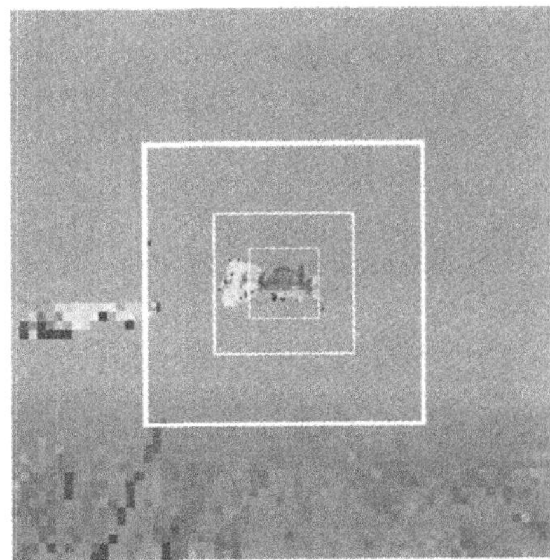

Fig. 4. Binocular retinal imaging. Central fovea and three peripheral regions indicated by white borders.

The first challenge that we are facing is to synthesize a prototype active vision system for the fish animat which is based solely on retinal image analysis. The vision system should be extensible so that it will eventually support a broad enough repertoire of individual and emergent group behaviors to make the artificial fish fully functional in its world. It is important to realize that we need not restrict ourselves to modeling the perceptual mechanisms of fishes. In fact the animat vision paradigm applies to any animat that models any animal—even a human being—to the level of fidelity that the artificial fish models a real fish.

The basic functionality of the animat vision system starts with binocular perspective projection of the color 3D world onto the animat's 2D retinas. This projection respects occlusion relationships among objects. It forms spatially variant visual fields with high resolution foveas and low resolution peripheries (Fig. 4). Based on the analysis of the incoming color retinal image stream, the visual center of the animat's brain supplies saccade control signals to the eyes to stabilize the visual fields during locomotion, attend to interesting targets, and keep a moving target fixated. The algorithms that are presently employed perform global optical flow computations and color histogram matching to identify and localize color objects. Using this relatively small set of active vision algorithms, the artificial fish is able to approach and track moving targets under visual guidance (Fig. 5). Eventually its strategic arsenal of active vision algorithms will enable it to forage for food, escape predators, rendezvous and socialize with mates, etc.

It should be possible to incorporate the animat vision system that we are developing for the fish animat into more complex animats that we intend to develop in the future, including a virtual mermaid and a terrestrial virtual human. It is my hope that the active vision systems that we will synthesize in this way will be relevant in whole or in part to physical robotics. I firmly believe that virtual animats in their dynamic world can serve as an useful proving ground for theories that profess sensorimotor competence in animal or robotic situated agents.

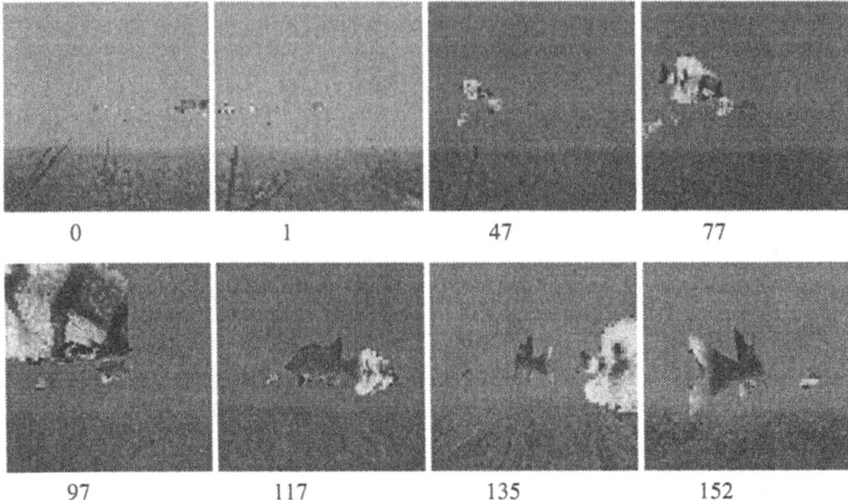

Fig. 5. Stabilization, fixation, and visually guided navigation in artificial fish. Retinal image sequence from the left eye of the observer fish with animat vision (b). Fish detects and foveates on a red fish target (frames 0–1). The red fish target remains fixated in the center of the fovea as the fish uses gaze directions to swim towards it (frames 47–117). The targeted fish turns and swims away with the observer fish in visually guided pursuit (frames 135–152).

Acknowledgements

I would like to thank my past and present collaborators, whose names are cited in the reference list, for their invaluable contributions to the research reviewed in this paper. Special thanks go to my PhD students whose work is highlighted in Sections 4 and 5; in particular, Yuencheng Lee implemented the realistic facial model, Xiaoyuan Tu implemented the artificial fish model, and Tamer Rabie implemented the animat vision system.

Funding for the research described herein was provided in part by the Natural Sciences and Engineering Research Council of Canada, the Information Technology Research Center of Ontario, the ARK Project (which receives its funding from PRECARN Associates Inc., Industry Canada, the NRC of Canada, Technology Ontario, Ontario Hydro Technologies, and Atomic Energy of Canada Limited), and Digital Equipment Corp. The author is a fellow of the Canadian Institute for Advanced Research.

References

1. D. Terzopoulos and K. Fleischer. Deformable models. *The Visual Computer*, 4(6):306–331, 1988.
2. D. Terzopoulos, A. Witkin, and M. Kass. Constraints on deformable models: Recovering 3D shape and nonrigid motion. *Artificial Intelligence*, 36(1):91–123, 1988.
3. D. Terzopoulos. The computation of visible-surface representations. *IEEE Transactions on Pattern Analysis and Machine Intelligence*, PAMI-10(4):417–438, 1988.

4. D. Terzopoulos, A. Witkin, and M. Kass. Symmetry-seeking models and 3D object reconstruction. *International Journal of Computer Vision*, 1(3):211–221, 1987.
5. D. Terzopoulos and D. Metaxas. Dynamic 3D models with local and global deformations: Deformable superquadrics. *IEEE Transactions on Pattern Analysis and Machine Intelligence*, 13(7):703–714, 1991.
6. D. Terzopoulos and M. Vasilescu. Sampling and reconstruction with adaptive meshes. In *IEEE Computer Society Conference on Computer Vision and Pattern Recognition (CVPR'91)*, pages 70–75, Maui, Hawaii, June 1991. IEEE Computer Society Press.
7. M. Vasilescu and D. Terzopoulos. Adaptive meshes and shells: Irrègular triangulation, discontinuities, and hierarchical subdivision. In *IEEE Computer Society Conference on Computer Vision and Pattern Recognition (CVPR'92)*, pages 829–832, Champaign, IL, June 1992. IEEE Computer Society Press.
8. D. Terzopoulos and R. Szeliski. Tracking with Kalman snakes. In *Active Vision*, pages 3–20. MIT Press, Cambridge, MA, 1992.
9. D. Terzopoulos and D. Metaxas. Tracking nonrigid 3D objects. In *Active Vision*, pages 75–89. MIT Press, Cambridge, MA, 1992.
10. D. Metaxas and D. Terzopoulos. Shape and nonrigid motion estimation through physics-based synthesis. *IEEE Transactions on Pattern Analysis and Machine Intelligence*, 15(6):580–591, 1993.
11. M. Kass, A. Witkin, and D. Terzopoulos. Snakes: Active contour models. *International Journal of Computer Vision*, 1(4):321–331, 1988.
12. I. Carlbom, D. Terzopoulos, and K. Harris. Computer-assisted registration, segmentation, and 3D reconstruction from images of neuronal tissue sections. *IEEE Transactions on Medical Imaging*, 13(2):351–362, 1994.
13. T. McInerney and D. Terzopoulos. A finite element model for 3D shape reconstruction and nonrigid motion tracking. In *Fourth International Conference on Computer Vision (ICCV'93)*, pages 33–37, Berlin, Germany, May 1993. IEEE Computer Society Press.
14. R. Szeliski, D. Tonnesen, and D. Terzopoulos. Modeling surfaces of arbitrary topology with dynamic particles. In *IEEE Computer Society Conference on Computer Vision and Pattern Recognition (CVPR'93)*, pages 140–152, New York, NY, June 1993.
15. T. McInerney and D. Terzopoulos. Medical image analysis with topologically adaptive snakes. In *First International Conference on Computer Vision, Virtual Reality, and Robotics in Medicine (CVRMed'95)*, Nice, France, April 1995. In press.
16. D. Terzopoulos and H. Qin. Dynamic NURBS with geometric constraints for interactive sculpting. *ACM Transactions on Graphics*, 13(2):103–136, 1994.
17. H. Qin and D. Terzopoulos. Dynamic NURBS swung surfaces for physics-based shape design. *Computer-Aided Design*, February 1995.
18. D. Terzopoulos and K. Waters. Analysis and synthesis of facial image sequences using physical and anatomical models. *IEEE Transactions on Pattern Analysis and Machine Intelligence*, 15(6):569–579, 1993.
19. Y. Lee, D. Terzopoulos, and K. Waters. Constructing physics-based facial models of individuals. In *Graphics Interface '93*, pages 1–8, Toronto, ON, May 1993.
20. D. Terzopoulos, X. Tu, and R. Grzeszczuk. Artificial fishes: Autonomous locomotion, perception, behavior, and learning in a simulated physical world. *Journal of Artificial Life*, 1(4), 1995.

Springer-Verlag and the Environment

We at Springer-Verlag firmly believe that an international science publisher has a special obligation to the environment, and our corporate policies consistently reflect this conviction.

We also expect our business partners – paper mills, printers, packaging manufacturers, etc. – to commit themselves to using environmentally friendly materials and production processes.

The paper in this book is made from low- or no-chlorine pulp and is acid free, in conformance with international standards for paper permanency.

Lecture Notes in Computer Science

For information about Vols. 1–918

please contact your bookseller or Springer-Verlag

Vol. 919: B. Hertzberger, G. Serazzi (Eds.), High-Performance Computing and Networking. Proceedings, 1995. XXIV, 957 pages. 1995.

Vol. 920: E. Balas, J. Clausen (Eds.), Integer Programming and Combinatorial Optimization. Proceedings, 1995. IX, 436 pages. 1995.

Vol. 921: L. C. Guillou, J.-J. Quisquater (Eds.), Advances in Cryptology – EUROCRYPT '95. Proceedings, 1995. XIV, 417 pages. 1995.

Vol. 922: H. Dörr, Efficient Graph Rewriting and Its Implementation. IX, 266 pages. 1995.

Vol. 923: M. Meyer (Ed.), Constraint Processing. IV, 289 pages. 1995.

Vol. 924: P. Ciancarini, O. Nierstrasz, A. Yonezawa (Eds.), Object-Based Models and Languages for Concurrent Systems. Proceedings, 1994. VII, 193 pages. 1995.

Vol. 925: J. Jeuring, E. Meijer (Eds.), Advanced Functional Programming. Proceedings, 1995. VII, 331 pages. 1995.

Vol. 926: P. Nesi (Ed.), Objective Software Quality. Proceedings, 1995. VIII, 249 pages. 1995.

Vol. 927: J. Dix, L. Moniz Pereira, T. C. Przymusinski (Eds.), Non-Monotonic Extensions of Logic Programming. Proceedings, 1994. IX, 229 pages. 1995. (Subseries LNAI).

Vol. 928: V.W. Marek, A. Nerode, M. Truszczynski (Eds.), Logic Programming and Nonmonotonic Reasoning. Proceedings, 1995. VIII, 417 pages. 1995. (Subseries LNAI).

Vol. 929: F. Morán, A. Moreno, J.J. Merelo, P. Chacón (Eds.), Advances in Artificial Life. Proceedings, 1995. XIII, 960 pages. 1995 (Subseries LNAI).

Vol. 930: J. Mira, F. Sandoval (Eds.), From Natural to Artificial Neural Computation. Proceedings, 1995. XVIII, 1150 pages. 1995.

Vol. 931: P.J. Braspenning, F. Thuijsman, A.J.M.M. Weijters (Eds.), Artificial Neural Networks. IX, 295 pages. 1995.

Vol. 932: J. Iivari, K. Lyytinen, M. Rossi (Eds.), Advanced Information Systems Engineering. Proceedings, 1995. XI, 388 pages. 1995.

Vol. 933: L. Pacholski, J. Tiuryn (Eds.), Computer Science Logic. Proceedings, 1994. IX, 543 pages. 1995.

Vol. 934: P. Barahona, M. Stefanelli, J. Wyatt (Eds.), Artificial Intelligence in Medicine. Proceedings, 1995. XI, 449 pages. 1995. (Subseries LNAI).

Vol. 935: G. De Michelis, M. Diaz (Eds.), Application and Theory of Petri Nets 1995. Proceedings, 1995. VIII, 511 pages. 1995.

Vol. 936: V.S. Alagar, M. Nivat (Eds.), Algebraic Methodology and Software Technology. Proceedings, 1995. XIV, 591 pages. 1995.

Vol. 937: Z. Galil, E. Ukkonen (Eds.), Combinatorial Pattern Matching. Proceedings, 1995. VIII, 409 pages. 1995.

Vol. 938: K.P. Birman, F. Mattern, A. Schiper (Eds.), Theory and Practice in Distributed Systems. Proceedings,1994. X, 263 pages. 1995.

Vol. 939: P. Wolper (Ed.), Computer Aided Verification. Proceedings, 1995. X, 451 pages. 1995.

Vol. 940: C. Goble, J. Keane (Eds.), Advances in Databases. Proceedings, 1995. X, 277 pages. 1995.

Vol. 941: M. Cadoli, Tractable Reasoning in Artificial Intelligence. XVII, 247 pages. 1995. (Subseries LNAI).

Vol. 942: G. Böckle, Exploitation of Fine-Grain Parallelism. IX, 188 pages. 1995.

Vol. 943: W. Klas, M. Schrefl, Metaclasses and Their Application. IX, 201 pages. 1995.

Vol. 944: Z. Fülöp, F. Gécseg (Eds.), Automata, Languages and Programming. Proceedings, 1995. XIII, 686 pages. 1995.

Vol. 945: B. Bouchon-Meunier, R.R. Yager, L.A. Zadeh (Eds.), Advances in Intelligent Computing - IPMU '94. Proceedings, 1994. XII, 628 pages.1995.

Vol. 946: C. Froidevaux, J. Kohlas (Eds.), Symbolic and Quantitative Approaches to Reasoning and Uncertainty. Proceedings, 1995. X, 420 pages. 1995. (Subseries LNAI).

Vol. 947: B. Möller (Ed.), Mathematics of Program Construction. Proceedings, 1995. VIII, 472 pages. 1995.

Vol. 948: G. Cohen, M. Giusti, T. Mora (Eds.), Applied Algebra, Algebraic Algorithms and Error-Correcting Codes. Proceedings, 1995. XI, 485 pages. 1995.

Vol. 949: D.G. Feitelson, L. Rudolph (Eds.), Job Scheduling Strategies for Parallel Processing. Proceedings, 1995. VIII, 361 pages. 1995.

Vol. 950: A. De Santis (Ed.), Advances in Cryptology - EUROCRYPT '94. Proceedings, 1994. XIII, 473 pages. 1995.

Vol. 951: M.J. Egenhofer, J.R. Herring (Eds.), Advances in Spatial Databases. Proceedings, 1995. XI, 405 pages. 1995.

Vol. 952: W. Olthoff (Ed.), ECOOP '95 - Object-Oriented Programming. Proceedings, 1995. XI, 471 pages. 1995.

Vol. 953: D. Pitt, D.E. Rydeheard, P. Johnstone (Eds.), Category Theory and Computer Science. Proceedings, 1995. VII, 252 pages. 1995.

Vol. 954: G. Ellis, R. Levinson, W. Rich. J.F. Sowa (Eds.), Conceptual Structures: Applications, Implementation and Theory. Proceedings, 1995. IX, 353 pages. 1995. (Subseries LNAI).

VOL. 955: S.G. Akl, F. Dehne, J.-R. Sack, N. Santoro (Eds.), Algorithms and Data Structures. Proceedings, 1995. IX, 519 pages. 1995.

Vol. 956: X. Yao (Ed.), Progress in Evolutionary Computation. Proceedings, 1993, 1994. VIII, 314 pages. 1995. (Subseries LNAI).

Vol. 957: C. Castelfranchi, J.-P. Müller (Eds.), From Reaction to Cognition. Proceedings, 1993. VI, 252 pages. 1995. (Subseries LNAI).

Vol. 958: J. Calmet, J.A. Campbell (Eds.), Integrating Symbolic Mathematical Computation and Artificial Intelligence. Proceedings, 1994. X, 275 pages. 1995.

Vol. 959: D.-Z. Du, M. Li (Eds.), Computing and Combinatorics. Proceedings, 1995. XIII, 654 pages. 1995.

Vol. 960: D. Leivant (Ed.), Logic and Computational Complexity. Proceedings, 1994. VIII, 514 pages. 1995.

Vol. 961: K.P. Jantke, S. Lange (Eds.), Algorithmic Learning for Knowledge-Based Systems. X, 511 pages. 1995. (Subseries LNAI).

Vol. 962: I. Lee, S.A. Smolka (Eds.), CONCUR '95: Concurrency Theory. Proceedings, 1995. X, 547 pages. 1995.

Vol. 963: D. Coppersmith (Ed.), Advances in Cryptology - CRYPTO '95. Proceedings, 1995. XII, 467 pages. 1995.

Vol. 964: V. Malyshkin (Ed.), Parallel Computing Technologies. Proceedings, 1995. XII, 497 pages. 1995.

Vol. 965: H. Reichel (Ed.), Fundamentals of Computation Theory. Proceedings, 1995. IX, 433 pages. 1995.

Vol. 966: S. Haridi, K. Ali, P. Magnusson (Eds.), EURO-PAR '95 Parallel Processing. Proceedings, 1995. XV, 734 pages. 1995.

Vol. 967: J.P. Bowen, M.G. Hinchey (Eds.), ZUM '95: The Z Formal Specification Notation. Proceedings, 1995. XI, 571 pages. 1995.

Vol. 968: N. Dershowitz, N. Lindenstrauss (Eds.), Conditional and Typed Rewriting Systems. Proceedings, 1994. VIII, 375 pages. 1995.

Vol. 969: J. Wiedermann, P. Hájek (Eds.), Mathematical Foundations of Computer Science 1995. Proceedings, 1995. XIII, 588 pages. 1995.

Vol. 970: V. Hlaváč, R. Šára (Eds.), Computer Analysis of Images and Patterns. Proceedings, 1995. XVIII, 960 pages. 1995.

Vol. 971: E.T. Schubert, P.J. Windley, J. Alves-Foss (Eds.), Higher Order Logic Theorem Proving and Its Applications. Proceedings, 1995. VIII, 400 pages. 1995.

Vol. 972: J.-M. Hélary, M. Raynal (Eds.), Distributed Algorithms. Proceedings, 1995. XI, 333 pages. 1995.

Vol. 973: H.H. Adelsberger, J. Lažanský, V. Mařík (Eds.), Information Management in Computer Integrated Manufacturing. IX, 665 pages. 1995.

Vol. 974: C. Braccini, L. DeFloriani, G. Vernazza (Eds.), Image Analysis and Processing. Proceedings, 1995. XIX, 757 pages. 1995.

Vol. 975: W. Moore, W. Luk (Eds.), Field-Programmable Logic and Applications. Proceedings, 1995. XI, 448 pages. 1995.

Vol. 976: U. Montanari, F. Rossi (Eds.), Principles and Practice of Constraint Programming — CP '95. Proceedings, 1995. XIII, 651 pages. 1995.

Vol. 977: H. Beilner, F. Bause (Eds.), Quantitative Evaluation of Computing and Communication Systems. Proceedings, 1995. X, 415 pages. 1995.

Vol. 978: N. Revell, A M. Tjoa (Eds.), Database and Expert Systems Applications. Proceedings, 1995. XV, 654 pages. 1995.

Vol. 979: P. Spirakis (Ed.), Algorithms — ESA '95. Proceedings, 1995. XII, 598 pages. 1995.

Vol. 980: A. Ferreira, J. Rolim (Eds.), Parallel Algorithms for Irregularly Structured Problems. Proceedings, 1995. IX, 409 pages. 1995.

Vol. 981: I. Wachsmuth, C.-R. Rollinger, W. Brauer (Eds.), KI-95: Advances in Artificial Intelligence. Proceedings, 1995. XII, 269 pages. (Subseries LNAI).

Vol. 982: S. Doaitse Swierstra, M. Hermenegildo (Eds.), Programming Languages: Implementations, Logics and Programs. Proceedings, 1995. XI, 467 pages. 1995.

Vol. 983: A. Mycroft (Ed.), Static Analysis. Proceedings, 1995. VIII, 423 pages. 1995.

Vol. 984: J.-M. Haton, M. Keane, M. Manago (Eds.), Advances in Case-Based Reasoning. Proceedings, 1994. VIII, 307 pages. 1995.

Vol. 985: T. Sellis (Ed.), Rules in Database Systems. Proceedings, 1995. VIII, 373 pages. 1995.

Vol. 986: Henry G. Baker (Ed.), Memory Management. Proceedings, 1995. XII, 417 pages. 1995.

Vol. 987: P.E. Camurati, H. Eveking (Eds.), Correct Hardware Design and Verification Methods. Proceedings, 1995. VIII, 342 pages. 1995.

Vol. 988: A.U. Frank, W. Kuhn (Eds.), Spatial Information Theory. Proceedings, 1995. XIII, 571 pages. 1995.

Vol. 989: W. Schäfer, P. Botella (Eds.), Software Engineering — ESEC '95. Proceedings, 1995. XII, 519 pages. 1995.

Vol. 990: C. Pinto-Ferreira, N.J. Mamede (Eds.), Progress in Artificial Intelligence. Proceedings, 1995. XIV, 487 pages. 1995. (Subseries LNAI).

Vol. 993: T.C. Fogarty (Ed.), Evolutionary Computing. Proceedings, 1995. VIII, 264 pages. 1995.

Vol. 994: M. Hebert, J. Ponce, T. Boult, A. Gross (Eds.), Object Representation in Computer Vision. Proceedings, 1994. VIII, 359 pages. 1995.

Vol. 997: K.P. Jantke, T. Shinohara, T. Zeugmann (Eds.), Algorithmic Learning Theory. Proceedings, 1995. XV, 319 pages. 1995.

Vol. 998: A. Clarke, M. Campolargo, N. Karatzas (Eds.), Bringing Telecommunication Services to the People – IS&N '95. Proceedings, 1995. XII, 510 pages. 1995.

Vol. 999: P. Antsaklis, W. Kohn, A. Nerode, S. Sastry (Eds.), Hybrid Systems II. VIII, 569 pages. 1995.